Biobased Materials

Ajay Kumar Mishra ·
Chaudhery Mustansar Hussain
Editors

Biobased Materials

Recent Developments and Industrial
Applications

 Springer

Editors
Ajay Kumar Mishra
Department of Chemistry
Durban University of Technology
Durban, South Africa

Chaudhery Mustansar Hussain
Department of Chemistry and EVSC
New Jersey Institute of Technology
Newark, NJ, USA

ISBN 978-981-19-6026-0   ISBN 978-981-19-6024-6 (eBook)
https://doi.org/10.1007/978-981-19-6024-6

This Springer imprint is published by the registered company Springer Nature Singapore Pte Ltd.
The registered company address is: 152 Beach Road, #21-01/04 Gateway East, Singapore 189721,
Singapore

Preface

The topic of bio-based materials is currently popular in the scientific community, working in such following areas as Recycled materials, Renewable materials, Materials for efficiency, Materials for waste treatment, Materials for reduction of environmental load, Materials for easy disposal or recycle, Hazardous free materials, Materials for reducing human health impact, Materials for energy efficiency, Materials for green energy, etc. This is a relatively hot topic in materials science and has strong demands for energy, material, and money savings, as well as heavy contamination problems, despite that the area of bio-based materials belongs to the most important fields of modern science and technology, no important encyclopedias have been published in the area of bio-based materials.

Biobased Materials: Recent Developments and Industrial Applications cover broadly the biomedical, pharmaceutical, construction, and other industries. It is an unique contribution from experts on hybrid biopolymers and bio-composites, bioactive and biodegradable materials, bio-inert polymers, natural polymers and composites, and metallic natural materials. Therefore, this book is suitable for useful reference for scientists, academicians, research scholars, and technologists.

Major challenges of bio-based materials are their efficient development, cost-effective, and green and environment-friendly production/applications. The content of the current book will be beneficial to the engineers and scientists for proper utlization of bio-based materials. The book brings the recent practices of the bio-based materials technology in different scientific and engineering domains. It helps the bounded industrial outcomes to reach the general readership of different domains. This book bridges the technological gaps between the industrial and academic professionals and the novice young students/scholars. The interdisciplinarity of this book is the main appeal to the huge number of readers.

Book consists of 14 chapters. Chapter "Biobased Material for Food Packaging" describes about the bio-based materials for different food packaging applications, whereas Chapter "Cytotoxicity and Biocompatibility of Biobased Materials" detailed about the cyto-toxicity and biocompatibility of bio-based materials. Chapter "Carbon Nanostructures, Nanomaterials and Energy Storage—A Critical Overview and the Visionary Future" consists of carbon nanostructures, nanomaterials and energy storage—a critical overview and the visionary future, whereas Chapter "Biobased Materials in Bioelectronics" focused on bio-based materials in bioelectronics. Chapter "Biobased Materials and the Vast Domain of Environmental Pollution Control—A Critical Overview" describes the bio-based materials and the vast domain of environmental pollution control—a critical overview while Chapter "Recent Trends in Eco-Friendly Materials for Agrochemical Pollutants Removal: Polysaccharide-Based Nanocomposite Materials" details the bio-based materials and the vast domain of environmental pollution control—a critical overview. Chapter "Protein-Based Biomaterials for Sustainable Remediation of Aquatic Environments" covers the recent trends in eco-friendly materials for agrochemical pollutants removal: polysaccharide-based nanocomposite materials, Protein-based biomaterials for sustainable remediation of aquatic environments and Chapter "Green Sustainability and Arsenic Groundwater Remediation in Developing countries—A Far-Reaching Review" consists of green sustainability and arsenic groundwater remediation in developing countries—a vast vision for the future. Chapter "Bio-Based Materials Used in Food Packaging to Increase the Shelf Life of Food Products" describes the bio-based materials used in food packaging for increasing the shelf-life of food products while Chapter "Bio Polymers and Sensors Used in Food Packaging—Present and Future Prospects" detailed about the biopolymers used in food packaging—present and future prospects. Chapter "Biofunctional Textiles: Functional Polymer-Carriers with Antiviral, Antibacterial, antifungal and Repellent Activity" detailed about the biofunctional textiles: functional polymer-carriers with antiviral, antibacterial, antifungal, and repellent activity while Chapter "Low Cost and Sustainable Treatment Options for Removal of Cd(II) from Drinking Water Using Indigenous Material for Rural Communities" describes the low cost and sustainable treatment options for removal of Cd(II) from drinking water using indigenous material for rural communities. Chapter "Green Synthesis of Zinc Oxide Nanoparticles Using *Citrus Sinensis* (Orange) Peel Extract for Achieving Ultraviolet Blocking Properties" consists of green synthesis of zinc oxide nanoparticles using citrus sinensis (orange) peel extract for achieving ultraviolet blocking properties and Chapter "Biohydrogen Production from Food and Beverage Processing: A Promising Strategy for Wastewater Management" summarizes the biohydrogen production from food and beverage processing: A promising strategy for wastewater management.

Overall, this book had provided a summary of the state-of-the-art knowledge to scientists, engineers, and research scholars about recent developments due to bio-based nanomaterials arena. The advances in *Biobased Materials: Recent Developments and Industrial Applications* in the context of modern society's interests have been considered preferably, that allow to identify grand challenges and directions for future research. The book contributors have been selected from all over the world and the essential functions of the *Biobased Materials* are presented rather than their anticipated applications. Moreover, up-to-date knowledge on economy, toxicity, and regulation related to *Biobased Materials* have been presented in detail.

Durban, South Africa   Ajay Kumar Mishra, Ph.D., FRSC
Newark, USA   Chaudhery Mustansar Hussain, Ph.D.

Contents

Biobased Material for Food Packaging

Dattatreya M. Kadam and Mrunal D. Barbhai

1 Introduction

Plastic usage is on the surge every year since its production started at the industry level. In 2020, the global plastic production was recorded as 368 million metric tons (https://www.statista.com/statistics/282732/global-production-of-plastics-since-1950/). It has been preferred for multiple usages in different sectors. Further, the application of plastics as packaging material in the food packaging industry is wide and growing, considering its ease and convenient usage. The plastics being manufactured using non-renewable resources like petroleum products make their degradation difficult (Chen et al. 2019). Along with this, the plastic waste ends up in landfills as there is no proper waste management system, causing plastic pollution, rising greenhouse gases, and leaving behind a large carbon footprint. Among the plastic waste generated, it is estimated that packaging contributes to almost 50% of global plastic waste produced, with food packaging contributing a major share (Ncube et al. 2021). Thus, food industries are motivated to adopt reducing, reusing, and recycling plastics; despite all these efforts, the accumulation of plastic waste still turns out to be a major problem having adverse environmental impacts (Gómez-Estaca et al. 2016). These synthetic plastics are not easily degradable and can require years together for complete degradation. Single-use plastic cannot be recycled the majority of the time and only 5% is recycled (Ncube et al. 2021). Considering this, many countries including India are banning plastics, especially single use plastics and researchers across the globe have also directed their efforts to finding sustainable alternatives like the development of biodegradable plastics in food packaging (Chen et al. 2019).

One such alternative solution being researched continuously, includes biobased material as a replacement for petrochemical-based plastics in food packaging.

D. M. Kadam (✉) · M. D. Barbhai
ICAR-Central Institute of Research on Cotton Technology, Matunga, Mumbai 400019, India
e-mail: dmkadam11k@gmail.com

© The Author(s), under exclusive license to Springer Nature Singapore Pte Ltd. 2023
A. K. Mishra and C. M. Hussain (eds.), *Biobased Materials*,
https://doi.org/10.1007/978-981-19-6024-6_1

Fig. 1 Types of plastics
based on their origin and
degradability. Adapted from:
Asgher et al. (2020)

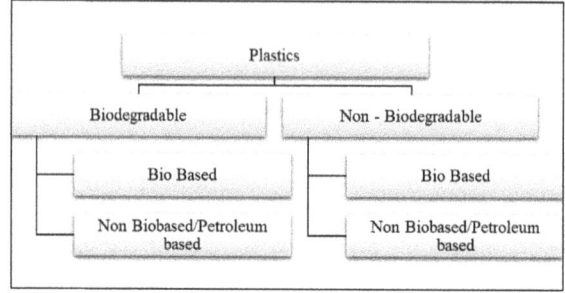

Bioplastics were defined by the European bioplastic organization as "plastics that are based on renewable resources that are biodegradable and/or compostable" (Peelman et al. 2013). There are different types of plastics as given in Fig. 1 viz., the ones that are biodegradable or non-biodegradable or the ones that are biobased and petroleum based (Asgher et al. 2020). Biobased materials are not always 100% renewable, however, there are some bioplastics that are 100% renewable while others are partially renewable. For example, polylactic acid (PLA), starch, polyhydrox-yalkanoates (PHAs) are biodegradable, while bio-polyethylene is nondegradable (Sudesh and Iwata 2008). Biopolymers can be obtained from three main sources, i.e., biomass, bioderived monomers or microorganisms. The biobased plastics derived from biomass include polymers extracted from polysaccharides, proteins, and lipids. Some of these raw materials originate from wastes generated by the agricultural industry. While the ones derived from microorganisms could be either synthesized naturally or through genetically modifying the microorganisms. In case of bioplastics derived from monomers, the chemicals are synthesized using biobased monomers, e.g., obtaining polylactic acid by polymerizing lactic acid (Weber 2000; Asgher et al. 2020). Based on their origin, Weber (2000) defined biobased material as "*one obtained from renewable resources and can find application in the food industry*". This suggests that the raw material required for developing biobased plastic is obtained from renewable sources like animals, plants, or microorganisms (Romaric et al. 2021).

In this chapter, the biobased polymer extracted using biomass and microorganism are discussed in detail along with their potential application in the food packaging industry (Fig. 2).

2 Polysaccharide-Based Material

The polysaccharide-based bioplastics can be either obtained from gums, lignin, cellu-lose, starch, and chitosan/chitin. These polysaccharides are crystalline in nature, have an affinity towards water, and are a good barrier for gases (Weber 2000). Among

Fig. 2 Extraction or production-based classification of biopolymers that are discussed in this chapter (Adapted from Chen et al. 2019)

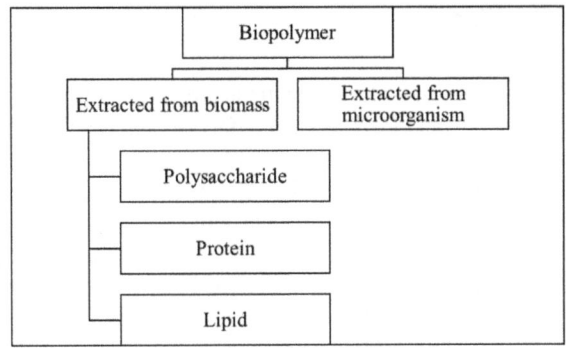

the polysaccharide-based polymers chitin/chitosan is of animal origin while gums, lignin, cellulose, and starch are of plant origin.

2.1 Cellulose and Hemicellulose

Cellulose is a linear carbohydrate polymer mainly obtained from the plant cell wall. It is acquired from the woody parts of plants or cotton linters, but in some cases can also be produced by microbes, i.e., bacteria, fungi, and algae (Abe et al. 2021). The different wood fibers have cellulose and hemicellulose ranging at 40–50% and 25–30%, respectively. Cellulose has 7000–15,000 D-glucose moiety linked with β-1,4-linkages. It has a degree of polymerization ranging from 1510 to 5500 (Yang et al. 2019). Cellulose xanthate is obtained from dissolving cellulose in sodium hydroxide and carbon disulfide mixture, which can be recast in an acid solution to produce cellophane film. Cellulose derivatives can be obtained by esterification or etherification of the hydroxyl group of solvated cellulose. The techniques involved in the preparation of cellulose polymers include extrusion or injection molding (Peelman et al. 2013). The structure of hemicellulose polysaccharides is simpler than that of cellulose and they are hydrophilic and brittle in nature (Asgher et al. 2020). Both the cellulose and hemicelluloses are mainly used in primary or secondary packaging, either as paper or cardboard boxes, bags, or also as wax- or plastic-coated paper packaging material. The difficulty in using cellulose and hemicellulose as materials for packaging liquid foods is due to their affinity toward water and crystallinity (Weber 2000). Thus, their polymer chains are broken using the dissolution method to obtain their derivatives that can be converted into bioplastics packaging material. Due to their low oxygen permeability, the bioplastic films made of cellulose and hemicellulose find usage in packaging.

2.2 Starch

Starch has been widely used in the preparation of various biodegradable plastic materials, as it is an easily available polysaccharide obtained from roots, tubers, cereals, and legumes. It is a storage polysaccharide and is made up of approximately 25–28% of amylose and 72–75% of amylopectin that have monomer of D-glucose (Abe et al. 2021). This amylose and amylopectin content of starch has influence over its properties. However, it is difficult to use starch for preparing edible coatings as it is brittle. It also has a higher melting point, closer to degradation temperature, thereby affecting the process of melting for preparing biofilms (Jariyasakoolroj et al. 2020). Thus, before its application as an ingredient in the bio-plastic making, starch is treated with some renewable plasticizers like glycerol, carboxylic acid, or sorbitol. To convert starch into thermoplastic material, it has to be extruded by applying thermal and mechanical energy. Marichelvam et al. (2019) reported extraction of starch-based bioplastic using corn and rice added with glycerol, citric acid, and gelatin. Further considering the hydrophilic properties of starch, hydrophobic polymers are blended in with the starch molecules along with some plasticizers while preparation of thermoplastic starch material (Weber 2000; Jariyasakoolroj et al. 2020). This not only helped in making the starch blends hydrophobic but also improved properties like thermal stability and mechanical strength (Yang et al. 2019). The addition of plasticizer helps to improve the flexibility or elongation of starch biofilms (Jariyasakoolroj et al. 2020).

2.3 Lignin

Lignin has aromatic rings and the chemical structure of lignin varies with the process of extraction used. Lignocellulosic fibers composed of cellulose (35–55% wt), hemicellulose (20–40% wt), and lignin (10–25% wt) are widely available biodegradable options for the synthetic polymers used in the preparation of bioplastics (Asgher et al. 2020). Different sources to obtain lignocellulosic fibers are straw of paddy, wheat, bagasse of sugarcane, flax, hemp, and so on. (Yang et al. 2019). Lignin has good moisture barrier properties and hence has been blended with different polymers like starch, cellulose, and proteins to produce bioplastics with better water resistance (Chantapet et al. 2013; Aqlil et al. 2017; Yang et al. 2019). For starch-lignin bioplastics, lignin is added through the interaction of hydrogen bonds. Aqlil et al. (2017) indicated that adding lignin to starch films improved their characteristics like water resistance. Further, the addition of graphene oxide to starch-lignin bioplastic improved the thermal stability, and reduced the water uptake and vapor permeability than only starch. Lignin can be also added to proteins like gluten to make bioplastic films. It was reported by Chantapet et al. (2013) that wheat gluten bioplastic when extruded with kraft lignin added a range of 10–30% weight improved the water resistance and mechanical properties.

2.4 Chitosan/Chitin

The source of obtaining chitin or chitosan is the backbone of the invertebrates. Some of the major sources of chitin include shell shrimps, crabs, and crawfishes (Kim et al. 2007). Chitosan—an extensively available, biodegradable polymer is obtained after deacetylation of chitin. It is employed in the manufacturing of edible coating and films. The films developed using chitin or chitosan possess excellent properties of gas (O_2 and CO_2) barrier (Souza et al. 2020). Chitin and chitosan also have good antimicrobial and cationic properties, thus making them excellent raw material for developing biobased films or coatings (Weber 2000; Peelman et al. 2013).

3 Protein-Based Material

Protein-based material can be used as edible coatings or films in the food packaging industry as their amino acid and side chains are easily modifiable for manufacturing the packaging material (Weber 2000). Properties like excellent gas barrier, mechanical properties, and an additional benefit of releasing nitrogen (that can perform as fertilizer) while degradation makes protein widely used biobased polymer in developing films or coatings (Chen et al. 2019). However, like polysaccharides and lipid-based polymers, protein polymers too have an affinity towards water. Also, they are brittle, making them difficult to process and use for packaging thus, biodegradable plasticizers (like polyols–glycerol, sorbitol; fatty acids–oleic acid, linoleic acid; water; monosaccharides–glucose, fructose, mannose; disaccharides and oligosaccharides) are added to improve their mechanical properties. Protein-based biopolymers can be obtained from two main sources, i.e., plant and animal sources (Fig. 3). The plant sources include zein, gluten, soy, and potato while animal sources include collagen, gelatin, casein, whey, and keratin (Weber 2000).

Fig. 3 Types of protein-based biopolymers

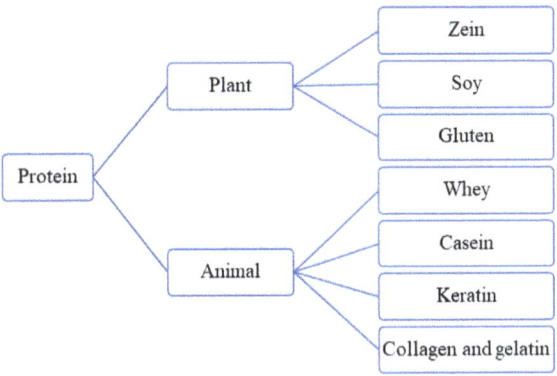

3.1   Plant Sources

3.1.1   Zein

Zein forms approximately 45 to 50% of corn protein and is obtained as a processing by-product. It has nonpolar amino acids and is insoluble in water thus exhibiting moderate quality water resistance; however, it is soluble in alcohol. It also exhibits oxygen or gas resistance and good mechanical properties (Peelman et al. 2013; Chen et al. 2019). One of the limitations of using only zein for film formation is its brittleness, which can be reduced by using appropriate plasticizers. Commonly used plasticizers are polyols like glycerol, or fatty acids (Weber 2000; Peelman et al. 2013; Kothai et al. 2014; Chen et al. 2019). The solvent extraction process or wet method is used for producing films using zein. In this method, zein is dissolved using an appropriate solvent, followed by casting the film on nonstick flat surfaces, that are allowed to stand till the solution evaporates and the film is peeled off. The solvents used include water, ethanol, methanol, acetone, or combinations of these solvents (Weber 2000; Chen et al. 2019). Kothai et al. (2014) also indicated the use of 50% aqueous ethanol for extraction of zein-based films along with the addition of glycerol as a plasticizer. Another method of preparing zein films is extrusion or stretching (Weber 2000; Chen et al. 2019).

3.1.2   Gluten

Gluten—a wheat storage protein, is made up of gliadin (soluble) and glutenin (insoluble) (Chen et al. 2019). The amino acid cysteine is present extensively in gluten (Weber 2000). Gluten films can be produced using different techniques such as cast films by casting a thin layer, followed by drying with aqueous alcohol, e.g., aqueous ethanol, or as thermo-pressed films where the films of proteins are collected from the surface of boiling protein solution (Chen et al. 2019). The manufacturing techniques affect the structure of protein thereby creating films having distinct characteristics. For example, cast films have better elongation and films obtained by thermos pressing have resistance to rupturing (Chen et al. 2019). The gluten films have a glossy surface, good O_2 barrier properties, and homogeneity, but are hydrophilic in nature, thus different hydrophobic plasticizers or oils are added to improve their resistance to water (Weber 2000; Peelman et al. 2013; Chen et al. 2019; Asgher et al. 2020).

3.1.3   Soy

Soy protein isolates are majorly used to prepare soy-protein-based films; other than protein fractions, soy flour and soy milk are also used. Different soy protein films can be obtained by techniques like extrusion, heating, casting, spinning, or compressing using thermal techniques. Like other protein films, soy films also show affinity toward water, limiting the resistance to H_2O, they also lack heat stability and mechanical strength. However, they have good gas barrier properties, transparency, cost-effectiveness, and flexibility (Weber 2000; Peelman et al. 2013; Chen et al. 2019; Asgher et al. 2020).

3.2 Animal Sources

3.2.1 Collagen and Gelatin

Animal protein collagen makes up approximately 20–25% of total body mass, and it is rich in methionine, glycine, and hydroxyproline/proline. Gelatin is produced after collagen is degraded partially with molecular weight ranging from 15 to 400 kDa based on the processing technique used for its production (Benjakula and Kittiphat-tanabawon 2019). Gelatin can be obtained from different sources like skin or bones of pigs or cows; fish skin, scales, or bones (Benjakula and Kittiphattanabawon 2019). Depending on the source, the concentrations of gelatin components may vary. For instance, proline, 4-hydroxyproline, and glycine content of gelatin from pig skin are 13%, 9%, and 33%, respectively. Different properties like degradability, cheaper cost, and easy availability make gelatin a good alternative raw material for preparing biobased films for packaging (Chen et al. 2019). Further, these films are transparent, have a good barrier to oxygen, tensile strength, and melting point at 37 °C; also, the addition of some essential oils can impart antioxidant properties to the films making it a suitable material for developing edible films (Chen et al. 2019; Asgher et al. 2020). These films can be made by hot (room temperature–35 °C) or cold ($\geq$room temperature) casting methods. Based on the type of casting used, the structure of gelatin film changes, like hot casts produce coil structure while cold cast produces spiral structure. The brittleness is more in the hot casted films than in the cold casted ones (Chen et al. 2019). Casting method results in thinner films than extruded films. Other methods of obtaining gelatin films are by using extrusion technology (Chen et al. 2019; Asgher et al. 2020).

3.2.2 Whey

It is a milk protein obtained as cheese processing byproducts. Whey protein isolates (50–80% protein content) and concentrates (90% protein) can be used to develop biobased films. The whey protein is rich in β-lactoglobulin, sulfur-containing amino acids, methionine, and cysteine. (Weber 2000; Chen et al. 2019). Despite their hydrophilic nature, the whey-based films are good packaging material, given their barrier properties for oil, gases, and aroma. The addition of plasticizers like lipids like oil and waxes can help in improving the water barrier characteristics of whey protein (Chen et al. 2019; Asgher et al. 2020). The whey needs to be thermally denatured in an aqueous solution to obtain the films. Maintaining the pH and temperature can help in developing uniform whey-based films (Chen et al. 2019). For improving the characteristics of whey-based biofilms, different treatments can be employed like ultrasound, UV (ultraviolet) radiations, and the use of alkalis (Chen et al. 2019). Application of whey-based films having antimicrobial properties are reported in the literature (Asgher et al. 2020).

3.2.3   Casein

It is a milk protein that is mainly composed of αS1-casein (38%), β-casein (36%), κ-casein (13%), and αS2-casein (10%) (Chen et al. 2019). Being easily available, rich in nutrients, nontoxic, biodegradable, emulsifiable, having good thermal stability, ability to form micelle, make casein excellent raw material for making biobased edible films (Chen et al. 2019). Despite these advantages, casein-based films have a cohesive matrix due to the presence of the polar and nonpolar amino acids, which causes brittleness during drying. However, this can be tackled by the addition of plasticizers like polyols while preparing the films (Weber 2000; Chen et al. 2019). Another major drawback of casein-based films is their affinity and solubility in water. It was reported that casein substances when immersed in water for a day can result in fifty percent weight gain (Weber 2000). Thus, it calls for treating the films using different techniques (like physical or chemical) that modify the polymer. Chemical treatment includes the use of waxes, tannic acids, and transglutaminase; while combining casein with lipids or polysaccharides is another solution to improve its thermal and mechanical characteristics (Chen et al. 2019).

3.2.4   Keratin

It is widely present and obtained from waste feathers, hairs, and nails and mainly consists of sulfur-containing amino acid cysteine (Weber 2000). Keratin can be processed into completely degradable bioplastic that is not soluble in water. However, it is affected by relative humidity thus while processing lamination or blending with other polymers or plasticizers can be used for improving its resistance to relative humidity (Weber 2000).

4   Lipid-Based Material

Lipids and lipid-based materials being nonpolar exhibit excellent property hydrophobicity, providing a barrier against the migration of moisture. This makes them excellent materials for developing biobased films or coatings. Lipids and waxes are blended with the protein or other polysaccharide-based polymers for improving the mechanical properties of bioplastics. These lipids and wax-based coating are often edible and cause no harm to humans (Reichert et al. 2020).

5 Microbial-Based Material

These bioplastics are synthesized by microbial fermentation and the major polymers include polyhydroxyalkanoates (PHA), polylactic acid (PLA), and exopolysaccharides (EPS).

5.1 Polyhydroxyalkanoates (PHA)

These are synthesized by microbial fermentation, where microbial cells produce the polymer. These polymers have to be garnered from microbial cells by use of solvents like methylene chloride, chloroform, and so on. The characteristics or properties of these polymers depend upon the microorganism involved and the substrate, i.e., the carbon source (Asgher et al. 2020). The most commonly used PHA is PHB (polyhydroxybutyrate), which is produced by bacteria and algae (Peelman et al. 2013; Jariyasakoolroj et al. 2020). The PHB has good gas barrier properties and is insoluble in water. The PHAs have a water vapor barrier similar to that of LDPE (low-density polyethylene) (Weber 2000). The major drawback is its high cost of production, which limits the employment of PHAs as bioplastics (Asgher et al. 2020). For reduction in cost and improving other bending strength and complex viscosity, PHAs are blended with different biodegradable polymers like polybutylene adipate-co-terephthalate (PBAT), polybutylene succinate (PBS), and PLA (polylactic acid). PHA films can be prepared using different techniques like a blow or injection molding and can be used in coating papers, films, adhesives, etc. (Nilsen-Nygaard et al. 2021). Under the trade name Biopol™, P(3HB-co-3HV) became the first commercialized PHA that finds application in packaging (Sudesh and Iwata 2008).

5.2 Polylactic Acid (PLA)

PLA can be obtained by fermentation of raw material, viz. sugarcane, wheat, rice, and corn, which leads to the synthesis of lactic acid and lactide (Rasal et al. 2010). It is sourced from renewable raw material and is thermoplastic polyester (Peelman et al. 2013). The properties of PLA are similar to poly(ethylene-terephthalate) (PET), so they are used for replacing PET in the packaging industry to a certain extent (Jariyasakoolroj et al. 2020; Asgher et al. 2020). The properties making PLA the most efficient biobased polymer for packaging include its biodegradable and transparent nature along with good resistance to breakage under tension. PLA is biodegradable; however, its melting point and glass point make it essential to be composted at the industrial level (Nilsen-Nygaard et al. 2021). Despite these advantages, some drawbacks of PLA that limits its usage as packaging material in the food industry include: brittleness, mediocre gas barrier properties, and high cost (Nilsen-Nygaard et al. 2021). The high cost of PLA can be attributed to its extraction and purification process. To synthesize PLA first, the raw material needs to be fermented, followed by purification and then polymerization (Peelman et al. 2013; Jariyasakoolroj et al. 2020; Asgher et al. 2020). After ring-opening polymerization of lactide or polycondensation of lactic acid, the pellets of PLA are produced. Depending upon the stereochemical compositions of lactide (L, L-lactide, D, D-lactide, and L, D-lactide), the polymer properties are determined. Various techniques can be used to process the polymer like extrusion over cast, injection molding, film extrusion, thermoforming, and blow molding (Peelman et al. 2013; Nilsen-Nygaard et al. 2021).

5.3   Exopolysaccharides (EPS)

These polymers are synthesized by gram-positive or gram-negative bacteria, blue-green algae, and fungi. The EPS contains mainly mannose, galactose, glucose, fructose, uronic acid, and other non-carbohydrates compounds such as pyruvate, succinate, acetate, and phosphate (Cottet et al. 2020). Some examples of EPS include alginate, dextrin, glucans, Levan, xanthan, Kefiran, bacterial cellulose, and so on (Asgher et al. 2020; Cottet et al. 2020). For instance, Kefiran—a biodegradable, water-soluble EPS is synthesized during milk fermentation in the preparation of Kefir. It has good anti-microbial properties, mechanical properties, and water vapor permeability. Plasticizers like glycerol can be used in the synthesis of kefiran-based bioplastic films (Asgher et al. 2020). Bacterial cellulose can be obtained from *Gluconoacetobacter xylinum, Acetobacter* sp., and *G. hansenii*, "pullulan" from *Aereobasidium pullullans*, "levan" from *Leuconostoc mesenteroides, Bacillus polymixa, Lactobacillus reuteri*, and others (Cottet et al. 2020).

6   Application of the Biobased Material in Food Packaging

For extending the shelf life of the food products especially the perishables like dairy foods, meat, fruits, and vegetables; films are being used during transportation and distribution. However, majorly synthetic materials are being used for preparing the films, causing a lot of environmental problems as these substances are mainly dependent on nonrenewable resources and are not degradable. This creates a toxic effect on the environment, thus extensive research is being conducted on developing biodegradable biobased plastics (Chen et al. 2019). Various bioplastics obtained from polysaccharides, lipids, proteins, or microorganisms (as discussed in the above sections) can find application as films, and coating in food packaging as part of active or intelligent packaging.

Polysaccharide-based bioplastics like chitosan are extensively used as biofilms and edible coatings to extend the shelf life of fruits and vegetables by delaying loss of water, ripening, and decaying. The food products are directly sprayed, brushed, and dipped in the molten compound for application of the chitosan films. These films can be applied to meat and meat products and their application helps in improving the appearance and delays peroxidation of lipids. Another advantage of using chitin or chitosan is its antimicrobial property which can enhance the shelf life of food by preventing the growth of microorganisms (Souza et al. 2020).

Various studies have indicated the application of chitosan-based films for extending the shelf life of food. Chien et al. (2007) indicated the use of chitosan coating at 0.5–2% (by dipping method) over the sliced mangoes stored at 6 °C, decreased the loss of water, helped in keeping the sensory qualities intact, and also repressed the microorganism's growth, thereby enhancing the shelf life of sliced mangoes. Another study also indicated the application of chitosan coating (0.5–2%) on guava stored at 11 °C, 90–95% relative humidity enhanced the shelf life of guavas. It was observed that coating with 2% chitosan maintained the firmness of guavas 14 times better than that of control fruits without the coating during the storage of 9 days. It also decreased weight loss as the transpiration and respiration were restricted by the chitosan film. During storage, the green color of guava (due to chlorophyll content) changes to yellow as a result of reduced chlorophyll content. This reduction was highest in the control guavas without chitosan coating while lowest in 2% chitosan-coated guava after the storage period. Also, the malondialdehyde formation was lower in chitosan-coated (0.5–2%) guavas (Hong et al. 2012).

Kim et al. (2007) have also specified that the use of chitosan films can improve the shelf stability of eggs. The authors studied the effect of different chitosan coatings prepared from α-chitosan having different molecular weights, viz. 282, 440, 746, and 1110 kDa and β-chitosan with a molecular weight of 577 kDa. The results evidently confirmed chitosan coating could protect the eggs during four weeks of storage at 25 °C, by preventing moisture and CO_2 transfer via eggs shells. This barrier for gasses provided by chitosan also prevented weight loss during storage thus, increasing the shelf stability. Amongst all the chitosan coatings, α-chitosan with 282 kDa molecular weight had the best bactericidal effects against *Salmonella enteritidis* than uncoated eggs (Kim et al. 2007).

Fakhouri et al. (2015) reported that after 21 days of post-harvest storage the grapes (Red Crimson) coated with starch-based edible film had a better appearance than the uncoated grapes. The enhancement of shelf life was due to the prevention of weight loss during storage. The starch-based film was prepared using different ratios of corn starch, gelatin, and glycerol (used as plasticizer). The addition of glycerol improved the flexibility and the increased gelatin content increased water vapor permeability, while as the starch content increased the thickness increased. Gelatin addition also decreased the opacity of starch-based films.

Nawab et al. (2017) reported that starch obtained from mango kernels can be utilized in producing biobased packaging material. The mango kernel starch films were prepared by gelatinizing starch, and adding glycerol and sorbitol as a plasticizer. This film was used to coat tomatoes post harvesting, and they were stored for 20 days. Coating with starch-based film extended the shelf life of tomatoes, without damaging any of the sensory properties.

The proteins have good film forming ability, better gas barrier, and mechanical properties than lipid and polysaccharide-based polymers, and are also rich in nutrition and thus are extensively used in manufacturing biobased films (Weber 2000; Chen et al. 2019). For example, it is reported that films made with soy protein exhibited approximately 670, 540, 500, and 260 times greater O_2 permeability than pectin, starch, polyethylene, and methylcellulose. The application of films can prevent flavor and moisture loss, and increase the shelf life by controlling the gas exchange (Chen et al. 2019). Kothai et al. (2014) reported the use of biobased protein films developed using potato peel powder, zein, and eggshell. It was reported that the water resistance of the paper box was improved by the application of the protein films developed using potato peel powder than the uncoated box. Çakmak et al. (2020) reported that edible film prepared using whey protein isolates at 8% w/v, essential oil (bergamot) at 4.5% and glycerol as a plasticizer at 39.2% had antimicrobial properties against S. aureus and E. coli.

PLA films were used for packing capsicums and it was observed that after 7 days of storage at 10 °C, the PLA films had a lower microbial load for total aerobic bacterial counts (4.40 log CFU (colony forming unit)/g), coliform bacterial counts (0.94 log CFU/g) and mold and yeast counts (2.96 log CFU/g) than low-density polyethylene (5.40, 3.04, and 2.35 log CFU/g respectively) and perforated low-density polyethylene (4.56, 1.61, 3.01 log CFU/g) film packaging, respectively. This was associated with higher water vapor permeability of the PLA biodegradable films (Koide and Shi 2007). Sudesh and Iwata (2008) reported that cups developed using PLA are used in some restaurants in Japan for serving chilled beverages. PLA with good tensile strength and mechanical properties has been reported to find applications in preparing films, jugs, bowls, and cups (Halonen et al. 2020). Further potential use of PHAs and PLAs is also reported as packing material for cheese (Srivastava et al. 2018).

Thus, as disused in this section, biobased material can be employed in active packaging, modified atmosphere packaging, and sensors as films or coatings given their antimicrobial properties (Halonen et al. 2020).

7 Conclusion and Future Prospects

Currently, the usage of bioplastic is limited, but is growing rapidly given the problems faced by the rising plastic pollution and toxic burden on the environment. The concept of sustainable use of resources and circular economy favors the use of biobased polymers in various sections of food packaging. As a result there is an increased market demand for biobased packaging material. Research is being conducted on manufacturing and applying biodegradable bioplastics in packaging food products that are perishable like fruits and vegetables, cheese, meat and meat products, etc. having a short shelf life. Given the beneficial properties of the biobased material, like biodegradability, composability, and antimicrobial properties, these polymers have the potential to replace the presently used conventional packaging material. Commercial usage of bioplastics is still limited, thus extensive trials are being done to gather all possible information that can aid in the commercial application of these biobased films and coating for various products. However, still, additional research needs to be conducted for understanding the functional and mechanical properties of these biobased films and their commercial application on a wider scale. Also, there are certain drawbacks associated with the application of some biobased materials (e.g., polyhydroxyalkanoates) in films and coating or food packaging, due to higher costs associated with their extraction and production, thus substantial research is required to understand different methods that can be employed to reduce the production cost of these biobased polymers. Advance research on the biodegradability and disposal of these biobased polymers can also assist in finding solutions to reduce the burden on landfills. The scope also lies in conducting experiments on the safety issues of bioplastic application in the food packaging industry. Overall, it can be concluded that biobased polymers have the potential to be used in food packaging that can improve the shelf stability of products and reduce the dependence on conventional plastic packaging material, thereby proving an alternative solution to reducing plastic pollution.

References

Abe MM, Martins JR, Sanvezzo PB, Macedo JV, Branciforti MC, Halley P, Botaro VR, Brienzo M (2021) Advantages and disadvantages of bioplastics production from starch and lignocellulosic components. Polymers 13(15):2484. https://doi.org/10.3390/polym13152484

Aqlil M, Moussemba Nzenguet A, Essamlali Y, Snik A, Larzek M, Zahouily M (2017) Graphene oxide filled lignin/starch polymer Bionanocomposite: structural, physical, and mechanical studies. J Agric Food Chem 65(48):10571–10581. https://doi.org/10.1021/acs.jafc.7b04155

Asgher M, Qamar SA, Bilal M, Iqbal HMN (2020) Bio-based active food packaging materials: Sustainable alternative to conventional petrochemical-based packaging materials. Food Res Int, 137.https://doi.org/10.1016/j.foodres.2020.109625

Benjakul S, Kittiphattanabawon P (2019) Gelatin. In: Melton L, Shahidi F, Varelis P (Eds), Encyclopedia of food chemistry, Academic Press, pp 121–127. https://doi.org/10.1016/B978-0-08-100596-5.21588-6

Çakmak H, Özselek Y, Turan OY, Fıratlıgil E, Karbancioğlu-Güler F (2020) Whey protein isolate edible films incorporated with essential oils: antimicrobial activity and barrier properties. Polym Degrad Stab 179:109285

Chantapet P, Kunanopparat T, Menut P, Siriwattanayotin S (2013) Extrusion processing of wheat gluten bioplastic: effect of the addition of Kraft lignin. J Polym Environ 21:864–873. https://doi.org/10.1007/s10924-012-0557-8

Chen H, Wang J, Cheng Y, Wang C, Liu H, Bian H, Pan Y, Sun J, Han W (2019) Application of protein-based films and coatings for food packaging: a review. Polymers 11(12):2039. MDPI AG. Retrieved from https://doi.org/10.3390/polym11122039

Chien PJ, Sheu F, Yang FH (2007) Effects of edible chitosan coating on quality and shelf life of sliced mango fruit. J Food Eng 78(1):225–229

Cottet C, Ramirez-Tapias YA, Delgado JF, de la Osa O, Salvay AG, Peltzer MA (2020) Biobased materials from microbial biomass and its derivatives. Materials (Basel, Switzerland) 13(6):1263. https://doi.org/10.3390/ma13061263

Fakhouri FM, Martelli SM, Caon T, Velasco JI, Mei LHI (2015) Edible films and coatings based on starch/gelatin: film properties and effect of coatings on quality of refrigerated Red Crimson grapes. Postharvest Biol Technol 109:57–64

Gómez-Estaca J, Gavara R, Catalá R, Hernández-Muñoz P (2016) The potential of proteins for producing food packaging materials: a review. Packag Technol Sci 29:203–224. https://doi.org/10.1002/pts.2198

Halonen N, Pálvölgyi PS, Bassani A, Fiorentini C, Nair R, Spigno G, Kordas K (2020) Bio-based smart materials for food packaging and sensors–a review. Frontiers in Materials 7:82

Hong K, Xie J, Zhang L, Sun D, Gong D (2012) Effects of chitosan coating on postharvest life and quality of guava (Psidium guajava L.) fruit during cold storage. Sci Hortic 144:172–178

https://www.statista.com/statistics/282732/global-production-of-plastics-since-1950/. Accessed 21 Dec 2021

Jariyasakoolroj P, Leelaphiwat P, Harnkarnsujarit N (2020) Advances in research and development of bioplastic for food packaging. J Sci Food Agric 100:5032–5045. https://doi.org/10.1002/jsfa.9497

Kim SH, No HK, Prinyawiwatkul W (2007) Effect of molecular weight, type of chitosan, and chitosan solution pH on the shelf-life and quality of coated eggs. J Food Sci 72(1):S044–S048. https://doi.org/10.1111/j.1750-3841.2006.00233.x

Koide S, Shi J (2007) Microbial and quality evaluation of green peppers stored in biodegradable film packaging. Food Control 18(9):1121–1125. https://doi.org/10.1016/j.foodcont.2006.07.013

Kothai S, Vadivu KS, Rajeswari N, Kavitha N (2014) Bio based coating materials in food packaging. Int J Eng Res Technol 03(02):1750–1752

Marichelvam MK, Jawaid M, Asim M (2019) Corn and rice starch-based bio-plastics as alternative packaging materials. Fibers 7(4):32

Nawab A, Alam F, Hasnain A (2017) Mango kernel starch as a novel edible coating for enhancing shelf-life of tomato (Solanum lycopersicum) fruit. Int J Biol Macromol 103:581–586

Ncube LK, Ude AU, Ogunmuyiwa EN, Zulkifli R, Beas IN (2021) An overview of plastic waste generation and management in food packaging industries. Recycling 6:12. https://doi.org/10.3390/recycling6010012

Nilsen-Nygaard J, Fernández EN, Radusin T, Rotabakk BT, Sarfraz J, Sharmin N, Sivertsvik M, Sone I, Pettersen MK (2021) Current status of biobased and biodegradable food packaging materials: impact on food quality and effect of innovative processing technologies. Compreh Rev Food Sci Food Safety 20(2):1333–1380

Peelman N, Ragaert P, De Meulenaer B, Adons D, Peeters R, Cardon L, Van Impe F, Devlieghere F (2013) Application of bioplastics for food packaging. Trends Food Sci Technol 32(2):128–141

Rasal RM, Janorkar AV, Hirt DE (2010) Poly(lactic acid) modifications. Prog Polym Sci 35:338–356. https://doi.org/10.1016/j.progpolymsci.2009.12.003

Reichert CL, Bugnicourt E, Coltelli M-B, Cinelli P, Lazzeri A, Canesi I, Braca F et al (2020) Bio-based packaging: materials, modifications, industrial applications and sustainability. Polymers 12(7):1558. MDPI AG. Retrieved https://doi.org/10.3390/polym12071558

Romaric O, Dabadé DS, Vieira-Dalode G, Sanoussi AF, Fagla-Amoussou AB, Hounhouigan MH, Hounhouigan DJ, Azokpota P (2021) Bio-based packaging used in food processing: a critical review. African J Food Sci 15(4):131–144

Souza V, Pires J, Rodrigues C, Coelhoso IM, Fernando AL (2020) Chitosan composites in packaging industry-current trends and future challenges. Polymers 12(2):417. https://doi.org/10.3390/polym12020417

Srivastava PR, Bano K, Zaheer MR, Kuddus M (2018) Biodegradable smart biopolymers for food packaging: sustainable approach toward green environment. In: Bio-based materials for food packaging. Springer, Singapore, pp 197–216

Sudesh K, Iwata T (2008) Sustainability of Biobased and Biodegradable plastics. Clean Soil Air Water 36:433–442. https://doi.org/10.1002/clen.200700183

Weber CJ (2000) Biobased packaging materials for the food industry status and perspectives. A European concerted action. Available at: http://www.biodeg.net/fichiers/Book%20on%20biopoly mers%20(Eng).pdf

Yang J, Ching Y, Chuah C (2019) Applications of lignocellulosic fibers and lignin in bioplastics: a review. Polymers 11(5):751. MDPI AG. Retrieved from https://doi.org/10.3390/polym11050751

Cytotoxicity and Biocompatibility of Biobased Materials

Serap Yalcin, Mehmethan Yıldırım, and Nadia İbrahim Kamil Kamil

1 Introduction

Small molecules and macromolecules generated directly or indirectly from biomass are referred to as bio-based molecules. They typically have high biocompatibility, low toxicity, degradability, a wide range of sources, and a low price. Furthermore, most bio-based compounds have distinct physical, chemical, and physiological features, such as optical activity, amphoteric, and hydrophilicity (Fang et al. 2021).

Bio-based polymers are byproducts of living creatures including plants, trees, microorganisms, and algae (Miao et al. 2014; Zhang et al. 2017; D'Souza et al. 2017; Yang et al. 2018. Including cellulose, chitosan, hemicellulose, cellulose-derived polymers, PLA, PHAs, soybean-based polymers, silk, kefir, and lignin, major biopolymers, and extractive, have been extensively researched for the production of high value-added products such as bio-based bio-medicinal/pharmaceutical products, chemicals, and functional materials (An et al. 2019; Dong et al. 2020; Si and Jiayun 2020). Furthermore, the increasing popularity of combination polymers bodes well for the development of novel biomaterials with desirable characteristics for biomedical and other applications. Novel copolymers, hydrogels, scaffolds, composite microparticles, nanoparticles, and nanofibers are being created for use in dental, ophthalmology, wound healing, cosmetics, pharmaceuticals, medicine delivery, food flavoring/preservative, and waste-water treatment (Birajdar et al. 2021) (Fig. 1).

The use of biomaterials in the medical fields has become widespread, considering their importance in many applications such as medical devices, drug delivery, gene transfer, tissue engineering, and biotechnology (Williams 2008). The progression

S. Yalcin (✉) · M. Yıldırım · N. İ. K. Kamil
Faculty of Art and Sciences, Department of Molecular Biology and Genetics, Kırsehir Ahi Evran University, 40100 Kırsehir, Turkey
e-mail: syalcin@ahievran.edu.tr

Fig. 1 Bio-based materials for medical applications

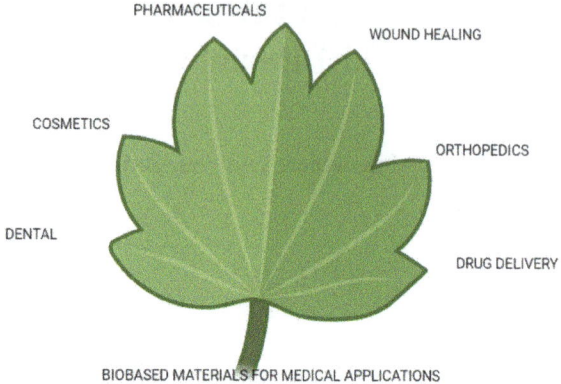

PHARMACEUTICALS

WOUND HEALING

COSMETICS

ORTHOPEDICS

DENTAL

DRUG DELIVERY

BIOBASED MATERIALS FOR MEDICAL APPLICATIONS

in pretreatment, separation, purification, structure determination, and depolymerization can be achieved by biopolymers (Si and Jiayun 2020). Biological materials undergo tests to ensure that they are not toxic and do not cause allergic reactions, genetic mutations, or affect the fertility levels of individuals. These tests are known as biocompatibility (Growth et al. 1995). The biocompatibility of biomedical materials is determined according to the international standard ISO-10993 (Growth et al. 1995). Medical devices go through an in vitro test called cytotoxicity that measures the toxicity on the cellular level and also ascertains if the inserted device causes cell death, these tests help to decide the use of biomedical in nanoparticles; cytotoxicity is measured by many methods including (ATP) test, (MTT) test, (LDH) test and the neutral red method (Peppas and Khare 1993; Schueller et. al. 1999; Seal et al. 2001). In this review, we covered the function of biomaterials their biocompatibility, biodegradability, and bioactivity, cytotoxicity/non-toxicity in especially medical applications.

2 Biomaterials

Over the ages, it has been observed that biomaterials have been benefited, as they were used to repair and treat tissues or organs that have experienced some sort of damage (Migonney 2014). They are commonly classified by their chemical structure (metal, ceramic, polymer, and composites) or based on their interaction with the biological environment (inert, bioactive, bioresorbable), the third way of classification is based on origin being synthetic or natural but it is considered to be less common (Davis 2003). It is believed that during the ancient civilizations such as Roman, Egyptian, and Greek, biomaterials were involved in dental applications as filling materials implants, and dental bridges; they were also perceived as prostheses to recompense for the loss of body parts (Migonney 2014; Hildebrand 2013).

3   Characteristics of Biomaterials

Some medical devices are supposed to remain in the patient's body for life. To achieve this, there must be a measure of the success of the interaction between biological materials and biological cells, therefore the acceptance or rejection of any device in the body can be defined as biocompatibility. It can be used to provide an understanding of toxicity, tissue compatibility, hemocompatibility, and functionality. There are several factors that affect biocompatibility, such as (Ramakrishna et al. 2001; Williams 1990).

1. Wettability is an important factor in the design of biomaterials, as the balance between hydrophilic and hydrophobic surfaces magnifies biocompatibility.
2. Chemical structure.
3. The electric charge of the surface; where it was found that polycations had more cytotoxicity than neutral polymers and polyanions.
4. The molecular weight of the polymer affects the biocompatibility as protein uptake, platelet adhesion, and the molecular weight of the polymer have a directly proportional relationship.
5. Polymer elasticity, surface topography, and roughness regulate the response of proteins and cells to material.
6. The interfacial free energy.

4   Types of Biomaterials

Biomaterials are divided into 3 different categories in the field of health;

1. Biomaterials of Natural (Herbal and Animal) Origin,
2. Synthetic (Polymer, Composite, Ceramic, Metal-Based) Biomaterials,
3. Hybrid (Semi-Synthetic) Biomaterials.

All these types of biomaterials have been used in the medical field for many years, in different generations (first, second, the third generation). With the latest developments in health and medical fields, it has been concluded that metals are one of the most widely used raw materials for load-bearing applications. Screws and wires used to break artificial joints, hip prostheses, implants, facial prostheses, etc. that come out with element combinations such as Co-Cr–Mo. It is widely used in material production (Davis 2003) (Fig. 2).

Polymers are of great surgical importance, especially in biomedical devices, tracheal tubes, facial prostheses, and kidney-liver treatments. Ultra-high molecular weight polyethylene is used in shoulder and hip joints and is one of the most frequently cited polymer examples.

One of the most preferred biocompatible materials in the twenty-first century is ceramics. They are especially preferred structures in areas such as ceramic materials, filling materials, and dental implants. However, they have a disadvantage in that

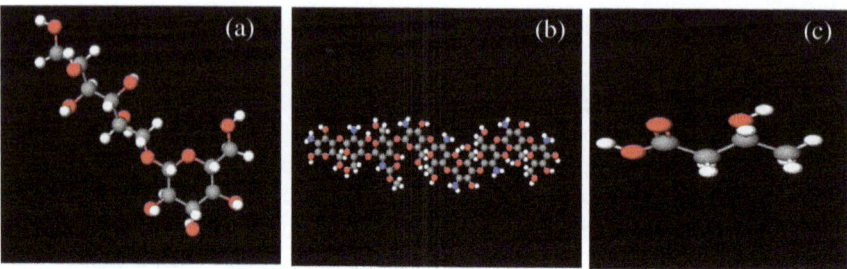

Fig. 2 **a** Kefiran, **b** Chitosan **c** PHB

they have poor fracture thresholds and are therefore used in limited applications as load-bearing materials (Davis 2003; Ramakrishna et al. 2001).

In addition, the type of composite material, which has been used especially with nanotechnology, has a low density and is widely used for prosthetic limbs due to its combination of high strength. To give a few examples, composite materials such as polymethyl methacrylate-glass filling are used for dental restorations. In addition, bisphenol A-glycidyl-quartz silica can be given as an example of the materials used in fillers.

In the field of regenerative medicine, biomaterials that follow the natural host tissue are used and polymers are often preferred. Naturally, derived polymers such as alginate, hyaluronic acid, gelatin, and collagen are used for the manufacture of three-dimensional scaffolds in cell growth, proliferation, and regeneration. A disadvantage in this section is that natural biomaterials have limited mechanical strength, thus limiting their application in load-bearing regions. These biomaterials can be applied in treatments, even if they are chemically modified. For example, when hyaluronic acid molecules are produced in certain amounts with different molecular weights, the efficiency of the treatment (especially in the field of dermatology and cosmetology) will increase. As another example, modified collagen chains such as hydroxylysine and PF are hybridized and applied in treatments (Williams 1990; Ramakrishna et al. 2001; Bhat and Kumar 2013).

Over the last decades, the use of biomaterials in medical science has been increasing. In the human body, many-body tissue like ligaments, teeth, bones tendons, and others have been replaced successfully by these biomaterials (Langer 1998). Nowadays, the various applications of these biomaterials include.

5 Bone Cement and Lenses

Bone cement or polymethyl methacrylate (PMMA) is widely used to fix bone injuries and in orthopedics, PMMA is considered an acrylic polymer that consists of a liquid MMA monomer and a powered MMA-styrene co-polymer; it acts as a filler that tightens the space that holds the implant against the bone. Bone cement types also

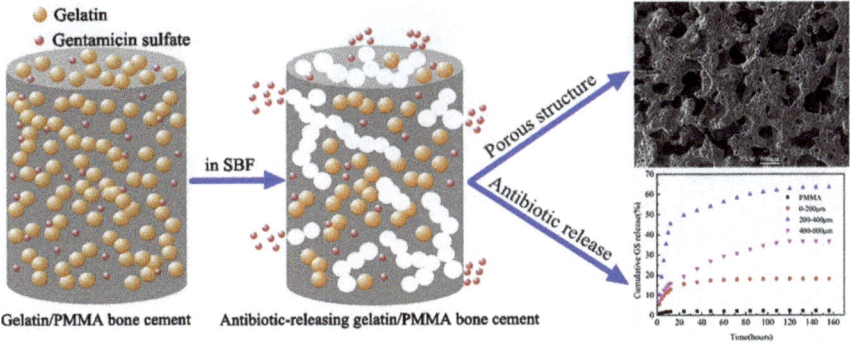

Fig. 3 Development of Gelatin/PMMA bone cement for antibiotic-releasing (Chen et al. 2019)

include: calcium phosphate cement (CPCs) and Glass polyalkenoate cement (GPCs); GPCs are bioresorbable and biocompatible, but with a low mechanical strength consequently, they are mainly used in maxilla-facial surgeries (Marin et al. 2020) (Fig. 3).

Due to the specific properties of the eyes, it is very important to choose the polymer carefully owing to the fact that an interaction between the materials used in the production of lenses and the eye can accrue. The anterior eye has several major aspects that are eyelid 'tear film and cornea, where it can control the requirements of surface properties, oxygen permeability, and modulus sequentially.

Glass, poly(methyl methacrylate) (PMMA), conventional hydrogels, silicone rubber, rigid gas permeable lenses (RGPs), and silicone hydrogels are used in the making of contact lenses. It has been proven that silicone hydrogel lenses have the greatest effectiveness in this field. Improvements continue to be made to this type of lens to ensure maximum eye compatibility and comfort. Silicone hydrogels are a safe option for extended use of the lens and the use of a lens pad, this is due to having higher Dk values and therefore greater oxygen transfer through extended wear (Langer and Peppas 1981).

The implantation of intraocular lenses in the posterior chamber (IOLs) in children has increased significantly recently. Currently, there are no intraocular lenses specifically designed for young children, so it is difficult to insert lenses designed for adults into children's eyes, as the crystalline lens in children is smaller than that of adults. Because the capsular bag does not continue to expand following the lensectomy, the materials used, designs, and sizes that can be implanted in children are more precisely established. The growth of crystalline lens mostly happens during the first two years of life. Capsular IOLs are described as a flexible open loop, one piece, made completely of poly(methyl methacrylate), modified C-loop designs to be in the bag placement (Bharadwaj 2021).

Although studies show that the PMMA material has biocompatibility, biocompatibility, reliability, relative ease of manipulation, and low toxicity, long-term studies are needed. Venancio et al. emphasize the long-term impact of the material in their

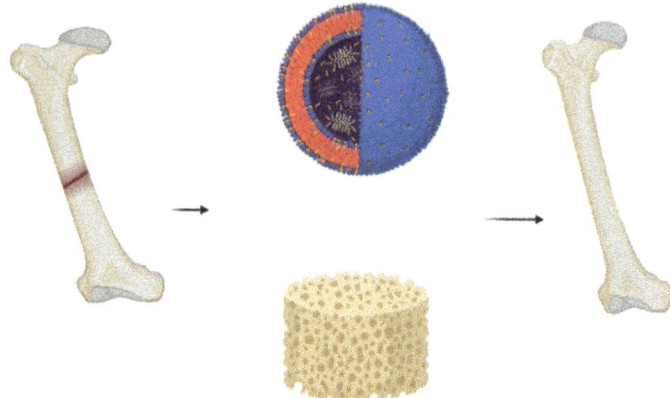

Fig. 4 Bone therapy with PMMA (https://biorender.com/)

work. Pikis et al. 2015 indicate that even small volumes of PMMA used for cranio-plasty may cause severe side effects related to thermal damage or exposure of neural tissue to methyl acrylate monomer. Although successful results have been obtained in the applications of these materials used in research, their toxic effects should not be ignored (Fig. 4).

6 Artificial Ligaments and Tendons

The forces are transferred between skeletal muscle tissues by means of tendons and ligaments, which are characterized by low requirements for oxygen and nutrition, low cell density, and weak regeneration capacity, in addition to their ability to lift high mechanical loads in the body. Excess pressure on these tissues causes permanent damage that affects function and movement. Surgical solutions are used to repair the damage, where tissue replacement with autologous or allogeneic grafts often results in the disease of the donor site, pain, and graft failure, Therefore, a biodegradable scaffold is being sought; Which when renewed will merge to become similar to the original tissue. Collagen, Alginate/CHT, PLLA, Silk, PLGA, PCL/CHT/CNC, PGA, PCL, PLCL/Collagen, and PCL/CHT are examples of biomaterials that have been used in studies about tendon and ligament (Santos et al. 2017; Bharadwaj 2021).

These materials do not cause a continuous inflammatory response, have a degradation time that corresponds to their function, have appropriate mechanical properties for their intended use, produce non-toxic biodegradation products that can be readily resorbed or excreted, and have appropriate permeability and processability for the intended application (Birajdar et al. 2021).

7 Breast Implantation

Generally, it's one of the most performed plastic surgical procedures in the US. For women who get diagnosed with breast cancer a lot of them eventually undergo mastectomy, so many will lean for the option of some form of breast reconstruction. Lastly, due to the increase in prophylactic mastectomies happening as a result of the better understanding of breast cancer genetics, the demand for better and safer breast reconstruction materials has increased; since 2006 silicone has been used as a breast implant but it was causing a lot of issues due to problems in biocompatibility that lead to capsular contracture, calcification, hematoma, necrosis and implant rupture. As a result, studies have been done to find a safer option which is the new generation polymer, TPE1 (Santos et al. 2017).

8 Drug Delivery with Polymeric Systems

A drug delivery system is a method or procedure for providing pharmaceutical compounds with high therapeutic benefits that employ the principles of chemistry, engineering, and biology. Modified release drug delivery systems are in high demand right now, to reduce dose frequency and boost medication efficiency at the needed place to reduce adverse effects (Wen et al. 2015; Tiwari et al. 2012). Until now, significant progress has been made in the development of several polymer-based sustained and regulated drug delivery systems (Sun et al. 2020; Liechty et al. 2010). Polymers have made substantial advances in drug delivery methods, allowing both hydrophilic and hydrophobic medicines to be transported to the site of action for longer periods of time (Wang et al. 2019). Polymer-based drug delivery systems can improve drug safety and efficacy, as well as patient compliance. These methods are intended to keep medication levels at therapeutic levels, reduce side effects and drug dosage, and make it easier to deliver pharmaceuticals with short in vivo half-lives (Harrison 2007; Borandeh et al. 2021).

Polymer-based systems have been studied by biologists, chemists, and engineers for more than 60 years on drug delivery systems. The specific formulations for use in in vitro and in vivo studies. Thus, development systems, which carry enzymes, antibodies, genes, and petite, can be administered to sick individuals with increased efficiency (Anselmo and Mitragotri 2014). Polymer structure can affect the distribution mechanism and release actions of different drugs. Under specific circumstances of polymer preparation and certain geometric shapes are achieved. In controlled release applications of biomaterials, some physicochemical, physical, toxicological, and diffusive tests are made (You-Xiong et al. 2004).

A certain working system is developed when using biomaterials. Polymer and lipid-based materials can be used for drug delivery. Current approaches to genetic engineering, molecular biology, bioengineering, and pharmaceutical technology are

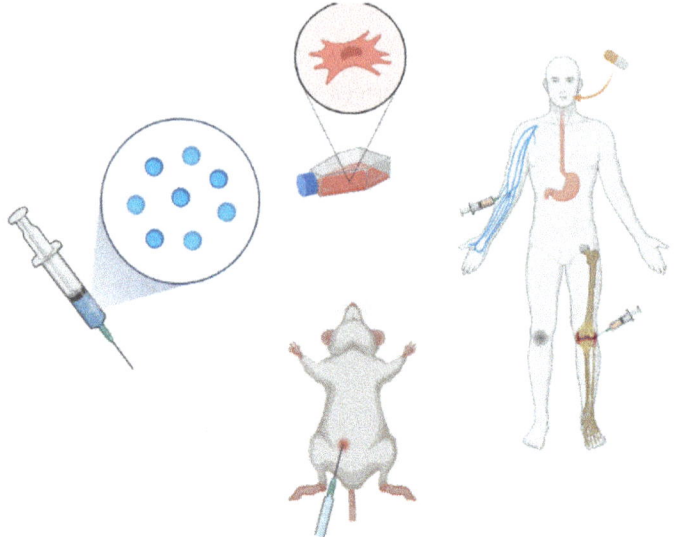

Fig. 5 Examples of polymer-based systems used in in vitro and in vivo studies (https://biorender. com/)

examined. Generally, it is desired to produce a formulation that provides biocompatibility with this material type and increases drug effectiveness. By targeting certain areas in the body and modifying those areas, it ensures that the drug is used in low but effective doses to obtain the therapeutic effect by minimizing toxicity (Yu et al. 2014; Langer 1998).

Researchers take advantage of their physicochemical and biochemical properties to ensure that bio-based materials or biomaterials progress towards the desired goal. Different ways are scanned to get the drug with biomaterial content such as pulmonary, ocular, nasal, transdermal, etc..... (Allen and Cullis 2004; Hoare et al. 2008; Fenton et al. 2018).

Pharmaceuticals, tissue regeneration scaffolds, drug delivery agents, and imaging agents are examples of the use of natural polymers in biomedical applications (Fig. 5).

Natural polymers can be found in a variety of places, including plants, animals, and microbes. Natural polymers–based scaffolds are intriguing for the medical area because of their resemblance to the extracellular matrix, mechanical tunability, high biocompatibility, non-toxic properties, and high water holding capacity (Fig. 6).

Table 1 summarizes some polymer-based drug delivery systems applications.

9   Surgical Sutures, Clips, and Staples to Close the Wound

Surgical sutures, staples, and clips have great importance in surgical procedures. But they also have some drawbacks, as they are unable to prevent fluid leakage from the

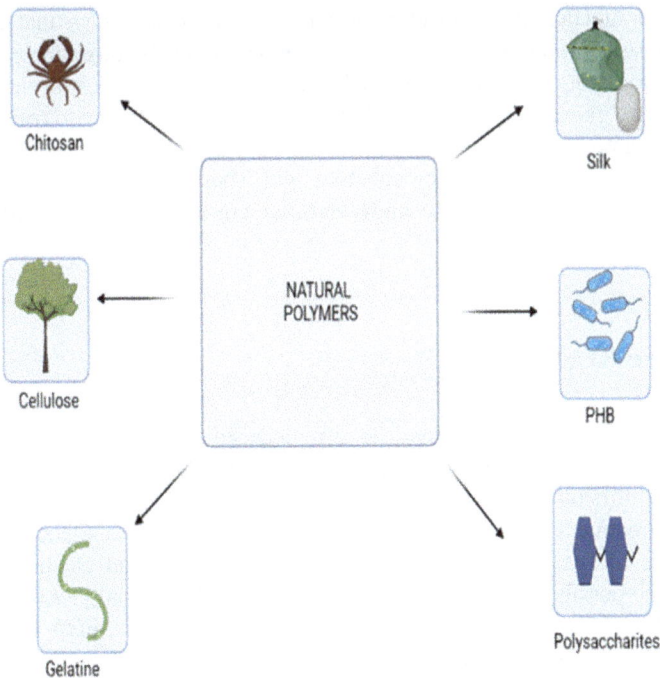

Fig. 6 Natural polymers (https://biorender.com/)

Table 1 Application of natural polymer-based drug delivery systems

Polymer	Aim	References
Sunflower	Drug delivery	Wang et al. (2016)
Silk	Drug delivery	Chambre et al. (2020)
Alginate	Drug delivery	Thomas et al. (2021)
PHB	Drug delivery	Yalcin et al. (2014)
Chitosan	Drug delivery	Del Gaudio et al. (2017)
Dextran	Drug delivery	Yalcin (2019)
Albumin	Drug delivery	Hyun et al. (2018)
Carrageen	Drug delivery	Yermak et al. (2018)
Guar gum	Drug delivery	Seeli and Phabaharan (2016)
Pullulan	Drug delivery	Ponrasu et al. (2021)
Zein and pectin	Drug delivery	Liu et al. (2006)
Lignin	Drug delivery	Cheng et al. (2020)
Gelatin	Drug delivery	Carvalho et al. (2019)

blood vessels and dura mater. In addition, surgical sutures cause scars that impair the healing process. To address this problem, biocompatible and biodegradable adhesives made of fibrin, polyethylene glycol (PEG), and cyanoacrylate have been designed. Cyanoacrylate glue is applied externally to close a skin wound due to residual cytotoxicity. Albumin, collagen, or chitosan are also used; Tissue adhesion is stimulated by irradiation with coherent or non-coherent light. These devices outperform sutures in being able to seal or repair wounds, and they are easy to apply (Vaishya et al. 2013).

10   Nerve Regeneration

It is known that mature neurons do not undergo cell division, which poses a serious threat to the body when it is injured. Therefore, nerve regeneration is a difficult and complex biological task. Scientists are working on providing treatment and solutions to this problem through conducting research focusing on the design of "nerve guide channels" or "neural channels", which increases the possibility of regeneration of axons and functional recovery. Biomaterials such as Poly(glycosaminoglycan-co-collagen), Collagen, PGA mesh coated with collagen, Gelatin, Poly(l-lactide-co-caprolactone), Polyglycolic acid (PGA), PLGA, Polyglactin, Poly 3-hydroxybutyrate, Poly(organo)phosphazene, hydrogels, Glycolide trimethylene carbonate, Polyurethane, and Chitosan are being used in nerve grafts. Building the perfect nerve conduit requires the following: Good biocompatibility, strong mechanical properties of the scaffold, transfer of bioactive factors, availability of a biodegradable and porous channel wall, possessing electrical activities and internal oriented matrix, having intraluminal channels that mimic the structure of nerve fascicles, and can be sterilized (Kou et al. 2010) (Fig. 7).

11   Biomaterials and Skin

Dermatological and cosmetological skin problems, one of the biggest problems of today, can be treated with products formulated within the framework of biocompatibility. Biomaterials such as collagen and hyaluronic acid are formulated and used not only in dental implants and fillings but also in skin problems. The main aim of the researchers and developers in the treatments applied to the skin is to renew the skin, eliminate the specific problem, and reveal the healthy tissue of the skin (Kim et al. 2003; Peppas et al. 2003).

Especially combinations of certain biomaterials are used in the cosmetological products produced today. One of the main materials at the forefront of these is hydrogels. Hydrogels are a three-dimensional polymeric network structure that is held together by weak cohesive forces including cross-linked covalent bonds, ionic bonds, and hydrogen bonds. Hydrogels, which fall under the hydrophilic polymeric

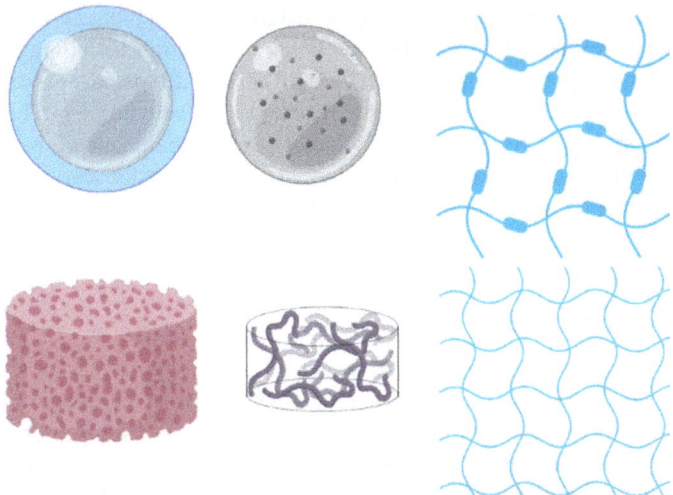

Fig. 7 The type of hydrogels (https://biorender.com/)

material class, show natural swelling ability in the water and are certainly suitable solvents, and their gel structures can reduce water by 10%. It actually functions as a humectant by keeping more than (Schueller et al. 1999).

Hydrogels can be applied to the skin as a result of a cosmetic formulation with other chemicals. The functions that can be realized by the emergence of such cosmetological products can be said as follows:

– It can provide water retention on the skin in multiple of its weight. This also applies to hyaluronic acid.
– Prevention of water loss occurs. In medicine, this problem is called transepidermal water loss.
– Provides a moisturizing effect as it acts as a humectant on the skin. Along with this effect, it functions in skin repair.
– -It can enable other biomolecules to be applied to the skin (ceramide, retinol derivatives, vitamin C derivatives, cannabidiol, etc.) to be better absorbed into the skin and have a greater effect (Seal et al. 2001).

The skin should not be perceived as only the face, neck, neck or feet, or hands, but it is known that these biomaterials are formulated to affect more difficult-to-reach skins such as the scalp. Since the absorption of high molecular size biomaterials by the body will be difficult, formulations with lower molecular size are being developed. The pH of the skin, its biochemical properties, the thickness of the dermis and epidermis layer, skin absorption, and non-toxicity are also the criteria that are not ignored while developing the formulation (Mitura et al. 2020).

Advances in regenerative skin tissue engineering have aided in skin repair as it works to provide faster wound healing. Scaffolding matrices such as porous, fibrous, microsphere, hydrogel, composite, and acellular scaffolds have been used in this field. These scaffolds are sometimes constructed from natural biomaterials including collagen and chitosan, and sometimes from synthetic materials including polycaprolactone (PCL) and polyethylene glycol (PEG). These scaffolds have several advantages that make them ideal for skin tissue regeneration, including High biocompatibility and high tensile strength. And by understanding the qualities, benefits, and drawbacks of different biomaterials and scaffolds new and better scaffolding will be applied to regenerate skin tissue (Lim et al. 2013).

12 Vascular Grafting

Cardiovascular devices have great importance for the treatment of many diseases, In the manufacture of artificial vascular grafts, the focus is mainly on the clotting factor and durability, and it has been proven that some materials such as metal, glass, ivory, silk, and nylon have failed to focus on these two factors. Thus, polyethylene terephthalate (PET, Da-cron) and polytetrafluoroethylene (ePTFE) are now used as large-diameter vascular grafts, as the interaction of these materials with blood and tissues is minimal. For vascular grafts with small diameters, the search for biomaterials is being stimulated by tissue engineering to create a new generation of vascular substitutes (Chaudhari et al. 2016).

Although polytetrafluoroethylene (PTFE) has been widely used as a vascular graft material, its bioinertness hinders endothelial cell attachment and endothelialization, limiting its use in small-diameter graft formation (Pashneh-Tala et al. 2016; Mi et al. 2018; Vijayan et al. 2020). Yu et al. (2021) studied modifying PTFE with a heparin-immobilized ECM coating, which is a viable way to functionalize biomaterials for producing small-diameter vascular grafts.

13 Prosthetic Meshes

Meshes are used to repair abdominal wall defects such as hernias. The ideal mesh has several characteristics as it must be biocompatible, able to resist mechanical infection and strains, sterilizable, and cost-effectively manufactured. When applying intra-abdominal support materials, polypropylene (PP), polyester (POL), and expanded polytetrafluoroethylene (ePTFE) are used. Polypropylene is more relied upon in hernia repair operations. Whereas ePTFE is mainly used in intra-abdominal sites as it is resistant to adhesion and its small structure reduces tissue growth. Sometimes collagen, polyglactin, polyglycolic acid, polydioxanone, hyaluronic acid, carboxymethyl cellulose, and even titanium are added to the base polymer materials to improve incorporation and temporarily change mechanical properties during the healing process. Biologically derived materials such as the dermis, pericardium, and intestines are also used, whether from cows, pigs, or humans. They also have other properties that differ from synthetic polymeric materials in terms of cost availability, and dimensional limitations and the mechanical properties of the biological materials are usually isotropic (Chirila and Harkin 2016).

Biomaterials are subject to certain properties that have great importance in ensuring that they do not cause undesirable reactions such as allergic reactions, rejection, or problems at the locus of the implant as a result, they should be biofunctional, bioactive, bioinert, and sterilizable (Ramakrishna et al. 2001).

14 Evolving Fast: Third Generation Biomaterials

First-generation biomaterials were used for medical implant applications in the 1960s and 70s. The main purpose of the use of this material was to minimize the toxicity in the host tissue and to ensure the biocompatibility of the applied treatment. In addition, there was an additional goal. Implant structures should not pose any mechanical or physical problems. The implant structures used had to resist corrosion in aqueous environments in a physical and mechanical balance, not to cause any carcinogenic or toxic side effects in living tissue, and to show appropriate mechanical properties without any problems. However, although first-generation biomaterials have been used for their purposes, second-generation biomaterials have entered our lives with the developing technology and new information (Hench et al. 2010).

Second-generation biomaterials were bioactive in other words, they try to affect physiological and cellular activities positively on health. Biomaterials have now gained a wider usage area as well as medical implants and this usage, did not pose any problem for dental and orthopedic treatments. Various materials such as bioactive ceramics, composites, and glasses have been developed for application in treatments. A type of material that can be tolerated by fracture fixation plates and screws for use in orthopedic surgeries can be briefly given as an example of second-generation biomaterials (Hench et al. 2002).

The third generation and most important class of biomaterials have been developed to specifically stimulate the cellular response. The first examples that come to mind are the third generation bioactive glasses and biocompatible foams that can be used for tissue regeneration (Hench et al. 1980).

Third generation biomaterials are making progress as medical and industrial technologies evolve. To give an example of this, the development of a scaffold with nanoscales that mimics the natural extracellular matrix of the host is still an ongoing project. The main goal of the researchers is the development and medical application of biomaterial-based artificial tissues with the same architectural structure as their natural counterparts. Therefore, biomaterials are of great importance for treatments. Prognostic methods are being developed for the use of affordable and innovative health services (Palakurthi et al. 2011) (Figs. 8 and 9).

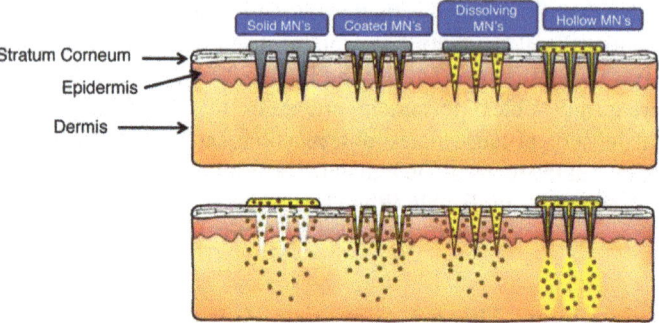

Fig. 8 The image shows schematic representations of different types of micro-needles (MANs) used for drug therapy (Fenton et al. 2018)

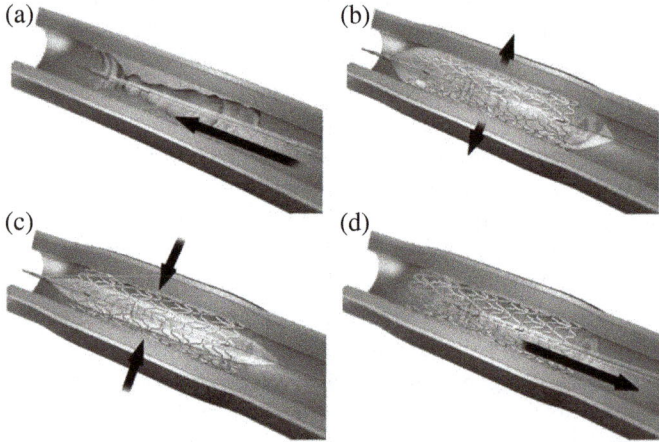

Fig. 9 Drug-eluting stents **a** The stent is attached to a catheter and placed in the diseased area. **b** The balloon that expands the stent is inflated. **c** The balloon is then deflated. After deflating, the drug-eluting stent is left as a scaffold. **d** The deflated catheter is removed, leaving the stent in the diseased area. He then releases the drug (Fenton et al. 2018)

15 Conclusion

The chapter emphasizes the wide range of uses for the aforementioned biomaterials. Hyaluronic acid, silk fibroin, chitosan, collagen, and other biomaterials have been successfully used in medical subfields. While there are a variety of connected benefits (such as biodegradability, biocompatibility, non-toxicity, nonantigenic, and amenability), the identified disadvantages of each unique polymer leave room for additional investigation.

References

Affatato S, Ruggiero A, Merola M (2015) Advanced biomaterials in hip joint arthroplasty. A review on polymer and ceramics composites as alternative bearings. Compos B Eng 83:276–283

Allen TM, Cullis PR (2004) Drug delivery systems: entering the mainstream. Science (New York, N.Y.), 303(5665), 1818–1822. https://doi.org/10.1126/science.1095833

An LL, Si CL, Wang GH, Sui WJ, Tao ZY (2019) Enhancing the solubility and antioxidant activity of high-molecular-weight lignin by moderate depolymerization via in situ ethanol/acid catalysis. Ind Crops Prod 128:177–185. https://doi.org/10.1016/j.indcrop.2018.11.009[5]

Anderson JM (2015) Exploiting the inflammatory response on biomaterials research and development. J Mater Sci: Mater Med 26:121. https://doi.org/10.1007/s10856-015-5423-5

Anselmo AC, Mitragotri S (2014) An overview of clinical and commercial impact of drug delivery systems. J Controlled Release Official J Controlled Release Soc 190:15–28. https://doi.org/10.1016/j.jconrel.2014.03.053

Bharadwaj A (2021) An overview on biomaterials and its applications in medical science. In: IOP conference series: materials science and engineering, Vol 1116, No 1. IOP Publishing, p 012178

Bhat S, Kumar A (2013) Biomaterials and Bioengineering Tomorrow's Healthcare. Biomatter 3(3):e24717

Birajdar MS, Joo H, Koh WG, Park H (2021) Natural bio-based monomers for biomedical applications: a review. Biomater Res. 2021;25(1):8. Published 2021 Apr 1. https://doi.org/10.1186/s40824-021-00208-8

Borandeh S, van Bochove B, Teotia A, Seppälä J (2021) Polymeric drug delivery systems by additive manufacturing. Adv Drug Deliv Rev 173:349–373. https://doi.org/10.1016/j.addr.2021.03.022 Epub 2021 Apr 6 PMID: 33831477

Carvalho JA, da Silva AA, Tedesco AC, Junior MB, Simioni AR (2019) Functionalized photosensitive gelatin nanoparticles for drug delivery application. J Biomater Sci Polym Ed 30(7):508–525

Chambre L, Martín-Moldes Z, Parker RN, Kaplan DL (2020) Bioengineered elastin- and silk-biomaterials for drug and gene delivery. Adv Drug Deliv Rev 160:186–198

Chaudhari AA, Vig K, Baganizi DR, Sahu R, Dixit S, Dennis V, Singh SR, Pillai SR (2016) Future prospects for scaffolding methods and biomaterials in skin tissue engineering: a review. Int J Mol Sci 17(12):1974. https://doi.org/10.3390/ijms17121974

Chen L et al (2019) Fabrication of the antibiotic-releasing gelatin/PMMA bone cement. Colloids and surfaces. B, Biointerfaces 183:110448. https://doi.org/10.1016/j.colsurfb.2019.110448

Cheng L, Deng B, Luo W, Nie S, Liu X, Yin Y, Liu S, Wu Z, Zhan P, Zhang L, Chen J (2020) pH-Responsive Lignin-based nanomicelles for oral drug delivery. J Agric Food Chem 68(18):5249–5258

Chirila TV, Harkin D (2016) Biomaterials and regenerative medicine in ophthalmology. Woodhead Publishing

Cobb WS, Peindl RM, Zerey M et al (2009) Mesh terminology 101. Hernia 13:1–6. https://doi.org/10.1007/s10029-008-0428-3

Davis JR (2003) Overview of biomaterials and their use in medical devices. Handbook of materials for medical devices, 1–11

Del Gaudio C, Crognale V, Serino G, Galloni P, Audenino A, Ribatti D, Morbiducci U (2017) Natural polymeric microspheres for modulated drug delivery. Mater Sci Eng C Mater Biol Appl 1(75):408–417

Dong H, Zheng L, Yu P, Jiang Q, Wu Y, Huang C, Yin B (2020) Characterization and application of lignin–carbohydrate complexes from lignocellulosic materials as antioxidants for scavenging in vitro and in vivo reactive oxygen species. ACS Sustain Chem Eng 8(1):256–266

Dos Santos V, Brandalise RN, Savaris M Biomaterials: Characteristics and properties. Topics in mining, metallurgy and materials engineering, 5–15. https://doi.org/10.1007/978-3-319-58607-6_2

D'Souza J, Camargo R, Yan N (2017) Biomass liquefaction and alkoxylation: a review of structural characterization methods for bio-based polyols. Polym Rev 57:668–694

Fang Y, Fan L, Bai H, Li B, Zhang H, Xin F, Ma J, Jiang M (2021) Bio-based molecules for biosynthesis of nano-metallic materials. Sheng wu Gong Cheng xue bao. Chinese J Biotechnol 37(2):541–560. https://doi.org/10.13345/j.cjb.200336

Fenton OS, Olafson KN, Pillai PS, Mitchell MJ, Langer R (2018) Advances in biomaterials for drug delivery. Advanced materials (Deerfield Beach, Fla.), e1705328. Advance online publication.https://doi.org/10.1002/adma.201705328

Groth T, Falck P, Miethke R-R (1995) Cytotoxicity of biomaterials—basic mechanisms and in vitro test methods: a review. Altern Lab Anim 23(6):790–799. https://doi.org/10.1177/026119299502 300609

Harrisson K (2007) Introduction to polymeric drug delivery systems. In: Jenkins MBT-BP (Ed), Biomed. Polym, Elsevier, pp 33–56. https://doi.org/10.1533/9781845693640.33

Hench LL (1980) Biomaterials. Science (New York, N.Y.), 208(4446), 826–831. https://doi.org/10. 1126/science.6246576

Hench LL, Polak JM (2002) Third-generation biomedical materials. Science (New York, N.Y.) 295(5557), 1014–1017. https://doi.org/10.1126/science.1067404

Hench LL, Thompson I (2010) Twenty-first century challenges for biomaterials. J Royal Soc Interface 7(Suppl) 4(Suppl 4):S379–S391. https://doi.org/10.1098/rsif.2010.0151.focus

Hildebrand H (2013) Biomaterials—a history of 7000 years. BioNanoMaterials 14(3–4):119–133. https://doi.org/10.1515/bnm-2013-0014

Hoare TR, Kohane DS (2008) Hydrogels in drug delivery: progress and challenges. Polymer 49(8):1993–2007

Huang Y-C, Huang Y-Y (2006) Biomaterials and strategies for nerve regeneration. Artif Organs 30:514–522. https://doi.org/10.1111/j.1525-1594.2006.00253.x

Hyun H, Park J, Willis K, Park JE, Lyle LT, Lee W, Yeo Y (2018) Surface modification of polymer nanoparticles with native albumin for enhancing drug delivery to solid tumors. Biomaterials 180:206–224

Kim SJ, Park SJ, Kim SI (2003) Swelling behavior of interpenetrating polymer network hydrogels composed of poly (vinyl alcohol) and chitosan. React Funct Polym 55(1):53–59

Kuo CK, Marturano JE, Tuan RS (2010) Novel strategies in tendon and ligament tissue engineering: Advanced biomaterials and regeneration motifs. BMC Sports Sci Med Rehabil 2:20. https://doi. org/10.1186/1758-2555-2-20

Kou Z, Wu Z, Tong KA, Holshouser B, Benson RR, Hu J, Haacke EM (2010) The role of advanced MR imaging findings as biomarkers of traumatic brain injury. J Head Trauma Rehabil (4):267–82. https://doi.org/10.1097/HTR.0b013e3181e54793

Langer R (1998) Drug delivery and targeting. Nature 392(6679 Suppl):5–10

Langer RS, Peppas NA (1981) Present and future applications of biomaterials in controlled drug delivery systems. Biomaterials 2(4):201–214

Lauto A, Mawad D, Foster LJR (2008) Adhesive biomaterials for tissue reconstruction. J Chem Technol Biotechnol 83:464–472. https://doi.org/10.1002/jctb.1771

Lee GH, Chang Y, Kim TJ (2014) 4-Characterization, Ultrasmall lanthanide oxide nanoparticules for biomedical imaging and therapy, pp 43–67. https://doi.org/10.1533/9780081000694.43

Liechty WB, Kryscio DR, Slaughter BV, Peppas NA (2010) Polymers for drug delivery systems. Annu Rev Chem Biomol Eng 1:149–173

Lim GT, Valente SA, Hart-Spicer CR, Evancho-Chapman MM, Puskas JE, Horne WI, Schmidt SP (2013) New biomaterial as a promising alternative to silicone breast implants. J Mech Behav Biomed Mater 21:47–56

Liu L, Fishman ML, Hicks KB, Kende M, Ruthel G. Pectin/zein beads for potential colon-specific drug delivery: synthesis and in vitro evaluation. Drug Deliv 13(6):417–23

Malekani J, Schmutz B, Gu YT, Schuetz M, Yarlagadda P (2011) Biomaterials in orthopedic bone plates: a review. In: Yarlagadda P (Ed) Proceedings of the annual international conference on materials science, metal and manufacturing. Global Science and Technology Forum, Singapore, 2011, pp 71–76

Marin E, Boschetto F, Pezzotti G (2020) Biomaterials and biocompatibility: an historical overview. J Biomed Mater Res 108:1617–1633. https://doi.org/10.1002/jbm.a.36930

Mater J, Chem B 6 (2018), pp 3475–3485; Vijayan VM, Tucker BS, Hwang PTJ, Bobba PS, Jun HW, Catledge SA, Vohra YK, Thomas V (2020) Non-equilibrium organosilane plasma polymerization for modulating the surface of PTFE towards potential blood contact applications. J Mater Chem B 8:2814–2825

Migonney V (2014) Biomaterials (Migonney/Biomaterials) ‖ History of Biomaterials, 1–10. https://doi.org/10.1002/9781119043553

Mi HY, Jing X, Thomsom JA, Turng LS, Promoting endothelial cell affinity and antithrombogenicity of polytetrafluoroethylene (PTFE) by mussel-inspired modification and RGD/heparin grafting

Miao S, Wang P, Su Z, Zhang S (2014) Vegetable-oil-based polymers as future polymeric biomaterials. Acta Biomater 10:1692–1704

Mitura S, Sionkowska A, Jaiswal A (2020) Biopolymers for hydrogels in cosmetics: review. Journal of materials science. Mater Med 31(6):50. https://doi.org/10.1007/s10856-020-06390-w

Mukherjee PK, Bahadur S, Chaudhary SK, Kar A, Mukherjee K (2015) Quality related safety issue-evidence-based validation of herbal medicine farm to pharma. In: Evidence-based validation of herbal medicine. Elsevier, pp 1–28

Muñoz-Bonilla A, Echeverria C, Sonseca Á, Arrieta MP, Fernández-García M (2019) Bio-based polymers with antimicrobial properties towards sustainable development. Materials (Basel) 12(4):641. Published 2019 Feb 20. https://doi.org/10.3390/ma12040641

Oliveira JM, Reis RL (2017) [Studies in Mechanobiology, Tissue Engineering and Biomaterials] Regenerative strategies for the treatment of knee joint disabilities 21‖Biomaterials as tendon and ligament substitutes: current developments. 8(Chapter 17), 349–371. https://doi.org/10.1007/978-3-319-44785-8_17

Palakurthi NK, Correa ZM, Augsburger JJ, Banerjee RK (2011) Toxicity of a biodegradable microneedle implant loaded with methotrexate as a sustained release device in normal rabbit eye: a pilot study. J Ocular Pharmacol Therapeutics: Official J Assoc Ocular Pharmacol Therapeutics 27(2):151–156. https://doi.org/10.1089/jop.2010.0037

Pashneh-Tala S, MacNeil S, Claeyssens F (2016) The tissue-engineered vascular graft-past, present, and future Tissue Eng. Part B-Rev 22:68–100

Peppas NA, Khare AR (1993) Preparation, structure and diffusional behavior of hydrogels in controlled release. Adv Drug Deliv Rev 11(1–2):1–35

Pikis S et al (2015) Potential neurotoxic effects of polymethylmethacrylate during cranioplasty. J Clin Neurosci Official J Neurosurg Soc Australasia 22(1):139–43

Ponrasu T, Chen BH, Chou TH, Wu JJ, Cheng YS (2021) Fast dissolving Electrospun nanofibers fabricated from jelly fig polysaccharide/pullulan for drug delivery applications. Polymers (basel) 13(2):241

Ramakrishna S, Mayer J, Wintermantel E, Leong KW (2001) Biomedical applications of polymer-composite materials: a review. Compos Sci Technol 61(9):1189–1224

Schueller R, Romanowski P (Eds) (1999) Conditioning agents for hair and skin. CRC Press

Si C, Jiayun X (2020) Recent advances in bio-medicinal and pharmaceutical applications of bio-based materials. Current Med Chem 27(28). https://doi.org/10.2174/092986732728200621 210700

Seal BL, Otero TC, Panitch A (2001) Polymeric biomaterials for tissue and organ regeneration. Mater Sci Eng R Rep 34(4–5):147–230

Seeli DS, Prabaharan M (2016) Guar gum succinate as a carrier for colon-specific drug delivery. Int J Biol Macromol 84:10–15

Si C, Xu J (2020) Recent advances in bio-medicinal and pharmaceutical applications of bio-based materials. Current Med Chem 27(28):4581–4583.https://doi.org/10.2174/092986732728200621 210700

Sung YK, Kim SW (2020) Recent advances in polymeric drug delivery systems. Biomater Res 24:12. https://doi.org/10.1186/s40824-020-00190-7

Sun Z, Song C, Wang C, Hu Y, Wu J (2020) Hydrogel-based controlled drug delivery for cancer treatment: a review. Mol Pharm 17(2):373–391. https://doi.org/10.1021/acs.molpharmaceut.9b01020

Thomas D, Mathew N, Nath MS (2021) Starch modified alginate nanoparticles for drug delivery application. Int J Biol Macromol 15(173):277–284

Tiwari G, Tiwari R, Bannerjee S, Bhati L, Pandey S, Pandey P, Sriwastawa B (2012) Drug delivery systems: an updated review. Int J Pharm Investig. https://doi.org/10.4103/2230-973x.96920

Vaishya R, Chauhan M, Vaish A (2013) Bone cement. J Clinical Orthopedics Trauma 4(4):157–163

Wang CE, Wei H, Tan N, Boydston AJ, Pun SH (2016) Sunflower polymers for folate-mediated drug delivery. Biomacromol 17(1):69–75. https://doi.org/10.1021/acs.biomac.5b01176

Wang B, Wang S, Zhang Q, Deng Y, Li X, Peng L, Zuo X, Piao M, Kuang X, Sheng S, Yu Y (2019) Recent advances in polymer-based drug delivery systems for local anesthetics. Acta Biomater 96:55–67. https://doi.org/10.1016/j.actbio.2019.05.044

Weiss M, Haufe J, Carus M, Brandão M, Bringezu S, Hermann B, Patel MK (2012) A review of the environmental impacts of biobased materials. J Ind Ecol 2012(16):S169–S181. https://doi.org/10.1111/j.1530-9290.2012.00468.x

Wen H, Jung H, Li X (2015) Drug delivery approaches in addressing clinical pharmacology-related issues: opportunities and challenges. AAPS J 17:1327–1340. https://doi.org/10.1208/s12248-015-9814-9

Wendels S, Avérous L (2020) Biobased polyurethanes for biomedical applications. Bioact Mater 6(4):1083–1106. Published 2020 Oct 15. https://doi.org/10.1016/j.bioactmat.2020.10.002

Williams D (1990) An introduction to medical and dental materials. Concise encyclopedia of medical & dental materials. Pergamon Press, Oxford and The MIT Press Cambridge

Williams D (2008) The relationship between biomaterials and nanotechnology. Biomaterials 29(12):1737–1738. https://doi.org/10.1016/j.biomaterials.2008.01.003

Wilson ME, Apple DJ, Bluestein EC, Wang XH (1994) Intraocular lenses for pediatric implantation: biomaterials, designs, and sizing. J Cataract Refract Surg 20(6):584–591

Xue L, Greisler HP (2003) Biomaterials in the development and future of vascular grafts. J Vasc Surg 37(2):472–480

Yalcin S (2019) Dextran-coated iron oxide nanoparticle for delivery of miR-29a to breast cancer cell line. Pharm Dev Technol 24(8):1032–1037

Yalcin S, Unsoy G, Mutlu P, Khodadust R, Gunduz U (2014) Polyhydroxybutyrate-coated magnetic nanoparticles for doxorubicin delivery: cytotoxic effect against doxorubicin-resistant breast cancer cell line. Am J Ther 21(6):453–461

Yang E et al (2018) Bio-based polymers for 3D printing of Bioscaffolds. Polymer reviews (Philadelphia, Pa.) 58(4):668–687. https://doi.org/10.1080/15583724.2018.1484761

Yermak IM, Gorbach VI, Glazunov VP, Kravchenko AO, Mishchenko NP, Pimenova EA, Davydova VN (2018) Liposomal form of the echinochrome-carrageenan complex. Mar Drugs 16(9):324

You-Xiong W, Robertson JL, Spillman WB, Claus RO (2004) Effects of the chemical structure and the surface properties of polymeric biomaterials on their biocompatibility 21(8):1362–1373. https://doi.org/10.1023/b:pham.0000036909.41843.1

Yu C, Yang H, Wang L, Thomson JA, Turng LS, Guan G (2021) Surface modification of poly-tetrafluoroethylene (PTFE) with a heparin-immobilized extracellular matrix (ECM) coating for small-diameter vascular grafts applications. Mater Sci Eng C Mater Biol Appl 128:112301

Yu J, Xu X, Yao F, Luo Z, Jin L, Xie B, ... Chen H (2014) In situ covalently cross-linked PEG hydrogel for ocular drug delivery applications. Int J Pharmaceutics 470(1–2):151–157

Zhang C, Garrison TF, Madbouly SA, Kessler MR (2017) Recent advances in vegetable oil-based polymers and their composites. Prog Polym Sci 71:91–143

Carbon Nanostructures, Nanomaterials and Energy Storage–A Critical Overview and the Visionary Future

Sukanchan Palit and Chaudhery Mustansar Hussain

1 Introduction

Carbon nanotubes, energy, electrical, electronics and environmental engineering applications are the needs of human civilization and human scientific progress today. Industrial wastewater treatment, drinking water treatment and management are the visionary aisles of science and engineering globally. Mankind's immense scientific and engineering prowess and scientific discernment are at a difficult situation as environmental protection concerns destroys the scientific conscience in the global scenario. Technology and engineering science of environmental remediation and water purification and separation science needs to be refurbished with the triumph of science globally. Application of carbon nanotubes in energy engineering and electrical engineering applications also needs to be reorganized as civilization treads forward. Provision of clean water and environmental and energy sustainability are the needs of civilization. Millions of people around the world are without pure drinking water and proper sanitation. The social situation of arsenic and heavy metal remediation needs to be envisioned in both developing and developed countries around the globe. Bangladesh and the state of West Bengal, India are fighting with a lesser vision against arsenic drinking water contamination. The challenges and limitations of arsenic groundwater remediation science are immense, scientifically inspiring and thought provoking. In this treatise, the author pointedly focuses on water resource management and wastewater management along with advances in nanotechnology

S. Palit (✉)
Department of Chemical Engineering, University of Petroleum and Energy Studies, Energy Acres, Post-Office-Bidholi via Premnagar, Dehradun, Uttarakhand 248007, India
e-mail: sukanchan68@gmail.com

C. M. Hussain (✉)
Department of Chemistry and Environmental Sciences, New Jersey Institute of Technology, University Heights, Newark, NJ 07102, USA
e-mail: chaudhery.m.hussain@njit.edu

© The Author(s), under exclusive license to Springer Nature Singapore Pte Ltd. 2023   35
A. K. Mishra and C. M. Hussain (eds.), *Biobased Materials*,
https://doi.org/10.1007/978-981-19-6024-6_3

applications. Carbon nanotubes and graphene are wonders of science and engineering and are the smart materials of tomorrow. Disinfection, decontamination, desalination and water reuse are the fundamental issues of global water science and technology. Here in this treatise the author pointedly focuses on the scientific success, the scientific vision and the scientific needs in the field of integrated wastewater management and integrated water resource management. Technological and engineering profundity in the field of industrial wastewater management needs to be refurbished and reorganized as civilization faces the ravaging environmental crisis of global climate change and frequent environmental disasters. This treatise widely pronounces the recent scientific advances in the field of environmental remediation and the applications of nanotechnology in mitigating environmental issues. A new dawn in the field of environmental engineering, chemical process engineering and energy engineering will surely emerge as man and mankind takes concerted efforts and research targets in global environmental protection science. Energy storage by carbon nanomaterials is a marvel of science today. Energy engineering and renewable energy are the needs of human civilization today. This treatise unfolds the greater vision of the alignment of energy engineering and nanotechnology/nanomaterials. Circular economy is the coinword of human scientific endeavors today. The author profoundly discusses these relevant scientific and engineering issues with minute details.

2   The Aim and Objective of This Study

The world of environmental engineering science, chemical process engineering, biotechnology, biological engineering and nanotechnology are today interlinked and interwoven with each other. The heart of chemical process engineering and the concepts of environmental engineering science are the needs of civilization and humankind today. Energy engineering and nanotechnology are the vast frontiers of the scientific landscape today. Industrial wastewater treatment and management are today aligned with the vast domain of nanotechnology, nano-engineering and applications of nanomaterials. The main aim and objective of this treatise is to elucidate the recent scientific and engineering advancements in the field of industrial wastewater treatment, drinking water treatment, groundwater remediation and the vast field of nanomaterials and engineered nanomaterials. Human factor engineering, industrial systems engineering, technology management and reliability engineering need to be addressed as global mitigation challenges in wastewater treatment, groundwater treatment and drinking water treatment surge forward. The author deeply elucidates the success of the science and engineering of nanotechnology and nanomaterials in the true scientific emancipation of integrated wastewater treatment and management. The sufferings of human civilization and humankind are tremendous today in developing and disadvantaged countries around the world. Sustainability whether it is social, economic, energy or environmental is in great need as civilization progresses forward. Today the challenges of conventional and non-conventional environmental engineering techniques are immense as scientific validation of environmental engineering and environmental science continues. Arsenic and heavy metal groundwater and drinking water contamination are monstrous issues in South Asia mainly in Bangladesh and the state of West Bengal, India. Entire South Asia is today plunged into this global engineering and technological issue. The face of human science, human technology and engineering science are all in a difficult situation globally. Countries around the world are gearing towards new innovations and new scientific profundity in applications of nanotechnology in environmental and water remediation. This treatise pronounces vehemently the needs and necessities of groundwater remediation and water purification science in the advancement of civilization. A new window of innovation and scientific affirmation will surely evolve in the search of scientific truth in environmental engineering science, energy engineering and nanotechnology. Man and mankind's immense knowledge prowess will then be surely envisioned. The application of carbon nanomaterials and carbon nanotubes in energy storage and energy engineering applications are the other visions of this well researched treatise. A new vista in the field of energy storage will emerge if the world of science and engineering takes concerted and affirmative steps and deep visionary direction in the field of carbon nanomaterials and energy storage. Man and mankind's immense scientific girth and determination will open new doors of scientific revelation in the field of both energy and environmental engineering. Carbon nanotubes and its applications in energy storage are the other cornerstones of this widely researched treatise. These are the salient features of this well researched treatise.

3 Industrialization, Urbanization and Application of Nanotechnology and Nano-engineering

Civilization, humankind, science and technology are in the avenues of newer scientific and technological rejuvenation and engineering steadfastness. The world of science and technology stands affirmative and envisioned as global climate change, global warming and frequent environmental catastrophes destroys the scientific landscape. Nanotechnology, nanomaterials and engineered nanomaterials are the wonders of science today. The applications of nanomaterials and engineered nanomaterials to the rapid development of human society, human scientific ingenuity and the vast domain of environmental protection are the utmost needs of the hour. Scientific provenance, deep scientific divination and vast scientific sagacity in the field of environmental engineering and nanotechnology will surely open new doors of innovation and instinct in decades to come. Rapid industrialization, steadfast urbanization and growing population are destroying the vast scientific and social fabric today. Industrialization and nano-science and nanotechnology are two opposite sides of the visionary coin. Technological vision in water purification science, drinking and groundwater treatment are in the phase of a new beginning. The triumph and ingenuity of science and technology in environmental remediation needs to be re-envisioned as civilization, man and mankind move forward. In this treatise, the author depicts profoundly the scientific needs, the scientific evolution and the scientific ingenuity in the application of nanotechnology and nanomaterials in industrial wastewater treatment and wastewater management. Globally, the shortage of pure drinking water is destroying the human habitat and public health engineering. Public health engineering in developed and developing nations around the world are today highly stressed and facing immense challenges. In a similar vision, basic, applied and fundamental sciences such as environmental engineering, chemical engineering and nanotechnology are on the path of newer scientific regeneration. Rapid industrialization and intense manufacturing scenario are in a similar vision challenging the human civilization. In such a situation, the ingenuity of applications of nanomaterials in pollution control whether it is air, solid or liquid will emerge as a harbinger of civilization. The author deeply ponders these intricate scientific issues. The world of energy science and engineering will surely be refurbished and science and technology will usher in a new era if positive directions in research are emancipated. The domain of integrated water resource management and integrated wastewater management needs to be reshaped and reinvented as global concerns for a circular economy surpasses one visionary boundary over another. Thus a newer generation in the field of nanotechnology and nano-engineering will emerge if they are aligned with the domain of circular economy.

4   What Do You Mean by Nanomaterials and Engineered Nanomaterials?

The triumph of science and engineering globally is unimaginable, visionary and far-reaching. Developed and developing nations around the world are in war-footing in providing basic human needs such as clean water, food, education, shelter, human habitat and the science of sustainability. Environmental and energy sustainability are the visionary vistas of science and technology globally. Thus also the need of the application of nanomaterials and engineered nanomaterials. Nanomaterials are chemical substances or materials that are highly manufactured and used at a very small scale. Nanomaterials are developed to exhibit novel characteristics compared to the same material without nanoscale features, such as increased strength, chemical reactivity or conductivity. ISO(2015) defines a nanomaterial as a material with any external dimension in the nanoscale (size range from approximately 1–100 nm) or having an internal structure or surface structure in the nanoscale. Nanomaterials in principle describe materials in which a single unit small sized (in at least one dimension) between 1 and 100 nm (the usual definition of nanoscale) measured in kelvin (Palit and Hussain 2018c; Palit 2018). Nanomaterials research takes a material science based approach to nanotechnology and has diverse applications in every branch of science and engineering. Engineered nanomaterials are chemical substances that are engineered with particle sizes between 1 and 100 nm in at least one dimension. It is well established that engineered nanomaterials derive many functional advantages from their unique and genuine chemical and physical properties. Graphenes, fullerenes and carbon nanotubes are examples of nanomaterials. The truth and abundance of science and engineering of nanomaterials research and nanoscale research needs to be envisioned and reorganized with the progress of global science and engineering. The novel properties of nanomaterials have resulted in tremendous interest in innovations across many industrial and commercial sectors. The technological and engineering revelation and the immense success of nanomaterials applications are slowly ushering in a new era in science and civilization. The immense scientific vision of nanomaterials, nanotechnology and nanoengineering will open newer vistas of knowledge management and scientific and engineering prowess in years to come.

5 The Scientific Doctrine of Industrial Wastewater Management and Integrated Water Resource Management

Industrial wastewater management and integrated water resource management are today aligned with diverse areas of applied science, fundamental science and engineering. Technological ardor and scientific fervor and validation in the field of water and wastewater treatment are today entering into a new phase in human civilization and the path towards human scientific progress. Technology management and integrated water resource management should be re-envisioned and aligned with each other in a global situation today. Loss of ecological biodiversity, ecosystem destruction and environmental catastrophes takes place when there is an immense stress on the environment and human habitat. Thus the need of a scientific doctrine of industrial wastewater management. Environmental or green sustainability is the other side of the visionary coin. Rigorous scientific advancements are the need of the hour as sustainable development in developed and developing countries around are at a state of immense distress. Industrial manufacturing, mass urbanization and rapid industrialization across the globe have veritably destroyed the scientific firmament and the scientific ardor. In this visionary treatise, the author deeply suggests the need of conventional and non-conventional environmental engineering tools. In the similar vision, nanotechnology and nanomaterials will surely be the forerunners towards a greater scientific emancipation and scientific and technological realization of environmental protection. Thus the need of a comprehensive treatise. The scientific doctrine of integrated water resource management will surely be implemented across all areas of human progress if Sustainable Development Goals are deeply implemented and successfully adhered.

6 Recent Scientific Advances in the Field of Carbon Nanotubes, Carbon Nanostructures and Energy Engineering

Carbon nanotubes and energy engineering are the coinwords of science and engineering today. A deep introspection is the utmost need of the hour. In this section, the author deeply unravels the recent advances in carbon nanotubes applications in energy storage and the vast world of energy engineering. Carbon nanostructures in energy applications are the other visionary areas of global research and development initiatives today. A newer genre will surely emerge in the distant scientific landscape.

Wang et al. (2015) deeply discussed with vision and cogent insight on multifunctional carbon nanostructures for advanced energy storage applications. Carbon nanostructures including graphenes, fullerenes etc. have found immense applications in a number of areas in alignment with a number of other materials. Technological validation and scientific verve and triumph are the needs of research pursuit in carbon nanotubes today. These multifunctional carbon nanostructures have recently attracted tremendous attention and interest for energy storage applications due to their large aspect ratios, specific surface areas and electrical conductivity (Wang et al. 2015). This well researched review (Wang et al. 2015) aims to report on the recent advances in energy storage applications involving multifunctional nanostructures. The challenges and limitations in energy storage applications are immense and far-reaching today. The advanced design and testing of multifunctional carbon nanostructures for energy storage applications specifically electrostatic capacitors, lithium batteries and fuel cells are elucidated in detail in this paper. Today sustainable energy from renewable sources such as wind, hydroelectric, geothermal, biological, nuclear and solar are in urgent demand globally. Carbon nanomaterials including Buckminster fullerenes, carbon nanotubes and graphene have attracted much attention in the last three decades. They have outstanding thermal, electrical, optical and mechanical properties. Thus the need of a detailed research and development initiative. Applications are sensors, photovoltaics, field emission transistors, fuel cells, composites, biomaterials and environmental protection. The authors in this paper (Wang et al. 2015) discussed in detail multifunctional carbon nanostructures for electrochemical capacitors, and multifunctional carbon nanostructures for lithium ion batteries. Also, multifunctional carbon nanostructures for fuel cells are the other cornerstones of this review paper. On the nanoscale, the physical and chemical properties of such multifunctional nanostructures are highly influenced by their structure and interfacial interactions with the surrounding nanomaterials. This well researched treatise redefines the applications of carbon nanostructures and opens up visionary aisles in the wide domain of nanotechnology (Wang et al. 2015).

Rai et al. (2011) discussed with vision and scientific far-sightedness modified carbon nanostructures for energy and display applications. Carbon nanostructures including fullerenes, nanotubes, nanocones, and graphene are gaining considerable attention for applications in memory based devices, electrical vehicles, and emergency power supplies. Advancements in science and engineering of carbon nanostructures are in the avenue of newer scientific regeneration. The scientific truth and success need to be envisioned at each step of academic rigor in the global scenario. Carbon nanostructures have been gaining considerable attention over the last three decades after the discovery of fullerenes in 1985 showing the possibility of stable curvatures of carbon firms (Rai et al. 2011). A new age of scientific regeneration is slowly unfolding in the field of carbon nanostructures. This paper with vision and vast scientific understanding elucidates the vast scientific intricacies of energy engineering applications of carbon nanostructures.

Renewable energy today is in the path of newer scientific ingenuity. Application of carbon nanostructures in wide areas of energy storage and energy engineering are the needs of science and engineering today. This entire chapter widely reviews the success of both carbon nanotubes and nanotechnology applications in energy and environmental engineering science. Renewable energy is today in the path of newer rejuvenation and there are today immense needs of carbon nanotubes, carbon nanostructures and graphenes. This article opens up newer visionary areas in the field of electrical engineering, electronics engineering and chemical process engineering. Also, areas of environmental engineering and science are unfolded as the integration of nanotechnology and diverse areas of applied science are envisioned.

7 Recent Scientific Advances in the Field of Industrial Wastewater Management and Water Purification and Separation

Scientific advancements in the field of industrial wastewater management are slowly entering a newer phase of might, divination and scientific determination. The challenges, the limitations, the prospects and the opportunities of industrial wastewater management and its treatments are vast and varied. In this treatise, the author with vision, scientific perseverance and scientific might is opening up new avenues of introspection in wastewater treatment and management. Integrated water resource management is today linked with vast application areas of nanotechnology. In the similar vein, wastewater management and nanomaterials are invariably linked with each other. A new dawn in human civilization is emerging as nations around the world confronts the wrath of climate change and ecological disaster.

Shannon et al. (2008) discussed with vision and scientific foresight science and technology for water purification in the coming decades. One of the most difficult and vexing problems afflicting people throughout the world is an inadequate supply of pure drinking water and proper sanitation. Human scientific rejuvenation in environmental engineering and water purification science are in a state of disaster (Shannon et al. 2008). Problems with water are expected to grow worse in the coming decades with water scarcity occurring globally even in regions considered to be water rich. In this paper (Shannon et al. 2008), the authors highlights some of the science and technology being developed to improve the disinfection and decontamination of water as well as to improve water supplies through desalination (Shannon et al. 2008). The many problems associated worldwide with the lack of clean and fresh water are widely known: 1.2 billion people lack access to safe drinking water, 2.6 billion have little or no sanitation, millions of people die annually and 3900 die in one day- from diseases transmitted through unsafe water and improper sanitation. Water strongly affects energy and food production, industrialization and the quality of the environment. Many freshwater aquifers are being contaminated and overdrawn in populous regions. A recent flurry of water research offers some hope in mitigating the impact of polluted water throughout the world. Civilization, science and engineering are in a state of immense distress and a burgeoning catastrophe (Shannon et al. 2008). The authors (Shannon et al. 2008) discussed in detail the areas of disinfection, decontamination, reuse and reclamation and desalination. Today there is an immense revolution in water purification science and drinking water treatment. Advancing the science of water purification will result in more innovations and scientific and engineering forays in water and wastewater treatment. This treatise widely opens up new arenas in the field of water purification science and drinking water treatment in decades to come. Developing nations around the world suffer a diversity of socio-economic-political-traditional constraints. Thus the need of a holistic sustainable development. Water purification will then be truly aligned with the science of green or environmental sustainability (Shannon et al. 2008).

Hamoda (1999) deeply elucidated with vision and scientific far-sightedness an integrated approach to wastewater management. Wastewater management focused for many years on the treatment and degradation of wastewater to protect the environment and human health. An integrated approach to wastewater management is today more extensive and vastly includes the cost of generation, collection and transport, treatment, disposal and reuse of wastewater in order to fulfill the guidelines of the global public health engineering scenario (Hamoda 1999). Circular economy and water treatment are the coinwords of scientific rejuvenation today. The role of public policy is increasing to establish goals and objectives of wastewater management. This paper (Hamoda 1999) outlines a systemic approach to be rigorously followed in wastewater management. This paper also examines the interrelationship between the functional areas of wastewater management and widely presents current trends in the generation, collection, treatment, disposal and reuse of water and wastewater in the Arab region (Hamoda 1999). Wastewater management is multi-dimensional. It embraces planning, design, construction, operation, maintenance, treatment, disposal and reuse systems. The author discusses major elements of wastewater management systems and the tasks associated with it (Hamoda 1999). The other hallmarks of this paper are wastewater management aspects, components of wastewater management systems, the functional elements and system analysis in waste generation. Today wastewater management programs must integrate the functional elements of the integrated waste management systems. Water and wastewater treatment are today in the vistas of new and novel environmental engineering science. The science and engineering of water and wastewater management according to this treatise will surely be the forerunners towards a new scientific order in environmental protection science (Hamoda 1999).

Shinde et al. (2009) discussed with vision, insight, foresight and scientific determination an integrated approach for wastewater treatment along with a focus on energy generation. Science, engineering and technology of wastewater treatment and management are today entering into a new visionary era (Shannon et al. 2008). Treatment of industrial and domestic wastewater poses many problems in the areas of treatment, cost, legislation, and adherence to pollution control norms set by governments and international organizations (Shannon et al. 2008). The authors discussed in minute details wastewater treatment technologies which are stabilization ponds, anaerobic ponds, facultative ponds, maturation ponds, aerated lagoons, and solar detoxification. In this paper (Shinde et al. 2009), the integrated wastewater treatment approach is elucidated in detail. A successful wastewater management and water resource management approach will generate biogas as renewable energy, reduce water, air and soil pollution, utilization of treated wastewater for algal cultivation, and reduce carbon dioxide emission (Shannon et al. 2008). Carbon sequestration is today's one visionary avenue of green or environmental sustainability. A deep scientific and technological introspection in wastewater management will veritably open new windows of innovation and scientific instinct in the wide arena of environmental engineering science (Shinde et al. 2009).

Integrated wastewater management and integrated water resource management are today aligned to each other. Futuristic vision of wastewater and water management are targeted towards greater realization of environmental protection science, energy engineering, petroleum engineering, fossil fuel science and chemical engineering science. In this entire treatise, the author deeply unfolds and unravels the doctrine of environmental engineering and industrial wastewater management and the linkages between them. A new era in the field of water purification science, water sustainability and environmental protection will surely emerge and science will veritably be enshrined in newer vistas.

8   Recent Scientific Pursuit in Nanomaterials and Water Remediation Science

Nanomaterials and engineered nanomaterials are the epitomes of science and engineering globally today. In this section, the author depicts profoundly the scientific needs of the application of nanomaterials in water purification, water sustainability, drinking water and industrial wastewater treatment. Today civilization, science and technology stands in the middle of deep scientific divination, scientific alacrity and an unending environmental crisis. The answers to these issues are research and development initiatives in the field of water purification science and environmental remediation. Man and mankind need to be envisioned and affirmed of global research and development initiatives in environmental engineering, chemical engineering, biological sciences, nano-biotechnology and biotechnology. In this section, the author deeply elucidates on the recent and significant research pursuit in the field of application areas of nanomaterials in water purification science. Civilization's scientific stance, revelation and profundity will surely open up new doors of innovation, invention and instinct in the vast world of environmental remediation.

Bhateria and Singh (2019) deeply elucidated with vast scientific determination nanotechnological applications of magnetic iron oxides for heavy metal removal. Heavy metal and arsenic removals are huge challenges and difficulties for scientists and engineers globally. Hamoda (1999) With the evergrowing increase in industrialization and mass manufacturing, heavy metals possess a great threat to the environment due to their discharge in water and industrial wastewater above permissible limits. The civilization and engineering science's immense scientific and academic rigor are facing vast difficulties and barriers today (Hamoda 1999). Environmental remediation and industrial pollution control are in the middle of immense scientific introspection and ingenuity. Heavy metals have serious toxic effects to human health, environment and public health engineering (Bhateria and Singh 2019). However new advancements in the budding area of nanotechnology provides immense assurance of mitigation. Development of cost-effective and novel 0D, 1D, 2D and 3D nanomaterials for environmental protection, industrial pollution detection and other vast and varied applications has attracted considerable attention. Scientific and engineering validation, verve and vision are the needs of civilization today. Zero valent iron and iron oxide particles are found to be the best candidates for heavy metal adsorption and removal (Bhateria and Singh 2019). Thus the need of a comprehensive treatise in this field. Various mechanical, optical and electrical properties of nanoparticles play a vital role in nanoparticle formation and vast interaction. Iron oxide nanoparticles (in a variety of chemical and structural forms) have already exhibited its vast diversity and immense potential in many frontier areas of environmental engineering. This review (Bhateria and Singh 2019) elucidates the application of iron and iron oxide nanoparticles in heavy metal removal in water and wastewater (Bhateria and Singh 2019). Heavy metals are metallic elements which have a relatively high density. Human scientific vision and vast scientific and engineering fortitude are the immediate needs of the hour. Heavy metals are capable of inducing toxicity even at low levels of human exposure. Due to heavy metal contamination of water and wastewater and the environment, there has been a considerable progress in ecology, public health and biodiversity (Bhateria and Singh 2019). In the last few decades, there has been an emerging progress in agricultural, industrial and urban activities leading to increased industrial pollution. Burning of fossil fuels, municipal wastes, fertilizers, mining and smelting of metallic ferrous ores, pesticides and sewage sludge are the primary sources of pollution whether it is air or water (Hamoda 1999). Metals such as arsenic, cadmium, chromium, lead and mercury are categorized under the prime concern metals to human and public health. Hence due to their high toxicity, they can induce multiple organ damage on exposure to also low concentrations. According to the United States Environmental Protection Agency and the International Agency for Research on Cancer, the metals like mercury, chromium, cadmium, arsenic and lead are highly categorized as carcinogens on the basis of scientific research pursuit (Bhateria and Singh 2019). The authors discussed in minute details nanomaterials as building blocks of nanotechnology, nanoparticles, iron nanoparticles, iron-oxide nanoparticles, synthesis of magnetic iron nanoparticles, and potential application of nanoparticles for heavy metal removal (Hamoda 1999). In this well researched treatise, the current progress of nanotechnology with a view on synthesis, characterization and applications of iron oxide nanoparticles has been vastly reviewed. Nanomaterials and engineered nanomaterials have great potential for the removal of recalcitrant contaminants due to their unique physical and chemical properties. In this review, the authors discusses with vision and purpose nanomaterials prioritization and its further application prospects. Iron oxide nanoparticles are found to be the efficient nanoadsorbents and their vast implications in heavy metal adsorption is one of the most promising applications. Application of nanomaterials in wastewater treatment is still in its early stages but needs to be envisioned as mankind moves forward. A new dawn in human scientific effort and scientific ingenuity will surely evolve if concerted efforts in environmental remediation are in the right direction (Bhateria and Singh 2019).

Yadav et al. (2017) deeply discussed with vision and lucidity nano-bioremediation technologies for environmental cleanup as a novel biological approach. Nanotechnology and biotechnology are today aligned with each other in the greater scientific realization and a larger emancipation of environmental protection. The most challenging task of the twenty-first century is to clean up the contaminants of the environment with the help of sustainable technologies. Nano-bioremediation is a new emerging technique for the remediation of pollutants using biosynthetic nanoparticles (Yadav et al. 2017). Civilization, science and engineering's immense stance, verve and validation in the field of nano-bioremediation are the utmost needs of the hour. The present review elucidates on the biosynthesis of nanoparticles from plants, bacteria, yeast and fungi which are surely emerging as nanofactories and potential applications in environmental cleanup (Shinde et al. 2009). The emergence of nanotechnology has been the subject of extensive research in recent decades by intersecting with various branches of science and technology and involving all forms of life on earth. This area is seen to be the next Industrial Revolution (Yadav et al. 2017). The authors (Yadav et al. 2017) discusses in detail nanoparticles, unique properties of nanoparticles, biogenic production of various nanoparticles, and the vast world of nanobioremediation. Soil and groundwater remediation with nanoparticles are the other main pillars of this well researched treatise. Nanoparticles can potentially be used for the remediation of soil and groundwater. In a visionary effect to combat the problem of water pollution and wastewater treatment, rapid and significant progress in wastewater treatment has been made which includes photocatalytic oxidation, adsorption, separation, disinfection, membrane processing and bioremediation. These unique properties of nanomaterials, for example, high reactivity and strong sorption are vastly explored for application in water/wastewater treatment based on their functions in unit operations in chemical engineering. Nanotechnology today has the immense potential to revolutionize existing technologies, particularly industrial pollution control. Futuristic vision and futuristic recommendations of the applications of nanoparticles in environmental and groundwater remediation are highly groundbreaking and overcome scientific frontiers. Today nanotechnology could provide scientific alternatives for environmental management and sustainable development without harming the environment. A newer vision and a newer scientific regeneration will surely evolve new vistas in scientific research pursuit in the field of both nanotechnology and bioremediation. The vast array of different opportunities, cross-disciplinary nature, vast potential for innovation and the potential impacts of nanotechnology leads to greater importance of nano-bioremediation. This treatise strongly focuses on the success of science, technology and engineering in these areas (Yadav et al. 2017).

Kumar et al. (2014) discussed with scientific vision, scientific foresight and scientific insight nanotechnology based water treatment technologies. The most important components of living beings on the earth are provision of pure drinking water and sustainability (Kumar et al. 2014). Global water scarcity is pervasive in water rich areas in the world as immense pressure has been created by the evergrowing population, industrialization, environmental changes and the loss of ecological biodiversity. This review highlights nanotechnology based water treatment technologies developed and used to improve desalination of sea and brackish water, disinfection and decontamination of water, nanophotocatalysis, and the vast domain of membrane separation processes. This review also delineates the fate and transport of engineered nanomaterials in water and wastewater treatment along with the risks associated with nanomaterials. Nanotechnology applications in water purification science are the cornerstones of scientific progress globally. The availability of clean fresh water has become a necessity of man and mankind. An estimated 1.2 billion people around the world do not have clean water to drink whereas 2.6 billion people do not have water for basic sanitation necessities (Kumar et al. 2014). The issue of public health engineering is in a state of immense scientific and social distress. Children under 5 years of age and elder groups are more prone to communicable diseases. Thus technology and engineering science are highly challenged (Bhateria and Singh 2019). Nanotechnology is the manipulation of matter at the nanometer scale to ultimately create novel structures, devices and systems. The authors (Kumar et al. 2014) discussed in minute details desalination, different energies and processes in desalination, membrane technologies for desalination, decontamination, disinfection, and the fate, transport and risks associated with engineered nanomaterials. Human health risk assessment is another cornerstone of this paper. All over the world, clean and safe water demand is rapidly rising with increased concerns and awareness about human health, public health engineering and the environment (Bhateria and Singh 2019). This paper deeply targets these scientific issues of water purification science (Kumar et al. 2014). Public health engineering, environmental engineering science and civil engineering are integrated to each other today. Thus a deep and broad scientific introspection and comprehension are the needs of the hour.

Science and technology in the global scenario are in the crucial rendezvous of introspection and futuristic vision. The challenges and the limitations of conventional and non-conventional environmental engineering tools are overcoming scientific frontiers. In this entire chapter, the prime focus is on the scientific success of nanotechnology and nanomaterials. A futuristic vision of risk assessment of nanomaterials stands as a huge pillar of this entire chapter. Application of advanced oxidation processes such as ozone-oxidation and photo-catalysis are today changing the face of human research pursuit. The author deeply envisions these areas of scientific truth and ingenuity.

9　Industrialization, Nanomaterials and the Visionary Road Ahead

Global advances in industrialization, urbanization and manufacturing are today in the path of new re-envisioning and regeneration. There is an unending disaster in environmental protection globally which is the provision of clean drinking water. Millions of people around the world are suffering terribly without clean drinking water and proper sanitation. Nanotechnology applications in water and wastewater treatment are the immediate needs of the hour. The challenges are huge and massive as civilization moves forward. Today there is a greater danger of lack of sustainable development in the global scenario. Human factor engineering, industrial systems engineering and reliability engineering need to be integrated with all domains of environmental engineering science. The pros and cons of industrial and manufacturing advances need to be readdressed with the passage of scientific vision, scientific determination and time. The limitations and the opportunities of nanotechnology applications in water and wastewater treatment needs also to be addressed with the progress of human vision. In this treatise, the author deeply unfolds the success of nanotechnology in mitigating civilization's environmental challenges. A new day will surely emerge in the field of nanotechnology, biotechnology and diverse areas of science and technology if proper concerted steps from scientists, engineers, civil society and the governments acts vociferously towards newer environmental engineering techniques. Nanomaterials and engineered nanomaterials are today aligned with the vast and varied domain of environmental protection. Industrialization and the application of nanomaterials will surely lead a long and visionary way in the true emancipation of science and technology globally (Wang et al. 2015; Yadav et al. 2017; Kumar et al. 2014; Hussain 2018; Palit et al. 2018; Palit and Hussain 2018b, 2018c; Palit 2018; www.wikipedia.com).

10　The Challenges and Difficulties of Arsenic and Heavy Metal Groundwater Remediation

The challenges of arsenic and heavy metal groundwater and drinking water contamination are today in the avenues of newer scientific reorganization and newer scientific revamping. Rapid industrialization and advances in industrial manufacturing are today destroying the social and scientific firmament globally. The world stands deeply appalled and mesmerized at the burgeoning issue of arsenic and heavy metal contamination in South Asia, particularly Bangladesh and India. Today the situations in Bangladesh and India are going out of control. The success and abundance of science and engineering of arsenic groundwater remediation need to be revamped as civilization moves forward. The world of science, engineering and human civilization stands in the middle of vast scientific introspection, scientific divination and scientific affirmation. Millions of people around the world are without pure water and proper sanitation. The situation in Bangladesh is absolutely alarming. Health issues due to contamination of drinking water by arsenic in South Asia needs to be reorganized and readdressed as mankind moves forward. Technological and engineering advancements in the global scenario are in the state of immense scientific regression and scientific catastrophe. The challenges and limitations of arsenic groundwater remediation are in the present day human civilization immense and groundbreaking. Thus the need of scientific and engineering instinct and innovations. In this treatise, the author deeply pronounces the scientific and technological needs of water purification science and the vast world of groundwater remediation. Application of fundamental science such as biological sciences, biotechnology, nano-biotechnology, biological engineering and applied geological sciences in water and wastewater treatment are the utmost need of the hour. A deep scientific introspection and contemplation in the application of nanotechnology in groundwater remediation will open new vistas of scientific endeavor (Palit and Hussain 2018a; Palit et al. 2018).

11 The Health Effects and Environmental Ethics of Nanomaterials Applications

Nanotechnology, nanomaterials and engineered nanomaterials have a severe health and safety issue. Public health engineering science in the application of nanomaterials in human society needs to be revamped and reinvented as civilization moves ahead. The industrial applications of nanomaterials also need to be reviewed as mankind tussles and deeply combats with environmental issues of nanomaterials. A new journey and a new scientific redeeming will surely enhance human's understanding in health and safety issues of nanomaterials and engineered nanomaterials. The success and forays of science and engineering will then be surely enhanced. Public health engineering and safety, health and environmental management today stands in the midst of deep scientific introspection and vast scientific vision. The health effects of nanomaterials applications in environmental protection needs to be reorganized and revamped as man and mankind are in the process of new regeneration (Yadav et al. 2017; Kumar et al. 2014; Hussain 2018; www.wikipedia.com; www.google.com).

12 Renewable Energy in the Global Scenario and Future of Human Civilization

Renewable energy scenario and energy engineering scenario in the global landscape stands in the middle of vision and ingenuity. The future of human civilization today depends on energy security and environmental integrity. Circular economy and green economy are the pillars of human scientific endeavors today. Recreate, regenerate and reuse are the coinwords of the circular economy today. Green sustainability and green nanotechnology are today aligned with each other. Energy security and renewable energy issues are plaguing the global scientific firmament today. Today human civilization is on the verge of environmental degradation and fossil fuel depletion. The needs of the United Nations Sustainable Development Goals are immense and far-reaching today. A new scientific regeneration and a new scientific beginning are in the process of emerging as environmental and energy sustainability needs to be implemented in every sphere of human progress. In this article, the author deeply describes the need of drinking water and industrial wastewater treatment in the proper scientific realization and proper engineering emancipation of arsenic and heavy metal groundwater remediation. The needs of integrated water resource management and wastewater management are vast and varied as humankind surpasses one visionary boundary over another. The future of human civilization will surely open a newer epoch in environmental engineering, petroleum engineering and chemical engineering as science and engineering progresses forward. The state of West Bengal, India and Bangladesh are in the difficult throes of an unending disaster of arsenic drinking water contamination. Thus the authors painstakingly discusses with vision and scientific and engineering fortitude these burning issues of energy and environmental engineering. Surely a new era in the field of renewable engineering will evolve as science moves forward.

13 Sustainable Development, Sustainable Engineering and the Visionary Road Ahead

Sustainability whether it is energy, economic or social are the absolute marvels and glories of human civilization today. Sustainable engineering and circular economy are the immense vision of today's human scientific progress. Energy engineering and renewable energy engineering are the newer aisles of science and technology today. Water sustainability and material sustainability are the coinwords of human scientific development. People, planet and profit are the pillars of sustainable development in the global scientific landscape today. Sustainable engineering and sustainable water and wastewater technology are today aligned to each other. These are the utmost needs of human progress today. The authors in this treatise deeply pronounces these fundamental areas of human scientific progress.

14 Nanostructures, Nano-engineering and the Scientific Sagacity of Human Progress

Nanostructures in energy engineering and renewable engineering will surely rule the vast scientific and engineering order in the future. The applications of nanomaterials and engineered nanomaterials in diverse areas of science and technology will veritably widen human scientific thoughts and scientific fortitude. Scientific benevolence, deep scientific knowledge and competence in the field of nanotechnology will usher in a new eon in the field of nanostructures. The scientific sagacity and the vast scientific doctrine in the field of nanotechnology and environmental engineering are the boons and marvels of global research and development initiatives. In this treatise, the authors deeply elucidates the scientific success and the vast scientific knowhow and profundity in the pursuit of nanomaterials and engineered nanomaterials in drinking water, groundwater and industrial wastewater treatment. Human civilization will surely be emboldened and restructured if academicians, researchers and policy makers across the world takes effective steps in combating global climate change and eco-biodiversity.

15 Future Flow of Scientific Thoughts and Futuristic Recommendations of This Comprehensive Study

Today nano-science, nanotechnology and nano-engineering are the harbingers of a new scientific order globally. Science and technology are veritably challenged as civilization surges forward towards a newer vision. Future scientific thoughts and future recommendations of the study are directed towards affordable and safety related nanomaterials application in water and wastewater management. Civilization, mankind and science are today in the process of a new beginning and newer scientific intellect. Engineering and scientific profundity of environmental protection science and industrial pollution control needs to be vehemently addressed with the mitigation of conundrums of science. Today safety, health and environmental management are the watchwords of science and civilization. In the areas of chemical process safety, a deep scientific and academic rigor are envisioned. Technology management applications in wastewater and water resource management will pave the newer avenues in environmental protection. The futuristic flow of scientific thoughts should follow sound academic rigor in environmental engineering and chemical process engineering. The world of science and engineering in water and wastewater treatment will surely usher in a new era if scientists, engineers and civil society are in the path of sound scientific rejuvenation and vision. Future of mankind lies in the hands of scientists and engineers. A multi-disciplinary approach involving environmental engineers, chemical engineers and geo-scientists will veritably open up a new order in environmental remediation science. Futuristic recommendations of this study should be targeted towards a new scientific and engineering beginning. Truly, water and wastewater treatment research and development initiatives globally will usher in a new visionary era. Millions of citizens around the world are without clean drinking water as concerns for arsenic and heavy metal groundwater contamination rises and science and mankind are highly challenged. Thus the need of a strong futuristic vision in water purification science and industrial pollution control. A concerted effort by scientists and engineers is the need of the hour. In the similar vein, energy and environmental sustainability are the success and marvels of science and need to be reorganized in the coming years. Strong and positive academic rigor will thus unabashedly open up new frontiers of research in water and wastewater treatment (Yadav et al. 2017; Kumar et al. 2014; Hussain 2018; Palit and Hussain 2018a, b, c) Energy storage and carbon nanostructures are the burgeoning areas of science and engineering today. Energy sustainability is the utmost need of global science and engineering today. The authors in this entire paper truly unfolds the success and triumph of engineering science and technology in vast and varied carbon nanostructures applications. Futuristic thoughts and futuristic recommendations are veritably directed in that visionary directions (www.wikipedia.com; www.google.com).

The scientific stance and the vast scientific profundity in the field of water sustainability and green sustainability need to be reinvented and unveiled as nano-science and nano-engineering move forward. Global concerns for wastewater management and water resource management also need to be restructured and the environmental engineering curriculum needs to be widened if water issues are mitigated. These are the salient features of this treatise.

16 Conclusion and Future Environmental Engineering Perspectives

Science, technology and engineering are today in the path of newer rejuvenation. Global warming, frequent environmental disasters and the loss of ecological biodiversity have urged scientists and engineers to gear forward towards new innovations and newer challenges. This treatise opens up newer thoughts and newer visions in the field of nanomaterials, nanotechnology and industrial wastewater management. Future environmental perspectives in the field of diverse areas of science and engineering are bright and groundbreaking. The world of nanotechnology and its applications are the wonders and marvels of science and engineering today. Today groundwater arsenic and heavy metal remediation and drinking water crisis are in the midst of scientific and engineering introspection. Human suffering and evergrowing poverty in developing and disadvantaged nations around the world are causes of immense concern. Scientific and technological challenges are immense in this century. Chemical process safety and risk management are the needs and scientific marvels of diverse areas of science and engineering. Today the situation is highly alarming as global climate change, depletion of fossil fuel resources and proliferation of nuclear weapons are destroying the vast scientific and engineering landscape. Thus the conundrums of science are absolutely appalling. Yet man and mankind will emerge successful if research and development initiatives in environmental protection move in the right direction. In this treatise, the author deeply elucidates the success and imagination of the vast world of nanotechnology and environmental engineering. Integrated wastewater management and integrated water resource management are the needs of civilization, science and engineering today. The author in this well researched chapter pointedly focuses on the application of nanomaterials and engineered nanomaterials in water and wastewater treatment. Today is a technology driven human society. The world of challenges and limitations in nanomaterials applications in water and wastewater treatment are immense and far-reaching. An interdisciplinary scientific approach in dealing with groundwater remediation and water purification will usher in a new era in the field of global science today. Integrated water resource management will also widen scientific thoughts and scientific ventures in decades to come. A new beginning and a newer scientific redeeming will change the global futuristic vision in water and wastewater treatment. Besides energy storage with the help of carbon nanostructures is a pillar of this paper. Today's technologically driven human society needs to be envisioned with respect to energy and environmental sustainability. A new age of scientific ingenuity will surely emerge as civilization moves forward. These vital issues are addressed in this treatise.

Success, innovation and inventions of environmental engineering are vast and varied today. Human habitat needs to be improved as United Nations Sustainable Development Goals are the objectives of human progress today. Integrated water resource management and wastewater management need to be integrated with Sustainable Development Goals. In the similar vision, nanotechnology, nanomaterials and engineered nanomaterials will surely widen scientific intellect, scientific perception and teaching and learning pedagogy in environmental engineering science in decades to come. Thus a newer global scientific and engineering initiative will emerge.

References

Bhateria R, Singh R (2019) A review on nanotechnological application of magnetic iron oxides for heavy metal removal. J Water Process Eng 31(2009):1–10

Hamoda MF (1999) An integrated approach to wastewater management. In: Fourth international water technology conference IWTC99. Alexandria, Egypt

Hussain CM (2018) Handbook of nanomaterials for industrial applications. Elsevier, Amsterdam, Netherlands

Kumar S, Ahlawat W, Bhanjana G, Heydarifard S, Nazhad MM, Dilbaghi N (2014) Nanotechnology-based water treatment strategies. J Nanosci Nanotechnol 14:1838–1858

Palit S (2018) Industrial vs food enzymes: application and future prospects, Chapter, In: Kuddus M (ed) Book–enzymes in food technology: Improvements and innovations. Springer, Singapore Pte. Ltd., Singapore, pp 319–345

Palit S, Hussain CM (2018a) Environmental management and sustainable development: a vision for the future, Chapter. In: Hussain CM (ed) Book–handbook of environmental materials management, Springer, Switzerland, pp 1–17

Palit S, Hussain CM (2018b) Remediation of industrial and automobile exhausts for environmental management, Chapter. In: Hussain CM (ed) Book–handbook of environmental materials management. Springer, Switzerland, pp 1–17

Palit S, Hussain CM (2018c) Sustainable biomedical waste management, Chapter. In Hussain CM (ed) Book–handbook of environmental materials management, Springer, Switzerland, pp 1–23

Palit S, Hussain CM (2018b) Nanomembranes for environment, Chapter. In: Hussain CM (ed) Book–handbook of environmental materials management. Springer, Switzerland, pp 1–24

Rai P, Pandey S, Arabale G, Nikolaev P, Arepalli S (2011) Modified carbon nanostructures for energy and display applications, In: 11th IEEE International conference on nanotechnology. Portland, Oregon, USA, pp 15–18

Shannon MA, Bohn PW, Elimelech M, Georgiadis JG, Marinas BJ, Mayes A (2008) Science and technology for water purification in the coming decades. Nature Publishing Group, London, United Kingdom, pp 301–310

Shinde GB, Vaidya RS, Govindarajan L, Raut NB (2009) Integrated approach for wastewater treatment–a focus on energy generation. In: Indo-Italian conference on emerging trends in waste management technologies. MAAER's MIT College of Engineering, Pune and Maharashtra Institute of Technology, Pune, pp 676–681

Wang Y, Wei H, Lu Y, Wei S, Wujcik EK, Guo Z (2015) Multifunctional carbon nanostructures for advanced energy storage applications. Nanomater 5:755–777. https://doi.org/10.3390/nano50 20755

Retrieved 2 March 2021, from www.google.com

Retrieved 2 March 2021, from www.wikipedia.com

Yadav KK, Singh JK, Gupta N, Kumar V (2017) A review of nanobioremediation technologies for environmental cleanup : a novel biological approach. J Mater Environ Sci 8(2):740–757

Bio-based Materials in Bioelectronics

Mayuri Kamble, Bhavna Kulsange, and Paresh H. Salame

1 Introduction to Bio-based Material and Bioelectronics

Bio-based materials are made up of substances obtained from biomass, and their organic characteristics indicate that they will degrade in the presence of natural organisms. They can be derived from natural sources or produced in the laboratory. Biomass is classified into three generations: (i) biomass from the first generation: this biomass is obtained from food crops such as sugar, starchy crops, vegetable oil, and animal fat. (ii) second-generation biomass includes wood, organic waste, food crop waste, and particular biomass crops. Most biomasses in this generation are classified as lignocellulosic; (iii) algae provide the third generation of biomass (Bardhan et al. 2015) 0.2 In an era where there is growing concern over the environment and green energy, bio-based materials are one of the most effective answers to the problem of environmental degradation. Bio-based materials are obtained or derived from various living or dead carbon sources, as well as other raw materials that have been processed through various processes such as biosynthesis, bioprocessing, and biological refining. Due to basic demands at the time, biomaterials have initially been unrecognized and the use of common metals was fairly adequate. Growing environmental concerns warranted that the materials possess not only desirable physical attributes such as tensile strength and flexibility but are also waste-free, efficient, and simple to construct. Thus, by recognizing nature as an efficient system, we direct our attention to bio-based materials. Materials derived from organisms are used virtually in every area of our daily life. Despite their easily modifiable structure, architecture, and physiochemical properties, synthetic polymeric biomaterials can be critical in a

M. Kamble · P. H. Salame (✉)
Department of Physics, Institute of Chemical Technology Mumbai, Mumbai, India
e-mail: ph.salame@ictmumbai.edu.in

B. Kulsange
Department of Zoology, RTM Nagpur University, Nagpur, India

variety of applications. These materials also present difficulties, as they are signif-icantly more complex to manufacture and rely substantially on fossil fuels such as coal and oil. As a result, there is a significant environmental impact, a detrimental effect on human health, and environmental degradation. In recent years, substan-tial research has been conducted on the production of bio-based materials using polyhydroxyalkanoates (PHA), alginate, and chitin. These sources can be used to create biocompatible, hypoallergenic, and non-immunogenic materials. Addition-ally, these materials can be employed as an extracellular matrix (ECM) and can be used to generate a variety of tissues, including cartilage, bone, and skin (Fig. 1).

The fields of biology, medicine, and security electronics have long held sway over the rest of the world's technological landscape. As a result of the convergence of biology and electronics, a new field of study known as Bioelectronics has emerged. With the use of an electronic component, researchers have created devices that might be used in various industries. The term "Bioelectronics" emerged due to its potential for combining life and electronics. One of the earliest investigations, which marked the beginning of the field of bioelectronics, was published in 1912 on the ECG (Elec-trocardiography) technology, which is based on measurements of body-generated electrical signals. The success and accuracy of this technique led to the develop-ment of many more devices, coupling electronics, and life sciences. Implantable electronic systems, such as the pacemakers, began to emerge in the 1960s with the discovery of transistors for stimulating different organs. Studies on electron transfer in electrochemical solutions in bio-based materials were also undertaken simultane-ously. These fields are now coming together to make it possible to record signals and stimulate cells at the cellular level. Bioelectronics can be used to solve problems in biology, medicine, and security by incorporating electronics. In addition to detecting

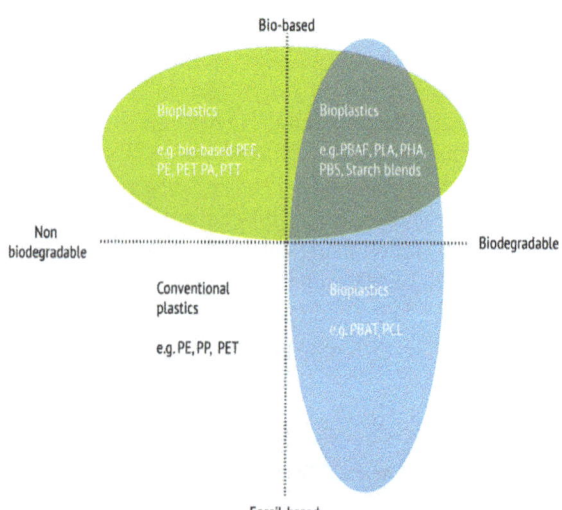

Fig. 1 Classification of Biobased polymer, bioplastics (after conrad et al. 2021)

and characterizing biological systems at the cell and sub-cellular level, the electrical system can also be used for other purposes. Understanding aberrant electrophysiological signals, as well as the identification of acceptable clinical symptoms, control of epileptic neuronal processes, and medications for a variety of complex neurological disorders, have been improved by bioelectronics technology. Despite these developments, ensuring biocompatibility, dependability, and stability will take time and effort. Mechanical incompatibilities between the biological tissue and materials employed in bioelectronic devices are blamed for these difficulties. The chemical and electrical properties of the gadgets and brain tissue are also significantly out of sync. Biosensors, biofuel cells (BFCs), and bioreactors are all examples of bioelectronic systems that can be used to generate new ideas.

2 The Interface of Bio-based Materials in Bioelectronics

It is possible to create medical devices that are both safe and cutting-edge through the use of bio-based materials integrated into bioelectronics. It has taken the researchers a long time to find ways to minimize or eliminate the environmental impact of dangerous elements. We can preserve the environment if we incorporate bio-based materials into our electrical equipment. Electronic transducers such as electrodes, field-effect transistors, and piezoelectric crystals can be used in conjunction with bio-based materials in the growing field of bioelectronics research. Customized bioelectronic devices are nothing without the electrical communication between the biomaterials and the corresponding transducers. The electrical characteristics of the biomaterial–transducer interface can be controlled by surface engineering of biomaterials like enzymes, antigen–antibodies, or DNA on electronic supports. This enables the electronic transmission of biorecognition events or biocatalytic change on transducers (Fig. 2).

3 Classification of Bio-based Materials

Bio-based material includes a wide variety of materials that are both derived from biomass and biodegradable content at the same time. However, not all bio-based materials are employed in the field of bioelectronics. Only a few materials require processing before they can be used in bioelectronics, yet, the applications for these materials are numerous in the field of bioelectronics. Here, we will discuss the synthesis of these bio-based materials and their application in Bioelectronics (Fig. 3).

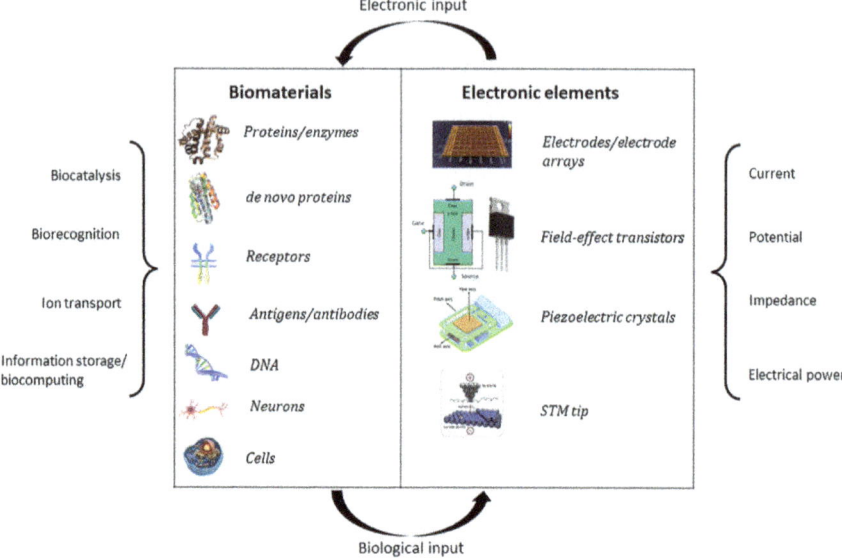

Fig. 2 Interface of biobased materials and electronic elements

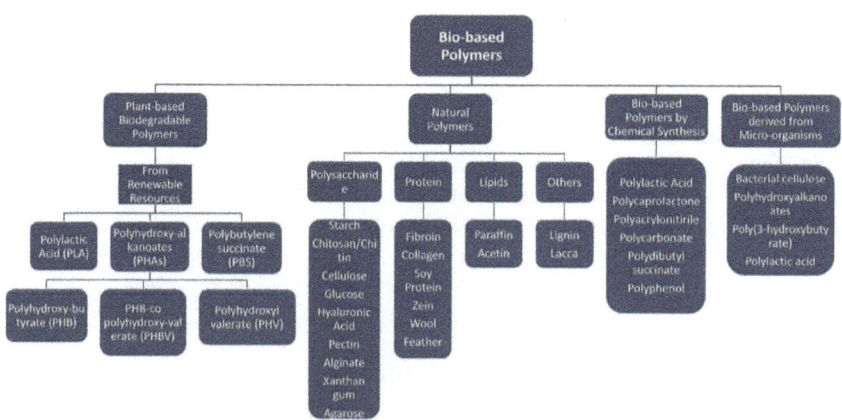

Fig. 3 Classification of Bio-based polymer

3.1 Bio-based Polymer: Synthesis and Application in Bioelectronics

3.1.1 Polylactic Acid (PLA)

PLA, or polylactic acid, is sometimes referred to as the "polymer of the twenty-first century" because of its potential in bioelectronics. PLA is derived from plants, is biocompatible, biodegradable, and successfully made on a large scale. Aliphatic bio-based polyester PLA is manufactured from 2-hydroxypropionic lactic acid (2-hydroxypropionic acid), which comes from animal or plant sources like cellulose and starch as well as fish waste and kitchen trash. PLA is an excellent substitute for traditional petroleum-based polymers because of its mechanical strength, biocompatibility, biodegradability, and, perhaps most importantly, its high compostability. Due to its superior thermal processability, PLA can be formed using a range of processes techniques such as melt extrusion, film casting, blow moulding, fibre spinning, etc.

I) Synthesis of PLA by microbial fermentation

In order to make PLA from biomass, three phases must be completed: fermentation, separation, and polymerization. For the synthesis of PLA using a ring-open polymerization process, lactide is used as a monomer, along with various catalysts and initiators. Also, various solvents such as chloroform, diphenyl ether, and toluene are used for synthesis. Ring-opening polymerization yields a polymer with variable molecular weights that may be tuned using a variety of process variables, including residence time, catalyst type, concentration, and temperature. Sequencing of the final polymer and the ratio of L- to D-lactic acid units are also programmable parameters. When it comes to biological applications like clippings and sutures, PLA is an excellent polymer because it decomposes mostly into water and CO_2 and is not carcinogenic or poisonous (Fig. 4).

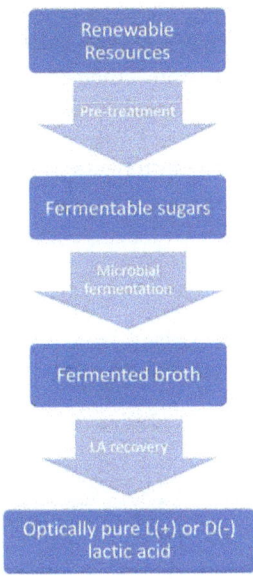

Fig. 4 Biosynthesis of PLA via microbial fermentation

II) Synthesis of L-lactic acid by a polycondensation reaction

There are three polylactic acid forms due to lactic acid's chirality, which include D- and L-type isomers, resulting in poly(L-lactic acid), poly(D-lactic acid), and poly(D, L-lactic acid). These can be synthesized via lactide production and azeotropic dehydrative condensation. The majority of commercially available high-molecular-weight PLA is produced via the lactide ring-opening polymerization method. While vacuum and temperature are steadily increased, lactic acid is directly polycondensed in bulk by distilling condensation water with or without a atalyst. The two reactions, such as esterification and depolymerization of PLA, result in various lactides. Direct polycondensation is a step-growth polymerization procedure in which water is eliminated as a byproduct. Direct polycondensation of LA yields only low-molecular-weight PLA because of the high viscosity of the reaction mixture and the low reaction rate, which is due to the unfavourable reaction equilibrium constant and the high reaction rate $(50,000 \text{ gmol}^{-1})$. High-molecular-weight PLA can be produced through direct polycondensation of LA or its oligomers. Several attempts have been made to tackle this challenge. Azeotropic polycondensation (AP) and solid-state polymerization (SSP) have been two of the most recent trends in polymer synthesis (SSP). Polycondensation, on the other hand, makes it impossible to regulate stereoregularity (Jain et al. 2019a).

I) Applications of bio-based Polylactic acid(PLA) in biomedicine (Fig. 5)

(i) Tissue engineering

Cellular scaffolds for tissue engineering are usually made from linear aliphatic polyesters such as PGA and PLA, as well as their copolymers (PLGA). When creating polymeric scaffolds, cell adherence to the polymer is an important consideration. The hydrophobicity of PLA and the absence of functional groups make it difficult for cells to adhere to it in culture. The slow hydrolytic breakdown is another drawback (Iwata

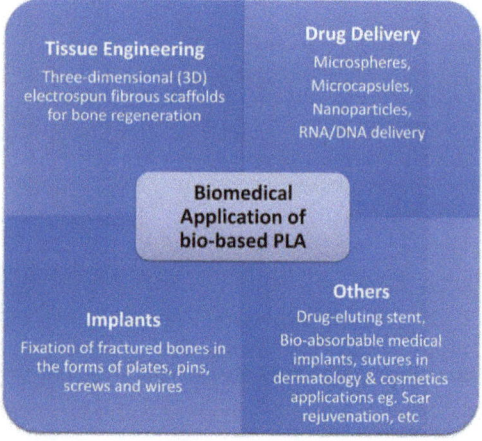

Fig. 5 Application of bio-based PLA in the biomedical field. (after Jain et al. 2019a)

and Doi 1998; Iwata et al. 2001). Techniques such as co-polymerization, mixing, and composite production were used to enhance PLA's mechanical capabilities. Various processes, such as electrospinning, particle leaching, and foaming have been used to examine PLA-polymer blends for the production of porous scaffolds. Recently, Nano-diamond (ND -ODA) functionalized PLA composites for tissue engineering were investigated. These ND-ODA composites were created by dissolving PLA in chloroform, which increased the mechanical characteristics of PLA when it was mixed with ND-ODA. These composites are used for tissue-engineering scaffolds. ND-ODA's biocompatibility, anti-thrombogenicity, and mechanical qualities match the mechanical properties of native blood vessels or tissues, further demonstrating their potential as an effective biomaterial. Printing porous scaffolds for bone tissue engineering via 3D printing is a unique technology. Also, electrospun fibrous scaffolds of PLA in 3D form have been proposed as viable tissue engineering tools in bone regeneration. 3D printed biodegradable scaffolds may be appropriate bone substitutes for clinical usage because of their quick vascularization and powerful osteoinduction bioactivity. PLA can also be mixed with other polymers to make tissue-specific scaffolds that show enhanced biodegradability and compatibility. Moreover, osteoblasts, osteoblast-like cells, and human umbilical vein endothelial cells were also able to grow, disseminate, and proliferate on PLA-printed discs.

(ii) **Drug delivery system (DDS)**

Biodegradable polyesters like PLA and PLGA [Poly(lactic-co-glycolic acid)] and liposomes, solid lipid nanoparticles, have all been studied as delivery techniques for biomedical uses. There have been advances in particle-based drug-delivery systems, such as micro-and nano-sized vesicles and polymerosomes, that can be used for therapeutics, diagnostics, and imaging. An alternative to the standard CremophorEL (now called KolliphorEL)-based Taxol® administration is nanoparticles of methoxy-poly(ethylene glycol)/PLA (mPEG/PLA). Mixture ratios can be varied to manage a wide range of features such as drug loading, particle size, and release patterns. If we need temporary fillings in the face/skin after reconstructive surgery, we can utilize PLLA injectable microspheres. PLLA microspheres have also been employed as an embolic material in transcatheter arterial embolization, which is an effective treatment for arterio-venous fistulas and malformations, extensive bleeding, and tumors. For a fiber-based transdermal delivery strategy, Siafaka et al. (2016) produced electrospun nonofibers of PLA/poly(butylene adipate) (PBA) blend. The effect of using EVA and PLA in combination is to develop paclitaxel-eluting stents in which the release rate and amount of drug may be controlled by adjusting PLA concentrations. An investigation of the adsorption of monoclonal antibodies (mAb) on the surface of PLGA nanoparticles was used to manufacture the nanoparticles.

(iii) **Bio-implant**

It is unnecessary to undergo a second surgery to remove biodegradable polymer implants after the problem site has been fixed. By controlling the L-LA/D-LA ratio (85:15)in the polymer, Haer et al. were able to increase PLA's mechanical capabilities (1998); this resulted in manufacturing screws and fixation plates for fracture fixation. These L-LA/D-LA plates were successful in healing fractures without the need for additional support. For orthopedic applications, the biocompatibility of four PLA films for the middle ear, ear canal, and nasal septum epithelial cells as well as fibroblasts and osteosarcoma cells was investigated using five cell types and found to be excellent. When a drug-releasing implant like a stent is made, the drug is either integrated into the matrix or adsorbed on the composite. Drug-eluting PLA/PCL nanofibers were created by inserting 5-fluorouracil into the polymer matrix. These nanofibers are one of a kind: PLA/PCL hybrid nanofibers (Gupta et al. 2014). Because of the synergism of two non-immiscible polymers, the release of carcinogenic medicines was larger in hybrid nanofibers than in pure polymers, suggesting that hybrid nanofibers could be used as therapeutic implants. Further, for reconstructive surgeries, such as ligament and tendon restoration and vascular and urological stents, long-term strength retention is required; for this PLA/PCL has been established as a worthy candidate.

3.1.2 Polyhydroxyl Acid (PHA)

PHAs are linear polyesters made up of hydroxyl acid monomers (HA) linked together by an ester linkage and assembled by various microorganisms as a carbon and energy reserve. PHAs have mechanical properties similar to those of petroleum-based polymers such as polyethylene and polypropylene. According to Chen and Patel (2012) at least 30% of soil-dwelling bacteria can synthesize PHA. PHA production potentials exist in the cell environment of bacteria in activated sludge, high seas, and severe settings. Both laboratory and industrial studies have shown that glucose fermentation can be used to synthesize polyhydroxyalkanoate (PHA). Hydroxylysates of hemicellulose, cellulose, and lignin can all be derived from plants. To synthesize PHA, pentoses can be employed as starting materials. PHAs can be completely degradable in three to nine months, which is one of their most attractive qualities. As a result of PHA depolymerization, several different types of bacteria and fungi can break down PHAs. The microbial population, the habitat, temperature, and the properties of the polymer to be destroyed all influence the rate of biodegradation (Adeleye et al. 2020).

I) Synthesis of PHA from sugar-rich feedstock/agricultural wastes

Alcaligenes latus, which produces up to 78% of the bioplastic monomer polybetahydroxybutyrate, uses maple sap as a carbon source (PHB). Sugar beet juice and A. latus were used together. The acquired data demonstrated a decreased PHA yield of 66% when compared to pure sucrose. For PHA production, soy molasses has shown an unexpected metamorphosis, with yields ranging from 5 to 17% PHA, due to Pseudomonas corrugata. A three-step fermentation process was used to produce PHA from cane molasses. An initial fermentation of molasses was followed by an accumulation of PHA and a batch fermentation to produce PHA, resulting in roughly 30% of cell dry weight (CDW) of PHA being produced. As a prelude to biopolymer-PHA manufacturing, sugarcane bagasse can be separated and utilized. Acid hydrolysis with 1% sulfuric acid at a reaction temperature 60 °C overnight was used to pretreat sugarcane bagasse, producing a low-furfural hemicellulose stream and a cellulose-lignin solid residue. By overcoming, neutralizing, filtering, and concentrating, the PHA fermentation feedstock was achieved. Xylose was isolated from soil samples, and bacteria were then evaluated for PHA synthesis from xylose and tolerance to inhibitors. To test the biorefinery concept, researchers looked into the generation of PHA from juice sugar sucrose, levulinic acid, and ethanol from enzyme-hydrolyzed cellulose hydrolysate (Adeleye et al. 2020).

(II)   Synthesis of PHA from waste oils

To make PHAs, the versatile organism Cupriavidus necator (formerly known as Ralstonia eutropha) was utilized. In small-scale batch fermentation studies, different concentrations of pure vegetable oil, heated vegetable oil, and waste frying oil were used. The results were promising. Rapeseed was used to make each of these oils. Rapeseed oil can be used by Cupriavidus necator to make the homopolymer polyhydroxybutyrate (PHB). Waste frying oil had a PHB concentration of 1.2 g/l, which was comparable to the glucose concentration. Pure oil and heating oil had PHB values of 0.62 and 0.9 g/l. However, due to rising costs, edible oils are no longer considered cost-effective for PHA manufacturing. Because of this, non-edible and waste cooking oils take precedence (Adeleye et al. 2020).

III)   Biosynthetic mechanism of PHA

The biosynthetic mechanism of PHA consists of three pathways(route) viz. (i) Pathway 1—The Acetyl-CoA to 3-hydroxybutyryl-CoA Pathway, (ii) Pathway 2—The β-Oxidation Pathway, (iii) Pathway 3—The in situ fatty acid synthesis pathway (Fig. 6).

(i)   Pathway 1—The Acetyl-CoA to 3-hydroxybutyryl-CoA Pathway

Synthetic PHA is made from precursors such as acetyl-CoA which is produced by breaking down sugars, fats, or amino acids. Condensation of two acetyl-CoA molecules into one acetoacetyl is the first step in the synthesis of P(3HB) by the',-ketoacylCoA thiolase (acetyl-CoA acetyltransferase; catalyses the first step in the creation of P(3HB). CoA. T. Adeleye and colleagues determined that acetoacetyl-CoA reductase is responsible for the second phase, in which acetoacetyl-CoA is converted to 3-hydroxybutyryl-CoA. PHB is produced by polymerizing 3-hydroxybutyryl-CoA monomers with P(3HB) polymerase (PHB synthase). PHA synthase (PhaCSCL) Cupriavidus necator is an example of this route, with a specificity for C_3–C_5.

(i)   Pathway 2—The β-Oxidation Pathway

The oxidation of fatty acids serves as a step in this PHA production process. Enoyl-CoA is the initial product of the oxidation of fatty acids. To produce MCL PHA polymerization, the enzyme R3-hydroxyacylCoA hydratase transforms the precursor

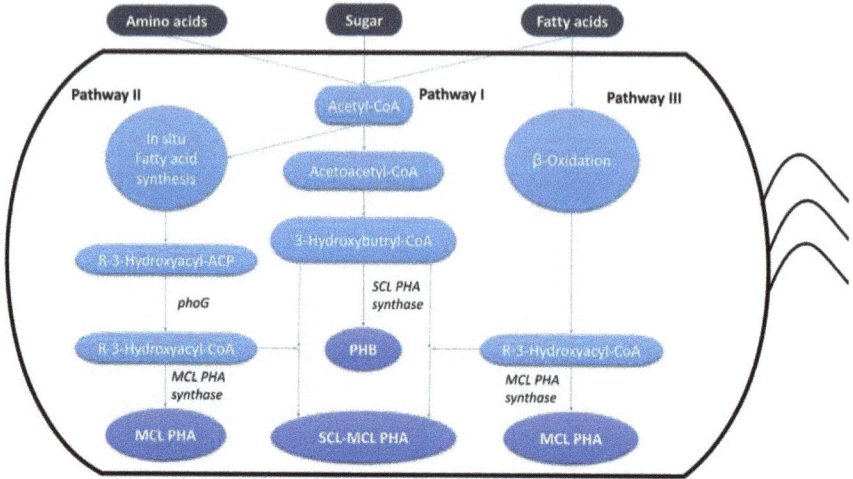

Fig. 6 Schematic representation of three pathways of the biosynthetic mechanism of PHA (after Chen et al. 2015)

enoyl-CoA to R-3-hydroxyacylCoA. The final stage of R-3-hydroxyacylCoA polymerization is carried out by MCL PHA synthase (PhaCMCL). Pseudomonas putida, Pseudomonas, P. oleovorans, and P.seudomonas aeruginosa are all part of this process.

(ii) Pathway 3—The in situ fatty acid synthesis pathway

For PHA production, R-3-hydroxyacyl-ACP must be synthesized in situ from fatty acids. The major enzyme that transforms 3-hydroxyacyl-ACP (acyl carrier protein) to 3-hydroxyacyl-CoA in this route is 3-hydroxyacyl-CoA transferase (PhaG). Metabolic engineering of MCL PHA production using PhaG led to the development of several custom biopolyesters. It is possible for the bacteria to simultaneously produce PHA precursors through the synthesis of fatty acids and—oxidation cycles.

I) **Applications of Bio-based Poly hydroxyl acid (PHA) in biomedicine**

(i) **Tissue Engineering**

Using a photochemical process, Stereolithography (SLA)technology is one of the 3D printing processes that may create components or products. The photocurable resin is polymerized using a UV laser to crosslink the chemical monomers. A major factor in their suitability for both hard and soft tissues is the wide variety of mechanical qualities that PHAs possess. Because they are highly biodegradable, they could be used to make bio-based products, including heart valves. Printing biomedical devices and tissue engineering scaffolds using SLA technology is a promising and precise method. Anatomical structures can be modelled with computed tomography scanning (CT-scan) or magnetic resonance imaging (MRI). The produced structures can be tested before surgery with this technology in orthopedic and reconstructive surgery (Mehrpouya et al. 2021).

(ii) **Bone scaffold**

A hybrid of additive manufacturing and wet spinning, CAWS (computer-aided wet-spinning) provides for fine control of fibre deposition through the use of layer-by-layer technology as well as 3D production of an external structure or internal architecture. Scaffolds are made of Poly(3-hydroxybutyrate-co-3-hydroxyhexanoate) (PHBHHx) Considering that this form encourages the correct cell attachment into the body, it is an excellent design for bone regeneration situations. CAWS has recently been utilized to build very small diameter stents made up of the PHBHHx biodegradable polymer using a spinning mandrel, which is a new development in the method (Mehrpouya et al. 2021).

(iii) **Bio-Implants**

While using polymers and natural fibres, PHAs are often mixed with them to create printing filaments with enhanced characteristics. Porous scaffolds with detailed designs may be produced using the Selective Laser Sintering (SLS) printing technology, which has a great deal of promise. A blend of PHAs can also benefit from this procedure, especially if they have been designed. Implants made of polyhydroxyapatite (PHA) are more stable and have a lower pH upon implantation. Polylactic acid (PLA) is more biodegradable than other therapeutically utilized polymers, including PLA, polyglycolide (PGA), and polylactic-co-glycolic acid (PLGA). Therefore, they're a great pick, and immune systems don't have problems with them. Therefore, 3D printing is an excellent and speedy solution for cells and systems (Mehrpouya et al. 2021).

3.1.3 Polyhydroxybutyrate (PHB)

Polyhydroxybutyrate (PHB) has a high stereo-regularity and crystallizes from the melt at high temperatures, which restricts its use in various applications. Due to its high crystallinity, PHB is brittle and can be broken easily. High processing temperatures are required because polymer breakdown and instability can occur at high crystallization temperatures. The thermomechanical performance of natural fibre reinforced poly(hydroxybutyrateco-valerate) PHBV biocomposites was good, allowing them to be used in semi-structural settings. Semi-structural adhesives have much less strength and stiffness than structural adhesives. It is possible to link a wider range of materials with these adhesives than with structural adhesives because they don't require heat to harden. Commercialization of composite polymers such as PHB,,PHBHHx, etc. as well as medium-chain length (MCL) PHAs, has been a success. In the case of polyhydroxybutyrate (PHB), examples include poly[(R)-3-hydroxybutyrate](P3HB), prolyl 4-hydroxylase (P4HB), and polyhydroxyvalerate (PHV). Conventional polymerization can yield homopolymers, random copolymers, and block copolymers of PHA depending on the bacterium species utilized and the growing circumstances. Many industries, including materials, fermentation, energy, and medicine, have made use of PHAs and their derivatives. Other bacteria that make PHB include Bacillus megaterium, Methylobacterium rhodesianum, Alcaligenes eutrophus, M. extorquens, and P. putida, as well as Escherichia coli (Rajan et al. 2018).

I) Biosynthesis of Polyhydroxybutyrate (PHB)

The shortage of critical nutrients like nitrogen, sulphate, potassium, magnesium, or phosphate can cause an increase in the creation of PHB in bacteria. Instead of synthesizing amino acids during the process, microorganisms create PHB and store it as distinct granules in the presence of abundant carbon. Other carbon sources like activated sludge and activated starch can be utilized to boost the PHB yield. The adoption of low-cost substrates will lead to lower production costs in the long run. One of the best microorganisms for collecting PHB is the gram-negative bacteria Alcaligenes eutrophus, which can collect up to 80% of its dry weight in PHB (Rajan et al. 2018).

I) Application of Polyhydroxybutyrate (PHB) in Bioelectronics

(i) Tissue engineering

There is a huge surface area, high porosity, biodegradability, cytocompatibility, and appreciability in the composite nanofiber membranes. Electrospinning was employed to fabricate a composite nanofiber scaffold from PHB, and carboxyl multiwalled carbon nanotubes grafted PHB. Carboxyl and hydroxyl groups of PHB were used in the condensation processes to attach carbon nanotubes. As a tissue engineering scaffold, a composite nanofiber membrane appears promising (Jirage et al. 2013) (Table 1).

(ii) Drug delivery system

The study's premise is that the degradation of PHB produces the compound 3-hydroxybutyric acid, which is present in humans, and that PHB-based systems have no cytotoxicity in general. Because of its low compatibility with medicinal agents, limited encapsulation effects, and poor drug delivery rates, the microbial PHB's high hydrophobicity and crystallinity restrict its application as a biomedical device. Since PHB has been chemically modified and synthesized in a variety of ways, its application as a drug delivery mechanism has increased (Jirage et al. 2013).

Table 1 PHB based blend polymer and their application in bioelectronics

PHB-based blend	Application
PHB/Chitosan	Drug delivery system (DDS)
PHB/polyethylene glycol	Tissue engineering
PHBV/Polylactic acid	Biomedical devices

3.1.4   Polybutylene succinate (PBS)

A)   Biosynthesis of Polybutylene succinate (PBS)

(i)   Synthesis of PBS by polycondensing succinic acid (or dimethyl succinate)

It is possible to make PBS by polycondensing succinic acid (or dimethyl succinate) and 1,4-butanediol, using either fossil or renewable monomers as starting materials. This technique of synthesis has the advantage of improving the thermal and mechanical properties, as well as thermoplastic processability. It is the most common process for manufacturing petrochemical succinic acid from maleic acid or its anhydride. Maleic acid is generated when one of the carbon–oxygen links is disrupted. In addition, the presence of hydrogen completes the process to form succinic acid by breaking the carbon–carbon double bond. The fermentation process can also result in the production of succinic acid. Various microorganisms have been examined and tested to make bio-based PBS since microorganisms are utilized in the biotechnological process of making succinic acid. Succinic acid is hydrogenated to 1,4-butanediol, which is then used to make PBS, as shown in this experiment. Using renewable resources rather than petroleum can be more expensive to generate succinic acid. Several bacteria, notably Anaerobiospirillum succiniciproducens, Actinogenes succinogenes, and Mannheimia succiniciciproducens, have been employed in a recent study to produce succinic acid. Organometallic catalysts were used at high reaction temperatures (190 °C) by Showa Highpolymer Co., Ltd. (Tokyo, Japan) to synthesize PBS, which resulted in a high molecular weight and complex removal of leftover metals or components due to solid metal–ester interactions. In reality, high-temperature reactions with organometallic compounds may be a factor in the degradation of monomers and polymers. Discoloration and reduced molecular weight would follow from this. PBS was synthesized with the help of 1,3-dichloro1,1,3,3-tetrabutyldistannoxane, a catalyst.

(ii)   Two-step synthesis of Polybutylene succinate (PBS)

Succinic acid and BDO are esterified to generate oligomers as the initial step in the production process. PBS with a higher molecular weight is created by oligomer polycondensation in the second step of this process. The PBS manufacturing flow diagram is depicted in Fig. 7. An inert gas intake, a mechanical stirrer, and a distillation column are commonly used in a reactor for the synthesis of the ester. Esterification can begin with stirring in this reactor at 160–190 °C and continued in a controlled atmosphere. Polycondensation begins as soon as no more water (or alcohol) is distilled out under normal pressure.

For the production of high-molecular-weight aliphatic polyesters, the simultaneous processes of condensation and degradation makes traditional polycondensation challenging. Because of this, in order to create polyesters with relevant mechanical characteristics, aromatic units, and chain extenders to raise the molecular weight or catalysts to speed up the kinetics are necessary. Both petrochemical and fermentation processes are used to produce benzoic acid. Catalytic hydrogenation of maleic acid

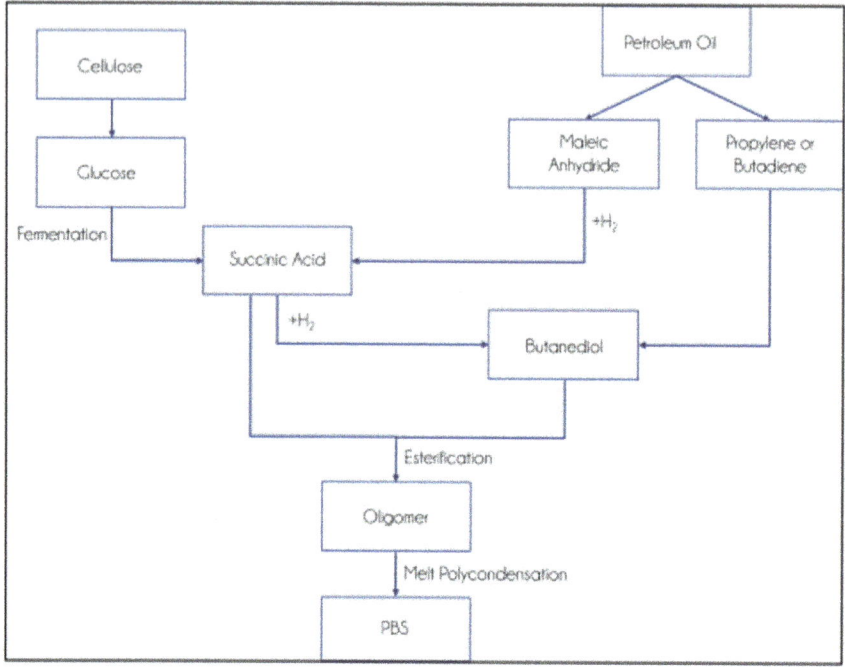

Fig. 7 Flow chart of two-step synthesis of PBS (after Xu et al. 2010)

and electrochemical synthesis of maleic anhydride in a bipolar membrane cell or a membrane-free cell has been used in the industrial production of petroleum-based SA over the past century. Fermentation produces benzoic acid because anaerobic and pro-fermentative microbes prefer it as an intermediary in their metabolism.

I) **Application of PBS in Bioelectronics**

(i) **Tissue engineering**

Bone marrow stem cells can be generated at a high rate by PBS. Compared to PLA and polyvinyl chloride, the results of PBS indicated a stronger tendency to (PVC). PBS can be employed in bone tissue engineering to induce new tissue growth, but its osteoblast compatibility and bioactivity are insufficient. During the process of creating a new tissue, physiological conditions have an impact on cell development, tissue regeneration, and the host reaction. As a biomaterial for tissue repair and tissue engineering, the rate of deterioration indicated its potential. PBS has been combined with inorganic materials by researchers in an effort to improve mechanical and thermal properties as well as spherulite size and gas permeability, all of which are critical for tissue engineering development. Along with improving cellular connections, such as selective endocytosis, adhesion, and orientation in injured tissue, it also aids in stimulating the tissue. To boost antibacterial and anticancer action, protein absorption, and cell proliferation, chitosan was added to pure PBS, rather than PBS on its own (Rafiqah et al. 2021).

(ii) **Bone Scaffold**

For tissue engineering research, a high compressive strength is required along with the ability to allow cell attachment, proliferation, and extracellular matrix (ECM) deposition, all of which lead to in-vivo bone regeneration with a suitable degradation rate. Because PBS is susceptible to bacterial infection and poor osteocompatibility after being implanted into the body, the development of tissue engineering from PBS confronts some hurdles. The use of a PBS copolymer in the creation of scaffolds to improve bone regeneration in dental sockets has also been the subject of extensive research. In addition to allowing bone cells to enter the construct, this substance also helps to keep them attached to it (Rafiqah et al. 2021).

Other than the previously noted polymers, PANI, Polypyrrole, and PEDOT: PSS are examples of synthetic polymers that can also be utilized in bioelectronics. As a result of the interaction of these polymers, bioelectronic devices that can be used to treat injuries as well as a biosensor are made possible.

3.2 Applications of Synthetic Bio-based Polymer in Bioelectronics

3.2.1 PEDOT: PSS

PEDOT: PSS, a conductive polymer, became the major candidate in the field of bioelectronics because of its transparency, conductivity, stability, and biostability. Bioelectronics as a biosensor relies on a precise contact between the conducting polymer and body tissue and the molecule. Numerous PEDOT functional group derivatives are available to improve the biocompatibility. Despite the fact that the structural composition of PEDOT has improved, the material's low fracture resistance and lack of attachment to the biological tissue mean that it cannot be used in bioelectronics. Ion gel and conductive hydrogel are key applications of this polymer because they improve tissue engineering by increasing the mechanical flexibility and molecular transport while also increasing hydration. Biosensors and medication delivery can both benefit from this polymer matrix. Polysaccharides (guar gum) have been combined with a PEDOT with an ionic liquid to promote mechanical flexibility. A flexible and water-resistant graphene-based electrode, a self-healing electronic implant, and other bioelectronics like Polyethylene glycol (PEDOT: PSS) are added to the mixture (Sadki and Chevrot 2003; Wei et al. 2015; Green et al. 2010). Fig. 8 shows the other applications of PEDOT:PSS in bioelectronics.

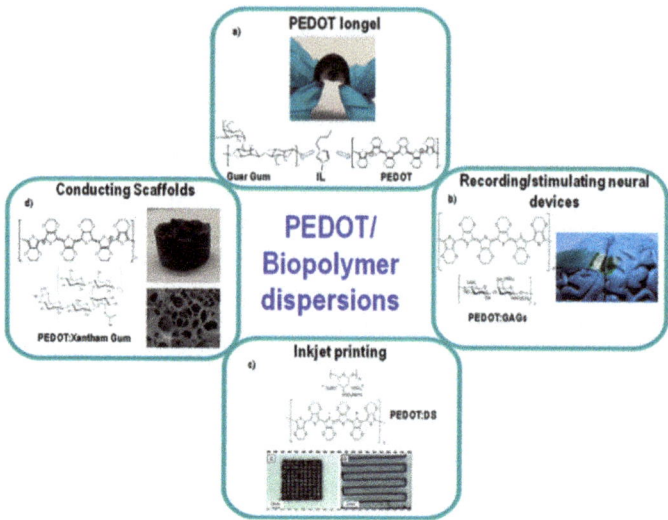

Fig. 8 Different applications of PEDOT biopolymer dispersion **a** PEDOT ion gel, **b** recording/simulating device, **c** inject printing, **d** preparation of scaffold. (after Mantione et al. 2017)

3.2.2 Polypyrrole (PPy)

In addition to their excellent biocompatibility, conducting polymers such as polypyrrole are also known for their versatility in a variety of applications such as polymer batteries, biosensors, actuators, and tissue engineering due to their ease of synthesis and processing with minimal modification. The use of polypyrrole in tissue engineering necessitates the modification of biological molecules in order to achieve a good contact with a tissue biomolecule. PPy not only aids in cell development but also aids in the adhesion of different types of cells. Endothelial cells, rat pheochromocytoma cells (PC12), glia cells from the dorsal root ganglia (DRG), and mesenchymal stem cells are among the types of cells used (MSCs) (Shirakawa et al. 1977). By conjugating diverse biomolecules to PPy scaffolds, such as neurotrophins and cell-adhesive molecules, it is possible to increase the conductivity of the scaffolds. The development of a biosensor for specific DNA recognition that is based on electroactive polypyrrole that functionalized with an oligonucleotide probe was demonstrated successfully (Korri-Youssoufi et al. xxxx). Polypyrrole electrodes combined with poly-dimethylsiloxane (PDMS) result in a highly stretchy electrode array, which results in a flexible, nonmetallic microelectrode array with a high degree of stretchability (MEA). The PPy electrode has been employed for electric stimulation of the ischiadic nerve as well as for in vivo electrocorticography recording in several studies. Langrnuir–Blodgett film of conductive PPy with alkyl-pyrroles linked to it is a type of conductive PPy. This film is deposited on the electrode surface, and it aids in the reoxidation of flavin enzymes that have been lowered in concentration (Singhal et al. 2002).

3.2.3 Polyaniline (PANI)

In bio-implants, polyaniline is used as an organic conducting polymer that allows for a better and more flexible interface to be created. As demonstrated by Tahir et al. (Tahir et al. 2007), polyaniline was synthesized with the assistance of protonic acid, a biosensor widely employed in detecting the bovine viral diarrhea virus. Polyaniline Polymer films with low cost inject printing, which is easy to manage in terms of physical parameters and their functioning, have recently been employed in biosensors, energy storage, display, and organic light-emitting applications diode that is capable of large-scale production. In bioelectronics, a carbon nanotube reinforced with a polyaniline bilayer soft hydrogel of bacterial cellulose is employed (Rebelo et al. 2019).

3.2.4 P3HT (3-hexylthiophene-2,5-diyl)

Organic semiconductor polymers, such as P3HT, have been widely used in bioelectronics applications for many years. When utilized as a bi-enhancer for biosensors, P3HT-Pb nanocomposites exhibit excellent performance. It has been demonstrated that P3HT-based artificial retina could restore light sensitivity in blind rats (Ming and QinChai 2022). P3HT can successfully mimic the native photoreceptor of the human eye (Ghezzi et al. 2013). When exposed to ultraviolet and visible light, the alkyl thiophene-derived polymer Rr-P3HT degrades, allowing for the identification of both reversible and irreversible steps in the process. Thin films coated with Rr-P3HT demonstrate the separation of ion complexes in a saline water solution (Bellani et al. 2014).

3.2.5 Hydrogel

As a crosslinked polymer network infiltrated with water, hydrogels have been intensively explored in tissue engineering due to their biological similarities to the body tissue. Hydrogel bioelectronics is progressing in tandem with achievements in material creation, with the two fields benefiting from each other's gains in fundamental understanding. In recent years, advances in hydrogel bioelectronics have been divided into four categories: (i) hydrogel coatings and encapsulations, (ii) ionically conductive hydrogels, (iii) conductive nanocomposite hydrogels, and (iv) conducting polymer hydrogels (Alba et al. 2010) (Fig. 9).

The introduction of hydrogel through hydrogel coating and encapsulation in bioelectronics is one of the simplest routes because of its high conformal contact with the skin and high degree of epidermis hydration. By improving biocompatibility via attenuating neuroinflammatory responses, we can widen the application of the hydrogel. The water-rich nature of hydrogels also allows controlled delivery of bifunctional substances (Winter et al. 2007). They are engineered to deliver nerve growth factors and anti-inflammatory drugs into the surrounding neural tissues, which

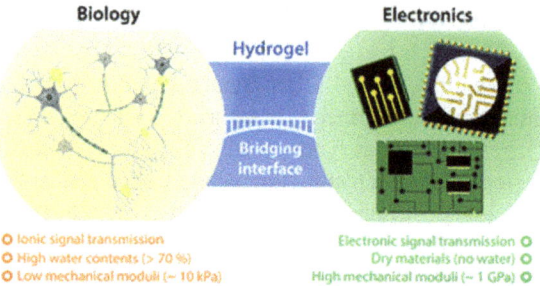

Fig. 9 Interface of Biology and electronics via hydrogel (after Yuk et al. 2018)

substantially alleviate neuroinflammatory responses. Bacterial cellulose is a soft hydrogel containing a large amount of water (90%) having a crystalline network are prominently used in carbon nanotubes in the introduction of CNT in bioelectronics (Rebelo et al. 2019).

3.3 Naturally Extracted Bio-based Polymer Used in Bioelectronics

As a result of their variability, polysaccharides have a complicated structure that comprises a variety of different monosaccharides. Ring-opening polymerization has shown to be an extremely effective approach for polysaccharides with high molecular weight. Polysaccharides appropriate stereoregularity can be produced using a suitable combination of the hydroxyl-protective group in anhydro-sugar, catalyst, and the polymerization temperature.

3.3.1 Cellulose

Cellulose is the foremost and well explored polysaccharide because of its biodegradable nature and abundance (plants). It consists of D-glucose through the 1,4- β glycosidic linkage. The chemical synthesis of cellulose is complex despite its simple structure. Natural synthesis, which includes plant photosynthesis, and microbial synthesis, are the two types of synthesis found in nature. Natural polymer bacterial cellulose (BC, or MC) can be produced by some bacteria and is also known as microbial cellulose (MC). Due to its unique structural and mechanical qualities compared to higher plant cellulose, BC is projected to be a desirable material for usage in a variety of fields. The diameter of BC fibres ranges from 20 nm to 100 nm, and they have a high aspect ratio, exhibiting a large surface area for a given mass. The combination of its unique characteristics and hydrophilicity led to a very high liquid loading capacity (Yoshida 2001).

(i) Synthesis of cellulose by the fermentation process of microorganisms

Microorganisms such as fungi, bacteria, and algae produce cellulose. All brown algae (Phaeophyta), most red algae (Rhodophyta), and golden algae (Chrysophyta) have cellulose at modest levels. Cellulose is also found in certain fungi in their inner cell wall layer in conjunction with -1-3/-1-6 linked glucan. Aerobacter, Agrobacterium, Azotobacter, Achromobacter, Rhizobium, Salmonella, and Sarcina are just a few of the bacteria that create microbial cellulose. The most efficient cellulose producers are A. xylinum, A. hansenii, and A. pasteurianus. A protofibril is formed when cellulose molecules are spun out of 'cellulose export components' or nozzles in the interior of a bacterial cell. Microfibrils with a diameter of 2–4 nm and bundles of protofibrils forming an 80 X 4 nm ribbon-shaped microfibril have been described. Both the ribbon-like polymer cellulose I and the amorphous, thermodynamically more stable cellulose II are produced by Acetobacter xylinum. In contrast, cellulose I is composed of parallel -1, 4 glucan chains, while cellulose II is comprised of randomly organized chains. They're more stable since they're antiparallel and contain many hydrogen bonds (Vandamme et al. 1998).

I) Applications of bio-based cellulose in Biomedicine and Bio-implants

(i) Wound healing/Tissue engineering

In Poland, researchers (Czaja et al. 2007) found that never-dried BC membranes outperformed conventional wound dressings in terms of (1) conforming to the wound surface (excellent moulding to all facial contours and a high degree of adherence even to contoured parts such as the nose, mouth, and other contoured parts were observed), (2) maintaining a moist environment within the wound, and (3) significantly reversing the damage caused by severe second-degree burns. Using these BC membranes, it is possible to treat huge and difficult-to-reach parts of the body. Silver nanoparticles can be infused into BC by soaking BC in silver nitrate solution, despite the fact that BC does not have any antibacterial action. Cellulose membrane coated with silver nanoparticles has been shown as an effective antimicrobial wound dressing material for burns and severe wounds because of its antibacterial properties (Mohite and Patil 2014).

(ii) Bio-implant

Biomaterials (scaffolds) that can be implanted into the body are also essential, and cellulose has been explored as a new material for use as a bone or skin graft. Moreover, bacterial cellulose has also been demonstrated as an injectible into a wound (Brown et al. 2006). To repair or replace human tissue, acetobacter-derived cellulose could be dehydrated with various organic solvents, such as methanol or ethanol. Yamanaka et al. (1989) used the moulding technique to produce hollow BC tubes with an inner diameter of between 2 and 6 mm. They utilized these hollow BC tubes to replace blood arteries and other tubular structures in the body, such as the trachea or the digestive tract bed. BC nanowhiskers (BCNW), scaffolds that may be adorned with nano-topographical features utilizing an ice microsphere templating approach as a new tool for decorating the pore walls of three-dimensional macroporous forms, were created by mixing BC and PLA (Mohite and Patil 2014).

(iii) **Drug delivery system**

Cellulose is largely employed in pharmaceutical coatings due to its superior film-forming characteristics. Sterilizing the BC ultrafine fibre network structure has no effect on its structure or characteristics. As a tablet film coating agent, BC offers great promise. Spray coating medication tablets can be improved by the addition of various ingredients, such as other plasticizers or polymers, to BC's aqueous dispersions. Because they are formed of nanoscale fibres randomly structured in 3D space, BCs can be employed in ferromagnetism medication delivery and serve as templates for the precipitation of magnetic inorganic nanoparticles. BC hydrogels appear to be a good candidate for controlled medication delivery systems (Mohite and Patil 2014).

(iv) **Electronics**

Conducting BC can be created by introducing multi-walled carbon nanotubes. A transparent flat-panel display composite made of BC nanofibre-reinforced acrylic resin with a low coefficient thermal expansion has been created. An organic LED was fabricated on this BC–acrylic resin combination, revealing its ability to be a flat-panel display substrate. Further, electroluminescence was also exhibited in this system, which again can be used in display technology. Metallic BC fibres are placed on the fibres of the fuel cell electrode to exchange protons in fuel cells. Development of low-cost technology for making highly electrically conducting composites employing BC pellicles as starting materials that can withstand sizeable elastic stretch and bending, particularly for flexible electronic devices, has been demonstrated (Mohite and Patil 2014).

3.3.2   Chitosan/Chitin

Several chitosan derivatives can be produced through the hydrolysis of chitin or chitosan, either chemically or enzymatically. Because of its unique biological impact, chitosan has garnered substantial interest as a biomedical substance. It can be used to treat obesity, hypertension, infections, tumors, inflammation, and free radicals. As a genetic material carrier in gene therapy, Chitosan has the potential to be employed as a non-viral cationic natural polymer in a wide range of biomedical, functional food, and pharmaceutical products.

I) **Applications of bio-based chitosan in Bioemedicine and Bio-imaging**

(i) **Bone tissue engineering**

There are a number of ways to make chitosan scaffolds, including freeze-drying, phase separation, gas foaming, and solvent casting. Chitosan-based scaffolds for bone tissue engineering using chitosan are best prepared by freeze-drying. Chitosan and its derivatives have been used in a variety of drug delivery methods in recent years. Bone morphogenetic protein-2 (BMP-2) and osteoblast development were promoted in rabbit models by water-soluble chitosan, according to a recent study (BMP-2) (Ruijin et al. 2016). Additionally, it can raise serum calcium levels and alkaline phosphatase activity in the bloodstream. Bone tissue regeneration techniques were developed using chitosan gels as a transporter medium to administer rhBMP-2. Bone healing can be boosted by using porous chitosan scaffolds (pore size 70 to 900 μm) combined with insulin-like growth factor-1 and BMP-2. Cell proliferation and adhesion to osteoblast cells were enhanced by the nonfibrous chitosan BMP-2 membrane that was used for bone tissue regeneration (Riaz Rajoka et al. 2019).

(ii) **Bio-imaging**

Bio-imaging is a technique that provides three-dimensional resolution with a real window, without any radiation exposure, and in vivo image acquisition. In this case, Telluride quantum dots embedded in chitosan nanoparticles have the potential to be an appealing bio-imaging agent. Because of their cytotoxicity, telluride quantum dots have some drawbacks as an imaging agent. However, once embedded in chitosan, it prevents the release of quantum dots for an extended period. As a result, it can reduce its long-term toxicity.Along with this Chitosan-embedded gadolinium nanoparticles can be synthesized and used for bio-imaging purposes because they can be located at the cancerous location for an extended period, which is beneficial for detailed imaging. Also, Chitosan can be used to reduce gold non-particles which function as photodynamic carriers and photothermal converters, as well as allowing photodynamic therapy (Riaz Rajoka et al. 2019).

(iii) **DNA delivery**

Chitosan has been shown to decrease the thermotropic enthalpy of the dipalmitoyl-sn-glycero-3-phosphocholine (DPPC) bilayer, a major component of the cell membrane and a standard experimental membrane model, during the gel to liquid crystalline transition and to promote the fusion of small DPPC vesicles to form large lamellar structures. This improves the interaction of OCMC and DPPC. The interaction between OCMC with DPPC may enhance the efficiency of OCMCS for gene or medication delivery (Inamdar et al. 2010). Moreover, Chitosan's amino groups are positively charged in an acidic environment. This charge results in a very strong electrostatic interaction with negatively charged macromolecules, such as DNA or RNA (Hejazi and Amiji 2003). Thus, during the gene therapy procedure, chitosan may be used as a genetic material delivery agent to protect the genetic material from nuclease activity and to enhance transfection efficacy, although the most critical factors to consider when using chitosan as a genetic material delivery agent are its molecular weight and degree of deacetylation (DA).Thus, chitosan with a high molecular weight (approximately 100 kDa) has a variety of applications in genetic material delivery processes, including delaying the release of the genetic material, forming stable physical shapes for polyplexes, and forming stable polyplexes with genetic materials (Köping-Höggård et al. 2004).

3.3.3 Agarose/Agarose Hydrogel

Agar, agarose, and agaropectin were extracted from the red alga Ahnfeltia plicata. An agarose-and-agaropectin-based agar has been used to define an agar. As a high-value product, agarose is frequently used in biotechnology and molecular biology. Agar and agarose can be used for a variety of purposes.

I) Synthesis of Bio-based Agarose

(i) Synthesis from A.Plicata

Natural polysaccharide agarose generates thermoreversible gels when cooled from 99 °C to a temperature below the ordering temperature, which is roughly 35 °C for conventional agarose. It was pre-treated using ultrasonication at 20 kHz in a citric acid–sodium citrate buffer solution (pH 5.0) in a 1:30 w/v ratio of the algal A. plicata. Cellulase concentrations of 10 U/kg were used to hydrolyze the alga for four hours at 50 °C. After that, the alga was exposed to 0.18% NaClO at a pH of 5.39 for ten minutes. After the solutions were discarded, the alga was rinsed with water to neutralize it. A 1.2% alkali solution was used in an autoclave to treat the alga for two hours at 121 °C in a 1:30 w/v ratio. We employed filtering to collect the solution and gel it at room temperature before cooling it to −20 °C and thawing it. The gel is then dried at 60 °C to get agarose gel (Zhang et al. 2019).

(ii) Synthesis of Agarose -HA Hydrogel

For this, agarose was dispersed in NaOH solution and was stirred for four hours and rinsed until the pH of the supernatant was neutral. By freezing and collecting the alkali-treated agarose, the agarose powder might be produced in the lab. Here, 3 g of agarose powder was dissolved in boiling deionized water and then cooled to 40 °C. Also, a 3 weight% solution of HA was prepared by dissolving it in deionized water at 40 °C. These (Agarose to HA solution) were mixed in three different weight ratios. The mixture was stirred at 40 °C for an hour to get a homogeneous solution. Epichlorohydrin (1 mL) and a 20% alkaline solution (1 mL) were added to 10 mL of the aforementioned composite agarose/HA solution. A hydrogel known as the composite agarose/HA hydrogel 1–3 was formed due to this reaction. As a final step, after careful removal of the composite (agarose/HAhydrogel), it was rinsed for one week with distilled water to remove any leftover ECH or impurities (Zhang et al. 2012).

I) Applications of bio-based Agarose in DNA electrophoresis and Tissue Engineering

(i) DNA electrophoresis

Agarose gel electrophoresis is a frequent technique in life sciences; biomolecular researchers utilize this method to separate biological components based on their sizes. Researchers can benefit from having the ability to segregate molecules by size, such as comparing unknown samples to known results or performing accuracy and quality check during other procedures. It is a common practice to employ agarose gel electrophoresis to separate DNA molecules in DNA modification techniques or investigations involving the identification of individuals by their unique DNA sequence (Lee et al. 2012).

(ii) **Tissue engineering**

Controlled self-gelling, water-solubility, adjustable mechanical, and nonimmuno-genic properties have made agarose a popular biomedical material. With its high stiffness and functional groups, agarose can support cell adhesion, proliferation, as well as cellular activity. Cells can thrive in a microenvironment that is tailored to their specific needs because of the changeable capacity of agarose to adsorb water. In the study of mechanical load reaction for chondrocytes and mesenchymal stem cells (MSCs), agarose-based hydrogels have been extensively used (Park et al. 2009). Drug delivery, cancer therapy, tissue engineering, and regenerative medicine, illness diagnosis, control, and treatment are just a few of the uses of agarose. For example, it has been used in the neurological system, bone creation, heart regeneration, and wound healing as well as adhesion to tissues such as skin, brain, and cornea. Agarose's characteristics can be altered via concentration, functionalization, and mixing in order to imitate the required tissue qualities (Salati et al. 2020).

3.3.4   Xanthan Gum

Xanthomonas campestris, a gram-negative bacterium, ferments glucose, sucrose, or other carbohydrate sources are used to yield xanthan gum. As a gelling agent when mixed with other gums, this biopolymer is employed as a thickener, stabilizer, or emulsifier in a wide range of industries, including food, cosmetics, pharmaceutical, and petrochemistry.

I)   Synthesis of bio-based Xanthan Gum

1.   Synthesis of Xanthan Gum from Shrimp Shell

By fermenting shrimp shells in an anaerobic batch bioreactor, xanthan gum is produced. Shrimp shells may be fermented into xanthan gums by three strains of Xanthomonas campestris, which are equivalent to sucrose fermentation. The shrimp shell was dried at 60 degrees celsius, ground for ten minutes at 50 rpm, and then stored for further study. There must be a determination of the amount of moisture, protein, fat, and ash. Calculating the carbohydrate content is as simple as removing 100% from the weight and then multiplying the result by 100. The shells of shrimp are ground into a fine powder and used in the subsequent fermenting process. To produce the gum, Xanthomonas campestris strains were incubated at 28 degrees celsius for 24 h at 180 rpm during the xanthan gum production process in yeast malt (YM) medium (3% malt extract, 3% yeast extract, 5% bacteriological peptone, and

1.0% glucose). 10 mL each of the inoculants were placed in 250 mL conical flasks and incubated at 28 °C for 120 h with agitation at 250 rpm and the xanthan gum gel was obtained (Sousa Costa et al. 2014).

I) Applications of Bio-based Xanthan Gum in Biomedicine

(i) Drug delivery service

As an excipient in tablet formulations, xanthan has also been used as a hydrogel support in drug release applications. When combined with other polymers, xanthan gum can assist in achieving specific desirable qualities. When xanthan and guar gums were employed to protect b-carotene-loaded liposomes, long-term stability was attained, while curcumin-loaded hydrogels derived from Schizolobium parahybae's xanthan and galactomannan were successfully used as an anti-inflammatory dermatological gel. Xanthan and lignin were cross-linked with polypropylene diglycidyl ether to produce superabsorbent hydrogels for use in medical devices. These aloe hydrogels can be used to treat skin illnesses or to restore damaged skin, as they contain chitosan and xanthan. Doxorubicin hydrochloride, an anticancer medicine used to treat a wide range of cancers such as haematological malignancies and soft tissue and solid tumors, was loaded into gold nanoparticles using xanthan, which was then utilized as a reducing agent (Petri 2015).

(ii) Tissue engineering

Using xanthan-based hydrogels, skin regeneration has been accomplished. Fibroblast proliferation was higher on xanthan-magnetite hybrid nanocomposites than on neat xanthan hydrogels when 0.4 T magnetic field was applied. In vitro neural differentiation of embryonic stem cells was also successfully accomplished using xanthan-nanomagnetite hybrid scaffolds. The differentiated cells demonstrated higher membrane potential amplitudes upon KCl depolarization, indicating successful synapse formation (Petri 2015).

3.3.5 Silk Nanofibre

Application of silk nanofiber in bio-implants

Bio-based polymers, such as silk nanofiber, play an essential role in developing next-generation wearable microelectronics implants. Nowadays, bioabsorbable conducting metal wires like Cu- or Al-based metallic wire, carbon nanotube wire, Ag nanowire fibre, silver/polyethylene terephthalate wire, and Zn-mg alloy wire are utilized to avoid implant inflammation, increase transmission, and reduce the weight of the implant. Because of its properties, such as conformal contact with the skin and the degree of epidermal moisture, the introduction of hydrogels into bioelectronics through hydrogel coating and encapsulation has shown to be one of the simplest approaches (Yang et al. 2020). It is possible to broaden the application of hydrogels by enhancing their biocompatibility and reducing neuroinflammatory responses. Hydrogels also enable the regulated delivery of bifunctional chemicals due to their water-rich nature. In order to significantly reduce neuroinflammatory responses, they are designed to transfer nerve growth factors and anti-inflammatory medications into the surrounding neural tissues. Bacterial cellulose is a soft hydrogel that contains a considerable quantity of water (90%), and it has a crystalline network. It is widely utilized in the production of carbon nanotubes and other materials. The incorporation of carbon nanotubes (CNTs) into bioelectronics has an impact on the functioning of silk-based materials and implants because it aids in the preservation of the original structure of silk in an implant (Li et al. 2020). The filtering method is the quickest and easiest approach for manufacturing silk nanofibers. It also has the added benefit of improving biocompatibility compared to the silk fibroin film (Niu et al. 2020). Silk nanofiber is an ideal natural polymer to use as a wire for an implant, because of its biodegradability, biocompatibility, low weight, high transmission, and high strength.

3.4   Bio-based Ceramics

Materials that are key constituents of vertebrate bone include polycrystalline aluminium oxide, hydroxyapatite, partly stabilized zirconium oxide, bioactive glass, and glass–ceramics. Polycrystalline aluminium oxide is a major constituent of the vertebrate bone. Plastic-hydroxyapatite composites are widely utilized to repair, reconstruct, and replace diseased or damaged human tissues and organs, such as bone. Bioceramics are employed in implants, and prostheses as porous or bulk materials with a defined shape, depending on the application (Heness and Besim 2003). Bioceramics can be classified into bioinert, bioactive, and bioresorbable. As a result, they are commonly employed in structural support body implants since they can retain their mechanical and physical properties while in the host body. Examples include bone screws, bone plates, and femoral heads, to name a few. It is made from the mineral alumina ($Al2O3$) and is primarily utilized in orthopedic joint replacement components, implant coating, and dental implants, among other applications (Heness and Besim 2003).

On the other hand, bioactive bioceramic, which retains a strong chemical interaction with the bodily tissue, increases the likelihood of successful tissue engineering. Bioceramics such as glass ceramics and dense nonporous glasses are examples of this type of material. Ceramics that are bioresorbable or chemically broken down by the body are the most promising ceramics in biotechnology since the body can degrade them. Bioceramics, both manufactured and natural, such as calcium phosphate, tricalcium phosphate, calcium sulfate, hydroxyapatite, and corals are included in this category because they are capable of replacing resorbable materials with endogenous tissue (Madsen and Kohn 2021).

Bioresorbable Bioceramics

3.4.1 Calcium Sulfate

For the past 90 years, calcium sulfate has been widely used as a biocompatible bone substitute. Among its many advantages are low cost, easy availability, and unlimited supply, lack of donor site morbidity, use as a delivery vehicle for numerous compounds (especially antibiotics), inherent osteoconductive characteristics (based on a structure similar to bone), and a proven safety record. Osteoconductive, osteoinductive, and osteogenic bone graft substitutes have the same or better biologic potential as autogenous bone grafts. In addition, it would be readily available and pose little risk to patients. Calcium sulfate fulfils many of these parameters. Capillaries, perivascular mesenchymal tissue, and osteoprogenitor cells are known to enter the matrix to produce new bone. The three crystalline phases of calcium sulfate (CS) are anhydrite ($CaSO_4$), hemihydrate ($CaSO_40.5H_2O$), and dihydrate ($CaSO_42H_2O$) (CSD). CSD is produced exothermically by mixing calcium sulfate hemihydrate (CSH) with water and allowing it to react (Beuerlein and McKee 2010).

It is vital to control the shape and size of CS because its performance is directly related to these uses. There are two forms of hemihydrates: α and β, with calcium sulfate hemihydrate (CSH) being the more frequently employed in medicine due to its high strength. The α-CSH form dissolves more slowly than the β-CSH form due to its density and stability, making it a popular choice for bone-filling applications. The primary disadvantage of calcium sulfate bone replacements is their quick resorption, which outpaces bone production/growth at the defect site, resulting in the creation of holes. Additionally, the CS bone cement must be coupled with a harder substance to produce greater mechanical performance despite its potential to rebuild the bone. By incorporating organic or polymeric components into CS, its strength and resorption rate could be increased (Hsu et al. 2017).

A) Synthesis of Calcium sulfate

(i) Method 1

As a starting point, the biochemistry-grade CSD powder (98%) was used. Microwave irradiation systems with precise temperature control were employed to synthesize high purity α-CSH in an air-conditioned (relative humidity 70%) cleanroom. 2 gms of CSD powder and 10 ml of deionized water were dissolved in a 55 mL Teflon pressure tube (350 psi) and magnetic string bar (diameter 6 mm 9 lengths 15 mm). This was followed by 10 min of microwave cooking at 100 °C, 130 °C, and 160 °C at 800 W on the Teflon pressure tube. To ensure a uniform reaction temperature, the mixture was constantly stirred while being heated with a microwave oven. In a filter flask with high-purity 100% ethanol, the synthesized samples were cleaned and filtered five times before being extensively dried in an electric dry oven at 60 °C for eight hours. This was followed by a further inspection of the powdered produced samples (Hsu et al. 2017).

(ii) **Method 2- From CaSO$_4$ powder**

Aqueous precipitation was used to create the CaSO$_4$ powder. To commence, distilled water was used to dissolve Ca(NO$_3$)$_2$ · 4H$_2$O, and to this Na$_2$SO$_4$ solution was added. The mixture was then agitated continuously for 120 min using a titration technique. Upon completing the mixing procedure, the solution had a pH of 5. The suspension was aged for an additional two hours after mixing was complete. The suspension was filtered through a Whattman 40 mm filtration system to remove the white precipitate. The filter paper (filter cake) was cleaned three times with distilled water to eliminate any remaining residue from the white precipitate. The filter cake was dried in an oven at 90 °C for 24 h to remove the wet precipitate's moisture content before being ground into a dry powder. This powder was then calcined at temperatures ranging from 300 to 1200 °C, with a one-hour soaking period (Zahari et al. 2020).

(iii) **Method -3 Synthesis of nanosized Calcium Sulfate**

In the first step, 3 mL of 5 wt% TritonX-114 was dissolved in 20 mL of cyclohexane and agitated for 30 min, following which 10 mL of 30 wt% H$_2$SO$_4$ aqueous solution was added to the mixture and spun continuously for 3 h. To this, 0.3 g of CaCO$_3$ was added and again stirred consistently for an additional 30 min. After centrifuging the suspension three times, it was washed with anhydrous ethanol to remove any remaining contaminants. This was kept in a vacuum oven at 60 °C for 4 h to get the desired solid products (Chen et al. 2019).

I) **Application of Bio-based Calcium sulfate in Bone Transplant**

With CaSO$_4$, the healing of the graft material is expedited, and its loss is minimized. As the body recovers, it will be replaced by the bone since it is tissue friendly. This material's high wettability and unusual surface roughness allow it to adhere to irregular bone surfaces. As a result, it is used in orthopedics as a bone filler. As a result, CaSO$_4$ has the potential to be used in biomedical treatment for bone regeneration because of its natural rate of absorption. In addition, it is regarded to be the fastest dissolving of all the currently available bone-repair components available.

Additionally, CaSO$_4$ can be used as a space-filler to allow the bone to heal itself passively and a bioactive chemical that can encourage new bone growth. As a bone transplant alternative, it can be utilized alone or in combination with other materials. For example, calcium phosphate precipitate can be produced on the surface of calcium sulfate and then deposited in a bone defect. CaSO$_4$ is a bioactive soluble substance that can stimulate new bone production and act as a filler to allow passive bone healing. It can be utilized as a bone graft substitute either alone or with other materials. For example, calcium phosphate precipitate can be formed on the surface of calcium sulfate and deposited in a bone defect to create a self-forming biologic apatite (Thomas and Puleo 2009).

3.4.2 Tri-calcium Phosphate

One of the most studied calcium phosphates is tricalcium phosphate (TCP; $Ca_3(PO_4)_2$). It's a calcium phosphate with a Ca/P ratio of 1.5 splits into two phases: α and β. α-TCP has a monoclinic space group crystal structure, while β-TCP has a rhombohedral space group crystal structure. The bioactivity of Tri-calcium phosphates is due to their physicochemical characteristics (Carrodeguas and De Aza 2011). TCP (Tricalcium phosphates) has three polymorphs: β-TCP at low temperatures and α- and α' TCP at high temperatures. Because it only occurs at temperatures over 1430 °C and reverts virtually instantly to α-TCP when cooled below the transition point, the α'-TCP is of little practical interest. β-TCP, on the other hand, is stable at normal temperature and converts reconstructively to α-TCP at 1125 °C, which can be kept for cooling at room temperature. β-TCP is a component of several commercial mono- or biphasic bioceramics and composites, while α-TCP is the principal constituent of the powder component of different hydraulic bone cements; both are used in dentistry, maxillo-facial surgery, and orthopedics (Jeong et al. 2019).

I) Synthesis of Bio-based Tri-calcium phosphate (TCP)

(i) Method 1

At 30 °C, *Serratia sp.* N14 bacteria were grown in a continuous carbon (lactose) limited culture in an airlift fermenter. Reticulated 1 cm^3 polyurethane cubes attached to cotton strings and cleaned with distilled water were used for the fermenter. Biofilm grew on the reticulated surfaces of the cubes in the fermenter for six days before they were transferred to the bioreactor for additional testing. For this experiment, 25 biofilm-coated cubes in a 25 cm glass column filled with calcium chloride ($CaCl_2$) and G-2-P solution were pumped through the column at 1 mlh^{-1}. The pH, temperature, and solution concentrations in the bacterial habitat can be precisely controlled using a bioreactor. More than 180 mg of the white crystalline substance was found in each biofilm-coated cube after 14 days of incubation at 30 °C (Thackray et al. 2004).

(ii) Method 2

Pure tricalcium phosphate powders were produced at 40 °C by maintaining a pH of 7 by the wet precipitation technique using the following precursors tetrachlorocalcium and diammonium hydrogen phosphate. In this method, diammonium hydrogen phosphate solution was added to the calcium nitrate tetrahydrate aqueous solution (1.2 M for TCP) in a double-walled jacket reactor with magnetic stirring. Vacuum filtering was carried out using an aqueous NH_4OH as a washing solution, followed by drying at 60 °C. The as-dried TCP precursor was calcined at 900 °C/2 h at 5 °C min-1 to enhance the synthesis of a single-phase TCP (Carrodeguas and Aza 2011).

Table 2 Application of TCP in Biomedicine (after Carrodeguas and Aza 2011)

Dentistry	Periodontology	Bony flaws on the second and third walls can be employed with or without membranes
	Implantology	In the event of rapid implant installation, defect augmentation following extraction to establish an implant base elevation of sinus floor gaps between the extraction socket and implant
	Cysts	Defects caused by cyst removal
	Dental and Maxillo-facial surgery	Defects caused by apicectomy Defects caused by the extraction of impacted teeth Corrective osteotomies result in defects All different types of bony craters and flaws in the face bones
Orthopedics	Bone faults and cysts are filled Autogenous bone replacement at the donor location In spinal fusion, vertebral body replacement, and joint replacement, it is used in addition to the cancellous bone	

I) Application of Bio-based Tri-calcium phosphate in biomedicine and orthopedics

1. Biomedicine

TCP has a variety of clinical applications. It is employed as a drug carrier in dental, craniofacial, maxillofacial surgery, and orthopedics, including vertebroplasty and kyphoplasty treatments (Table 2).

(i) Coating for Cell adhesion

Calcium phosphate coatings can be applied to a range of materials in order to increase their bioactivity. The sol–gel and electrodeposition processes are the most frequently used for coating calcium phosphate. Most calcium phosphate coating research focuses on metal implant applications to reduce implant corrosion while simultaneously improving the bioactivity. The surface reactivity and cell adhesion of calcium phosphate coatings have been enhanced due to research on these coatings. Ti-6Al-4 V alloys with porous surfaces were used in the human body because they were biocompatible. A thin HAP surface was used to construct a sol–gel coating technique, which was used to boost post-implantation bone ingrowth and osteoconductivity while decreasing post-implantation bone loss. HAP was applied to the porous surface of cylindrical implants to seal them in place. This alloy was used in in vivo testing of the rabbit bone, and osteoconductivity was increased by increasing the amount of desired protein adsorption. Antibacterial drugs and growth factors have been the subject of numerous studies to improve their potency by encapsulating them. An AgNO3 and TCP coating was applied to the material's surface using a laser-engineered net shaping technique to reduce infection, boost cell-material interaction, and enhance antibacterial activity (Thackray et al. 2004).

(ii) **Bone cement**

Bone deformities can be filled and repaired using calcium phosphate cement. Polyethylene glycol (PEG), poly (lactic-co-glycolic acid) (PLGA), cellulose, gelatin, collagen, and synthetic polymers such as PEG, polycaprolactone (PCL), poly (L-lactic acid) are often used in cement (PLLA). Combining these polymers allowed calcium phosphate cement to control injectability, porosity, mechanical properties, and degradation rate. A calcium phosphate cement with improved injectability and flow has been developed for use in the urethra in vesicoureteral reflux illness and minimally invasive surgery for bone defect repair. Beta-TCP pastes were mixed with hyaluronic acid or PEG to create the calcium phosphate cement. In addition to the slow rate of bone regeneration and disintegration, a pore size limit on new growth, a lack of mechanical strength, and an inflammatory reaction to synthetic polymers, calcium phosphate cement has several other disadvantages (Eliaz and Metoki 2017).

(iii) **Bone Scaffold**

Calcium phosphate-based scaffold are frequently used in bone tissue engineering and bone implants. The properties of calcium phosphate scaffolds are stable, allowing for precise control over porosity and biocompatibility. Biocompatibility and revascularization are enhanced because of the pore size of the scaffold. This makes it suitable for implant application. Scaffolds can be made from a variety of materials, including collagen, gelatin, PCL, PLGA, and PLLA. Because the hydrogel lacks mechanical strength for load-bearing, calcium phosphate microbeads encapsulating human umbilical cord mesenchymal stem cells were combined with the alginate hydrogel to make up for it. Cells can be seeded deep into a scaffold and injected with minimally invasive procedures by using this combination. The mechanical properties of this injectable alginate hydrogel scaffold were superior to those of traditional hydrogels (Sun and Yang 2015).

3.4.3 Bio-based Hydroxyapatite

Phosphorite is a type of rock containing phosphorus as its primary element; the translucent phosphate crystals found in phosphorite are apatite. In the case of those with a high concentration of hydroxide, they are known as Hydroxyapatite (HAp), which is a calcium phosphate with the formula $Ca_{10}(PO_4)_6(OH)_2$, primarily found in bones. Its structure and composition are very comparable to that of hard tissues of humans. Even though there are other calcium phosphates available, such as fluorine-rich hydroxide (called fluorapatite), and chlorapatite, which contain chlorine as a key ingredient, hydroxyapatite has been discovered to be the most stable of all calcium phosphates studied. This material is used to create synthetic bones and teeth because it does not alter in composition or structure when exposed to varying pH and physical conditions such as pressure, temperature changes, and the presence of different fluids in the human body. Even for bone healing, it is employed because it stimulates bone ingrowth (Sun and Yang 2015).

I) Synthesis of Hydroxyapatite

(i)   **Method 1**—Using titanium dioxide-coated anodes and stainless-steel cathodes, dangerous chemical-containing water from factories can be treated using the electrolysis method. It can be seen that the phosphate component of the unclean water precipitates on the cathode in the form of hydroxyapatite, which is a calcium phosphate compound. No further chemical treatment was required for this precipitation, and it is also simple to collect the precipitate so that it can be reused in other applications later on (Legeros 2008).

(ii) **Method 2**—The hydrothermal approach can be used to create the desired result when the synthesis temperature is low, and the conditions that must be maintained for the reaction are modest. The crystallinity quotient of the yield obtained from this method was relatively high. Calcium ions comprising oxides are employed as precursors, and amines are used as chelating agents in conjunction with these ions. For the production of HAp, a variety of precursors and chelating agents can be utilized. The selection of the appropriate precursor and the chelating agent is based on the desired form and structure of the final product (Legeros 2008).

(iii) **Method 3**—The sol-gel procedure, also known as wet chemical synthesis, is a very popular way of synthesizing HAp, and it is described below. The initial stage is the hydrolysis of the metal alkoxide, followed by the polymerization of the solute that is created, and finally, the drying and roasting of the gel to obtain the final product. Calcium citrate or calcium acetate, as well as phosphoric acid, are the precursors in this process. It is possible to fabricate HAp nanorods or nanowires (Legeros 2008).

I) Applications of Bio-based Hydroxyapatite as A Synthetic Bone and Tissue Engineering

Electronic engineering and management of a biological system, such as the nervous system or brain mechanism, is what bioelectronics is all about. As a result, it has made it possible to treat previously untreatable diseases and has opened the door to curing long-term problems. Tissue engineering uses hydroxyapatite, which is employed to correct any flaws in the tissues that still exist. Many natural and synthetic procedures can be used to replace the bone tissue that has been ruptured. When healing tissues in the past, bone grafts were used because of their capacity to stimulate bone growth or osteogenicity. It is critical that the substitute we use for bone repair has this quality. Grafts have a significant drawback in that they are hard to come by and carry a high risk of infection; this is where HAp comes in. The bone's primary component, HAp, has a high tissue affinity, so the problem of rejection is eliminated. It has a low risk of infection and can be utilized to regenerate the bone tissue successfully. Compared to the grafts previously employed, the osteogenic property of the synthetic bone is excellent (Szcześ et al. 2017).

It is possible to make a dense version of porous HAp by employing certain sintering conditions, and this may also be done without pressure. With excellent thermal stability and good mechanical qualities, fine microstructures are created. It is possible to create a porous HAp using chemicals like naphthalene, paraffin, or hydrogen peroxide. These porous HAp granules are utilized for local drug delivery because they are pre-filled with medication. A further benefit of HAp is that it prolongs the time for the medicine to stay in the affected area, allowing for complete healing. Porous HAp granules, custom-made for this purpose, are employed as bone gap fillers. Regeneration of bone and the correction of underlying abnormalities are all possible with the help of the HAp. Since it is biocompatible, it quickly dissolves in collagen and works on healing the affected areas. Since it has such high mechanical strength, the HAp can be used to both improve dental hygiene and repair worn-out enamel. We can use tissue engineering to fight the disease and improve our lives by using HAp as a medium.

4 Classification Of Bioelectronics

4.1 Bioelectromagnetics

Bioelectromagnetics is a subfield of bioelectronics that deals with the explanation and investigation of electromagnetic, magnetic, and electric phenomena occurring in biological tissue. It is a multidisciplinary field that integrates biology, physics, and engineering science, as well as simulation methodology. This phenomenon is characterized by the following events (Fig. 10):

1. The behavior of excitable tissue.
2. The electric currents and potentials in the volume conductor.
3. The magnetic field at and beyond the body.
4. The response of excitable cells to electric and magnetic field stimulation.
5. The intrinsic electric and magnetic properties of the tissue (Fig. 11).

First division: A) bioelectricity B) bioelectromagnetism (biomagnetism), and C) biomagnetism. This horizontal division is tied by Maxwell's equation

Then the division is made vertically for: I) measurement of fields, II) stimulation and magnetization, and III) measurement of intrinsic electric and magnetic properties of tissue.

(1) Measurements of field—Detection of the magnetic and electric signals generated by the diverse actions of live tissues in the body. The amount of electromagnetic energy produced by an active tissue can be measured electrically or magnetically, either within or outside the region of the body where the source is located.

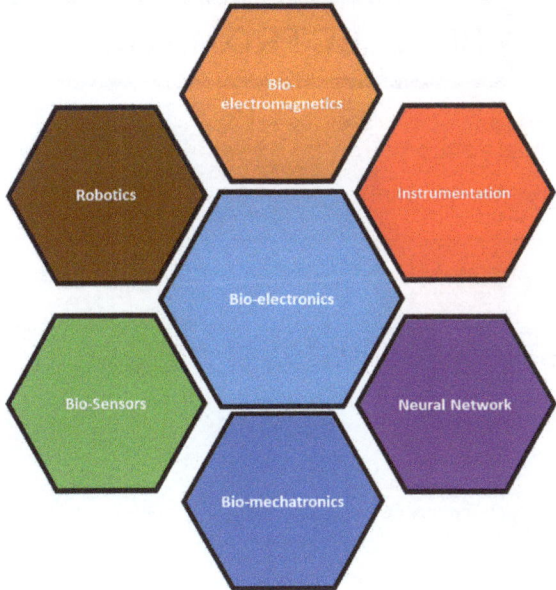

Fig. 10 Classification of Bioelectronics

(2) Simulation and magnetization—In the presence of an electric or magnetic field, an electric simulation describes the effect of applied electric and magnetic fields on the active tissue; in order to activate the tissue, electric and magnetic energy is applied. As previously stated, electric and magnetic energy is generated outside the biological tissue itself. When magnetic energy is supplied to a piece of tissue containing a ferromagnetic material, the substance becomes magnetized and becomes magnetic.

(3) Measurement of intrinsic properties—electric and magnetic energy generated by an electric device outside the biological tissue, is then applied to the tissue. While measurements of properties are listed in Table 3.

4.1.1 Application of Bioelectromagnetics

Bioelectromagnetism has many applications in different fields such as physics, molecular, and cellular biology. We can also find it in biomedical engineering, for the production of medical equipment and instruments and its use in the treatment of various diseases.

I) **Bioelectromagnetic devices**

There is yet another contrast between bioelectromagnetic (BEM) devices: the device might be either thermal or non-thermal in nature. Low-frequency devices are typically utilized in bioelectromagnetism, and low-frequency radiation has a variety

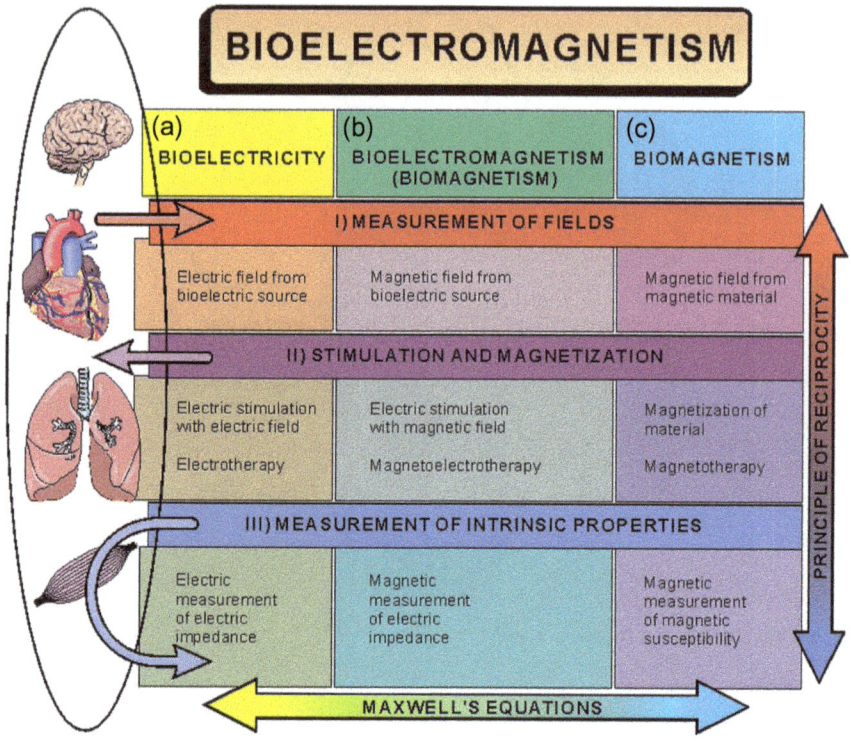

Fig. 11 The taxonomy of organization of bio-electromagnetism into its subdivisions (after Malmivuo and Plonsey 1995) (Vieira et al. 2018)

Table 3 Measurement of intrinsic properties in a different field (after Malmivuo and Plonsey 1995) (Vieira et al. 2018)

A. Bioelectricity	B. Bioelectromagnetism	C. Biomagnetism
Impedance cardiography		Magnetic susceptibility
Impedance pneumography		Plethysmography
Impedance tomography	Impedance tomography	Magnetic remanence measurement
Electrodermal response (EDR)		Magnetic resonance imaging(MRI)

of uses, including radiofrequency hyperthermia, surgery, and lasers, among others. Some processes generate heat in tissues, whereas others do not. Biologic nonthermal refers to a modality that does not generate gross tissue heating, whereas physically nonthermal refers to devices that have physiologic temperatures that are below the thermal noise. Due to the fact that thermal noise is significantly lower than the level required to generate tissue heating, any physically nonthermal application is also automatically and physiologically nonthermal in nature. Several traditional applications that use electromagnetic radiation are prominently used in a wide range, including the complete family of medicines known as electrophysical agents and their derivatives. They are employed to alleviate pain, muscle spasms, and inflammation and enhance superficial and deep circulation status. Electrical energy is frequently used to aid in the diagnosis process. This is true for tests such as electromyography, biofeedback, electroencephalography, electro-retinography, and imaging tests such as magnetic resonance imaging positron emission tomography, computed tomography (CT), ultrasound, and radiography. The use of electric and magnetic energy varies depending on the instrument in question, with certain devices emitting ionizing radiation (x-ray/CT) (Malmivuo and Robert 1995).

II) Bioelectromagnetic therapy

Bioelectromagnetic treatment (BMT) is described as the application of electromagnetic field to predict the disease and to maintain the health of patients. It will create an electric current in this treatment as a result of the variation in the electromagnetic field. It is equipped with static magnets that are incapable of generating an electrical impulse unless and until the device is physically moved. Detectors and magnets are capable of detecting bioelectronic fields of sufficient intensity (i.e., having an intensity within a certain range). The electromagnetic fields that can be applied usually range from less than the earth's magnetic field to 103 times the earth's magnetic field in strength and frequency. It can also aid in the improvement of outcomes by relaxing muscles, lowering stress, and allowing channels and blood vessels to be opened before the administration of needles. The use of bioelectromagnetic therapy is particularly prevalent in treating arthritis, osteoporosis, Alzheimer's disease, Parkinson's disease, multiple sclerosis, atherosclerosis, hypertension, and cardiovascular disease.

(i) Osteoporosis and Arthritis

Bioelectromagnetic (BT) therapy is extremely helpful in the healing process of non-union fractures and stimulating the production of new bone tissue. It is one of the most effective treatments available for building strong bones. Significant gains in bone density have been documented in a short period of time after receiving appropriate treatment. Due to the fragility of their bone, persons with osteoporosis frequently experience "compression." Bioelectromagnetics aid in the restoration of normal bone density, which helps to avoid future fractures and the alleviation of pain and healing of existing fractures. During a three-month period, one study (Rindge 2008) found that bone density increased by 5.1%, indicating an increase in bone density (Malmivuo 1999).

Materials used in Bioelectronic therapy for Osteoporosis and arthritis

Because all organisms on Earth are geomagnetic, magnetic materials can generate magnetic fields, including SMFs and PEMFs. These magnetic fields are directly associated with the survival and evolution of life on Earth. Furthermore, while magnetism results from the movement of electrons, cells, tissues, and organs all have their own magnetic fields, and various magnetic fields may interact with one another in a variety of ways. As a result, various scientific studies have been conducted to investigate the potential health consequences of magnetic materials.

The application of magnetic material films in the treatment of bone ailments has potential (Meng et al. 2013; Rindge 2008). It was discovered in 2013 that New Zealand white rabbits suffering from lumbar deficiency showed enhancement in osteogenesis after the implant of superparamagnetic nano scaffolds with external SMF. Using New Zealand white rabbits' tibias as test subjects, Bambini et al. (2017) (Meng et al. 2013) placed dental implants containing NdFeB magnets into their tibias and discovered that the SMF generated by the magnetic implants assisted bone repair. The clinical application of magnetic materials for bone regeneration is still in its early stages. As technology progresses, there are three key paths in which magnetic materials for therapeutic purposes could emerge. The first category includes interventional and auxiliary devices used in minimally or non-invasive treatment. Biodegradable materials will be developed, as will the micromachining of implanted tools, surface anticoagulation, and tissue proliferation modulation to accomplish this.

The usage of biologically generated materials will focus on the second path. Although collagen, sodium hyaluronate, chitosan, and silk fibroin are commonly utilized in the clinic, the quality and variety of materials must be enhanced and expanded to be effective. In addition to improving biocompatibility, quality, and stability, such alterations will enable the material to be used in conjunction with bioderived components. The third direction is the development of nanomaterials, such as nano-coatings and carriers for the controlled release of nano-drugs, among other things. Furthermore, computer-aided three-dimensional printing could be used for precise machining, quick prototyping, and the customization of magnetic materials, among other things. Researchers have devised various designs for applying magnetic materials to help bone regeneration; however, the ideal dose, adverse effects, and long-term stability have not yet been determined by the researchers (Peng et al. 2019).

(ii) **Alzheimer's Disease, Parkinson's Disease**, multiple sclerosis

In disorders affecting the central and peripheral nervous system such as Parkinson's disease, Alzheimer's disease, multiple sclerosis, and stroke the positive output has been recorded due to the application of bioelectromagnetics. In one case study of an old Parkinson's disease patient, a sudden increase in muscular control, and energy has been reported due to 35 days of daily bioelectromagnetic therapy (Malmivuo 1999).

Materials used in the treatment of Alzheimer's disease and Parkinson's disease

The two major neurodegenerative diseases, Alzheimer's disease (AD) and Parkinson's disease (PD) are characterized by low levels of the neurotransmitters acetylcholine (ACh) and dopamine (DA), in the brain. Hence, we need to focus on natural inhibitors that help elevate both neurotransmitters' level (Peng et al. 2019) (Table 4).

(iii) **Atherosclerosis, Hypertension, Heart Disease**

With the combination of laser therapy and Bioelectromagnetic therapy, a positive output in cardiovascular disease has been reported. In one case study of 66 elderly

Table 4 Natural compounds and their function in AD and PD

Disease	Material (Natural compound used)	Specialization	Function
Alzheimer's disease(AD)	Cholinergic Inhibitor	Agonists of the nicotinic cholinergic receptor	Compensate for the low levels of ACh (anticholinesterase)
	Alkaloids	Source of anticholinesterase	
	Terpenoid		
	Nicotinic compound		
Parkinson's disease (PD)	Dopaminergic agonists(DA)	Formed within the brain by conversion of its precursor L-DOPA	Activates dopamine receptors
	Monoamine Oxidase inhibitor	Antidepressant	A fast breakdown of DA & Net increase in DA level

patients suffering from heart diseases, improvement in microcirculation, myocardial reactivity, central hemodynamics reducing the biological age of the cardiovascular system, and a geroprotective effect were recorded (Malmivuo 1999).

III) Bioelectromagnetic medicine

It is possible to detect endogenous magnetic fields and use them for diagnostic purposes. Superconducting quantum interference devices (SQUID) can detect extremely weak magnetic fields emitted by organs, which may be analyzed to learn a great deal about how they function. The use of low-frequency magnetic fields to cure bone non-unions is an active use of bioelectromagnetics in therapeutics. Non-invasively accelerating the pace of bone growth and mechanical integrity of knitted fractures can be accomplished by wrapping a coil around the affected area and administering a milli-Tesla alternating current or square-wave magnetic field through the coil. A portable version of this device is used to treat the majority of fractures that do not heal on their own. Applications are now being explored for the therapy of ligament injury, osteoporosis, chronic skin ulcers, and tendenitis, among other conditions (Houghton and Howes 2005).

4.1.2 Polymers and Ceramics Used in Bioelectromagnetics

See Table 5.

4.1.3 Future Application of Bioelectromagnetics

A diverse range of electric fields and a diverse range of magnetic fields can be found in living creatures. The electromagnetic waves and ultra-weak photons emitted by animals, tissues, and cells are theorized to act as messengers of information between

Table 5 Application of bio-based materials in bioelectromagnetic and their fabrication process. (after Winkler et al. (2018)

Class of materials	Bio-Based materials	Fabrication method	Application in Bioelectromagnetism
Polymer	Chitosan	Electrospinning technique	Facilitates the osteogenesis in MG-63 human osteoblast-like cells
	PVA	Ultrasonication, conventional co-precipitation, Chemically cross-linked, and dialysis	Accelerate bone regeneration at the fracture site. In novel drug delivery system that releases a cytokine-like PTH to promote bone regeneration
	PLA	Electrospinning technique	Increases proliferation rate and faster differentiation of osteoblasts
	Silk fibroin	Electrospinning technique	Wearable electronics and implantable bioelectronics
Ceramics	HA	Dip-coating magnetization technique	Attracts more growth factors and other biomolecules

and inside cells. Understanding the significance of these endogenous fields is likely to result in considerable improvements in our ability to collect information from living systems and to impact living systems in a good manner. Other developments in cancer research appear to be quite promising at this time. According to the current evidence, malignant tissues posses electromagnetic properties distinct from those of normal tissue. On many occasions, cancer has been linked to alterations in the endogenous electric fields of the organism. Tumor tissue also appears to respond differently to applied fields, according to the research findings. All of this has the potential to be beneficial in the identification of cancer in its earliest and least invasive stages. Furthermore, research revealing that applied fields might preferentially damage the malignant tissue, as well as studies proving that applied field effects can lengthen the lives of cancer patients, lend credence to the potential of such an outcome. Additionally, research revealing that applied fields might preferentially damage the malignant tissue and studies proving that applied field effects can lengthen the lives of animals with cancer raise the possibility of cancer treatments based on bioelectromagnetics. There is some evidence suggesting that magnetic fields can help reduce the detrimental effects of ionizing radiation on the body. Magnetic radiation trauma treatment is now a possibility as a result of this. Others have discovered higher damage caused by pulsed magnetic field application after exposing mice to X-rays, so we will have to wait until we have a better knowledge of the impacts before moving forward with this research (Winkler et al. 2018).

4.2 Bioi Instrumentation

Bioinstrumentation refers to the development of technology for the measurement and manipulation of parameters within biological systems, with a particular emphasis on the application of engineering techniques for scientific discovery as well as the detection and treatment of illness. Bioinstrumentation is composed of five major components, which are as follows:

(i)   **Measurand**—The measurand is a physical quantity measured using an instrumentation equipment. The human body is a source of the measurand and produces bio-signals, for instance, the surface of the body or the heart's blood pressure (Jaganathan et al. 2014).

(ii) **Sensor/Transducer**—The transducer is a device that converts one type of energy into another, typically electricity. The piezoelectric signal, for example, turns mechanical vibrations into electrical signals. Depending on the measurement, the transducer produces a usable output. The sensor senses the signal from the source. It is what connects the signal to the human (Jaganathan et al. 2014).

(iii) **Signal Conditioner**—The output from the transducer is converted into an electrical value using signal conditioning circuits. This value is sent to the display or the recording system by the instrument system. Amplification, filtering, analog-to-digital, and digital-to-analog conversions are all part of the signal conditioning process. Instrument sensitivity is increased using signal conditioning (Hang and "(PDF) 2002).

(iv) **Display**—t is used to visually depict the parameter or quantity being monitored, for instance, a Chart recorder, Cathode Ray Oscilloscope (CRO). Alarms are often used to detect audio signals. Viz. Signals are generated, detected, and displayed by a Doppler Ultrasound Scanner for Fetal Monitoring (Hang and "(PDF) 2002).

(v)   **Data storage and transmission**—Data storage is a method of storing information that can be retrieved later. In recent years, hospitals have begun to use electronic health records. Telemetric systems use data transfer to send data from one point to another over a long distance (Hang and "(PDF) 2002).

4.2.1   Application of Bioinstrumentation

(i) Scientific method

As part of the scientific method, researchers frequently need to deploy an apparatus to collect data. For example, we might hypothesize that exercise decreases blood

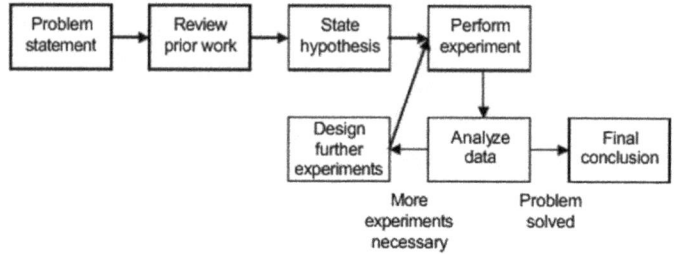

Fig. 12 Schematic representation of working of the scientific method. (after Hang et al. 2002) (Jaganathan et al. 2014)

pressure, but we need to test and analyze the hypothesis to confirm or deny it. Experiments are frequently repeated numerous times. The results can then be statistically evaluated to estimate the possibility that the results were created by chance. The findings are published in scholarly publications with sufficient detail so that others can duplicate the experiment and validate the findings (Fig. 12).

(ii) **Clinical diagnoses**

Instrumentation is widely used by physicians to collect data as part of a scientific approach to treat the disease. For example, a physician observing a patient's history and discovering that the patient has complained of blurry vision would consider diabetes as a possible differential diagnosis for the patient.

Pressure, flow, and temperature are examples of biomedical measurement systems that can be categorized based on the level of the sensation perceived. It is possible to compare multiple methods for measuring any quantity using this classification system, which is a significant advantage. The principle of transduction is utilized when applying a second classification system, such as resistive, inductive, capacitive, ultrasonic, or electrochemical. Different applications of each principle can assist in better comprehension of each concept; in addition, new applications of each principle may become apparent. Each organ system, such as the cardiovascular, pulmonary, neurological, and endocrine systems, can be explored separately by employing various measurement techniques to gather information. However, because this method isolates all the key metrics for specialists who need to know only about a specific area, it results in much overlap in terms of the amounts felt and transduction principles.

4.2.2 Biomermristor

Non-linear two-terminal electrical components, initially described and named by Leon Chua in 1971, are known as memristors (Chua 1971). Typically, a bio-memristive device is created by sandwiching a functional layer consisting of a bio-derived material between two conducting electrodes. Conductive filaments, Schottky excitation, the tunnelling impact of charge, the traps and de-trapped electrons, and the charge transfer mechanisms all play a role in determining the function of the memory (Fu et al. 2020). The presynaptic membrane, synaptic gap, and postsynaptic membrane resemble the top electrode, functional layer, and bottom electrode in the memory device. A pulse signal is supplied to the top electrode of the memristor and then transported down to the bottom electrode, causing the flow of conductive ions in the active layer (Sun et al. 2020).

Researchers are drawn to biomaterials because of their low cost, environmental friendliness, renewable nature, long-term viability, lightweight, ease of fabrication, high storage density, and quick reading/writing speeds. Overuse of metal oxides and compounds has resulted in environmental degradation and wasteful use of natural resources. As a result, using organic materials that are good for the environment and long-lasting is necessary. Since biopolymers can be made from living creatures, they don't need to be synthesized and processed in a sophisticated chemical manner. The current environmental issues and the cycles of sustainable energy can be efficiently mitigated by trash recycling. Because of its outstanding scalability, high flexibility, ease of processing, and low fabrication cost, the biomemristor has attracted much interest. Because of their biodegradability, natural biomaterials and polymers can be used in electrical devices to help achieve electronic device recycling for a more sustainable electronics sector. Biomemristor development hinges on the notions of reusability and biodegradability (Fu et al. 2020; Sun et al. 2020).

4.2.3   Bio-based Material Used in Biomemristor

Materials used as functional layers in memristive devices fall into three categories: inorganic materials (compounds and metal oxides), organic materials (natural biomaterials and polymers), and organic–inorganic hybrid materials. Inorganic materials are the most common s used as functional layers in memristive devices. Mesoscale memristive devices based on natural biomaterials are constructed from biopolymers created by living organisms. This group of biopolymers can be further subdivided into two groups: carbohydrate-based and protein-based. The memristor effect has been demonstrated in various biomaterials, including dead leaves, silk fibroin, ferritin(protein), and fruit peel, among other things (Sun et al. 2020).

A biomemristor with ag/WS/ITO structure was successfully fabricated using WS(walnut skin) which exhibits a rectangular hysteresis loop of current and voltage (Fu et al. 2020). Along with this bio-memristor made up of egg albumen fabricated with hydrogen-peroxide has been obtained (Sun et al. 2020). Moreover, plat leaves are used for the production of biomemristors such as Ag/dried leaves /Ti/PET structure fabricated on polyethylene terephthalate (PET) substrate (Fu et al. 2020). However Ag/Lotus leaves/ ITO are also derived from lotus leaves (Mao et al. 2019).

Biomaterials-based memristive devices have a wide range of potential applications, such as artificial synapses for neuromorphic computing, wearable electronics, and implantable device for personalized health, Resistive random-access memory

(ReRAM)devices for artificial intelligence. Moreover, biomaterial-prepared memristive devices have many other advantages, including implantability, wearability, renewability, degradability etc. (Fu et al. 2020).

Recently, the memristive device, which is prepared using biomaterials as basic components, has become the research focus due to its potential application in medical diagnosis in the biomedical field (Sun et al. 2020).

4.3   Biomechatronics

Biomechatronics is a branch of engineering that combines biology, electronics, mechanical engineering, and mechanics to create devices that can compensate and replace physiological functions in humans. These devices are used in therapeutic and diagnostic procedures and are manufactured using the union of biology, electronics, mechanical engineering, and mechanics (Fig. 13).

Optimizing the interaction between human physiology and electrochemical devices is the primary goal of biomechatronics research and development. This technology, which has the ability to mimic the human body, is used in industries such as robotics and neuroscience, among others. The feedback loop can be improved by transmitting and receiving information from a device, which allows for more precise control. The information from the devices is collected by the mechanical sensor and sent to the controller/biosensor via the interface in this configuration.

A biosensor is used to detect the intended movement of humans, which can be put within or outside of biomechatronic devices. The information collected by the biosensor is interpreted by the controller and transmitted to an actuator. A user's original muscle can be replaced by the actuator, which can either aid or completely replace the user's movement.

Fig. 13 Schematic representation of biomechatronic system integration (after Mazzolai et al. 2010) (Qi et al. 2019)

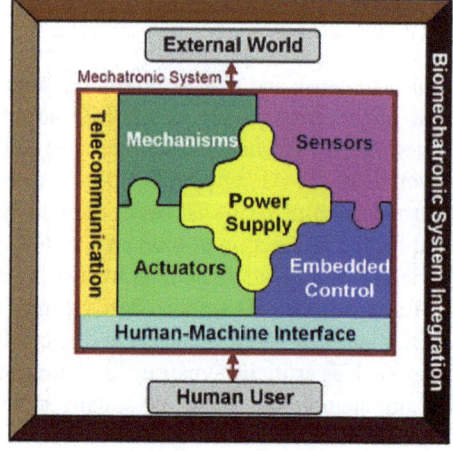

4.3.1 Application of Biomechatronics

Biomechatronics, as the name implies, has a wide range of applications in the medical, biological, mechanical, and electronic fields, among others. Robotics-prosthetics, biosensors, and neural networks are the most important and most researched applications (Fig. 14).

1. Robotic- Prosthetics

It is critical to understand the involuntary and voluntary movement of the body and the function of the body's muscles to comprehend the fundamental requirements of prosthetic and orthotic devices. Not only is the robotic prosthetic being employed, but also robotics in conjunction with virtual reality (VR) approaches to help patients with walking difficulties after a stroke. Traditional ways for controlling robots, such as pulling on the cables of body-powered gripper systems or reading binary myoelectric signals from residual native muscles, have advanced beyond standard care. Medical bionics techniques utilizing surgical, implanted, and surface signal detection strategies can directly access the intended motor control signals generated by the brain and nerves. It has also been established and is beginning to mature possibilities for prosthetic feeling, allowing the activities of previously insensitive prosthetic limbs to be sensed through suitable feedback channels. Advanced control and feedback combined with bone integration were recently demonstrated to be advantageous, highlighting the significance of building more inclusive and clinically beneficial systems (Naidu et al. 2012) (See Fig. 15).

(i) Robotic leg and arm prosthetics

Traditional prosthetics made using a specially designed compression cup, bionic arms and limbs are attached to the body and brought into touch with the skin with the help of the sensor. Body-powered harnesses are now being used to ease muscle movement due to technological advancements. Bionic prosthetics can recognize specific electric

Fig. 14 Integration of various disciplines in biomechatronics (after Naidu et al. 2012) (Mazzolai et al. 2010)

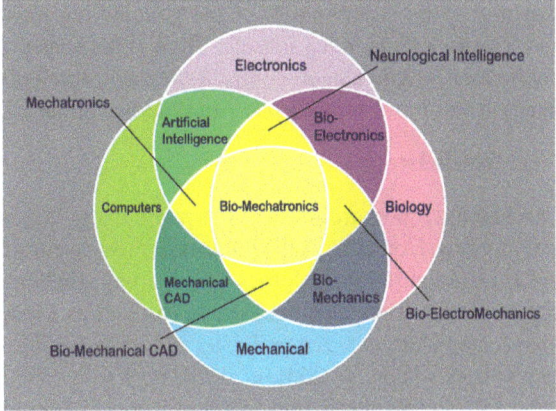

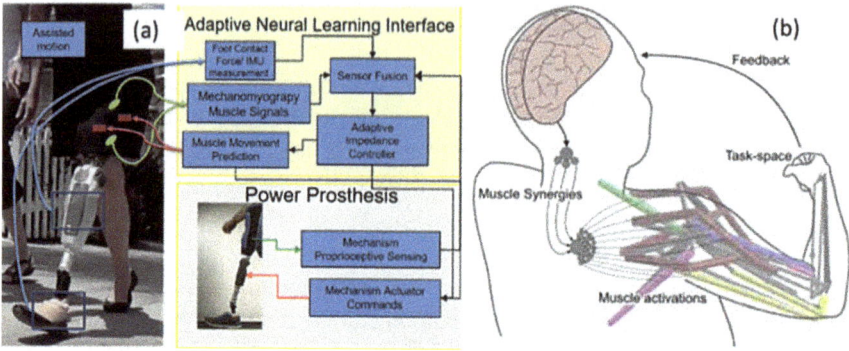

Fig. 15 **a** working of sensory-motor interface for lower extremity robots(SMILER) (after Filip et al. 2018), **b** schematic of working of robotic arm (after McPhee et al. 2018)

impulses from muscle and translate them into action; this movement is related to an actual muscle which performed that movement prior to the amputation, allowing the user to resume their normal activities. A procedure known as 'electromyography' is used to measure and record the activity in bionic limb and arm prosthesis sensors, which are electrodes that make contact with the skin and record the activity. Because they are wearable, these bionic prostheses are simple to use; we may effortlessly remove or attach the prosthetics. While we can modify the functionality of the prosthetic with the use of a single button or switch, we cannot change the appearance of the prosthetic (Zhang et al. 2018; Marasco et al. 2021).

(ii) **Knee brace**

A controlled knee brace has been designed to assist stroke patients who have partial paralysis in the rehabilitation of their leg muscles after the stroke. It is possible to increase the functionality of existing knee braces by using the newly built assistive bionic joint, which features a control algorithm that makes use of sensor signals collected from the patient's leg muscles to operate the joint. Several fluidic muscles that mirror the functionality of the genuine leg muscles are used to replicate the activity of each main leg muscle. To read the sensor data and change the muscular contraction length, a microcontroller is used, which provides the wearer with greater strength and movement (Wilson and Vaidyanathan 2017; Nagel et al. 2017).

(iii) **Exoskeleton**

Exoskeleton robots are able to move particular joints and accomplish various anatomical motions. However, as the number of moving parts and device modules increases, system configuration becomes increasingly challenging. Mechanical design and control methods are also made more difficult as a result of the shoulder's changing joint center. Currently, ArmeoPower and armeoSpring are the main two commercially available exoskeletons (Valle et al. 2021). A fully functional soft robotic elbow sleeve made up of elastomeric and fabric-powered pneumatic actuators with intent-based control, using a combination of elastic fabric and soft actuators increases the force required to do some action. According to the researchers (Valle et al. 2021), there was a substantial difference in control methods between the two groups, which could aid rehabilitation patients in completing ideal motions. Fang et al. (Elliott et al. 2013) have discovered that in a normal gait, people's arms swing in unison with their legs. A rotational orthosis for gait training with arm swing was created as a result of this research. Gait rehabilitation could benefit from the use of an exoskeleton.

4.3.2 Bio-based Polymers Used in Biomechatronic

A prosthetic is composed of two pieces: a prosthesis and a socket. The prosthesis is the largest of the two elements. Metal or polymer can be used to construct the prosthesis that will replace the missing arm or leg, while the polymer is used to construct the socket, which provides for greater flexibility and durability. The choice of the material is critical in the field of prosthetics because the weight and cost of the prosthetics are directly related to the materials used. Previously, lightweight metals such as titanium and aluminium were utilized in prosthetics, but nowadays, polymers had replaced them.

As prostheses should be lightweight, polymers are primarily used for their construction. Researchers from Fraunhofer institute (Fang and Hunt 2021) discovered a biopolymer compound that was utilized to create functioning models. The production process makes use of biopolymers and natural fibres with a few additions, while the fabrication technique makes use of the plate extrusion process; on the other hand, knee braces are manufactured utilizing biodegradable polymers through the injection moulding method. Polypropylene-based materials including polyethylene, polypropylene, acrylics, and polyurethane have largely supplanted leather and wood in the production of a lightweight pylon, and carbon fibres are now used to make lightweight pylon instead of leather and wood. Other than these, computer-aided design and manufacturing (CAD/CAM) tools, as well as a mixture of plaster of paris and other materials, are used to create advanced smart prosthetics for patients. Construction of the socket of the prosthesis is accomplished by the use of a lamination or sandwich layering process. Initially the material is cast between two layers of polyvinyl alcohol (PVA) and polyvinyl chloride (PVC), resin such as epoxy resin with carbon fibres, and hybrid matt, is injected with the help of a vacuum into the gap between the two layers. Now, natural-based bio-composites are widely employed, and natural fibre can be weaved into layers to create a variety of different effects. Hence it is preferable to use the biopolymer, or natural thermoplastic, which is created within the cells of living organisms, rather than traditional petroleum-based plastics. Polyhydoxyalkanoate (PHA) is a naturally occurring compound generated by a wide variety of microorganisms for use as a carbon storage source. Polyhydroxyvalerate (PHV), in addition to polyhydroxybutyrate (PHB), is also used in the fabrication of prostheses, as previously stated (Krombholz and Ganster 2013; Me and Rosalam 2012).

4.3.3   Future Scope of Biomechatronic

The high price of biomechatronic devices in the medical field warranted further investigations in the areas of research and development and commercialization, especially regard to battery power and usability. In addition, a brain link between the human body and a prosthetic is also in need of advanced technology. As a result of advancements in this sector, it is now possible to construct intelligent prostheses and orthotic systems that can adapt to the needs and conditions of their users. Smart pancreas pacemakers for diabetes patients can be developed which can release insulin in a controlled manner, allowing the patient's blood sugar (glucose) level to be maintained. The construction of a mentally controlled digital camera can be accomplished through the use of neural connections and intelligent robotic systems in order to provide a vision for blind individuals (Krombholz and Ganster 2013).

4.4   Robotics

Robots are machines that are able to do complex tasks on their own without human intervention. Some field of rorbotics such as soft-human-friendly robots, complex minimally invasive medical testing like colonoscopy, angiography, angioplasty, localized medicine delivery, and other medical operating procedures are dependent on bio-based materials penetration (Khan et al. 2015). If not surpass, bio-inspired soft robots have the potential to match natural organisms' flexibility and multi-functionality. For this, all the components of soft robots must be woven together in the design of bio-inspired robots. For example, integrating soft and stiff materials into structures with global deformability can achieve the desired capabilities. The progress in this field is also dependent on the advances in the area of 3D printing technology and developement of new multifunctional materials (Zhang et al. 2018).

4.4.1   Soft Robotics and Soft Computer

Traditional rigid-bodied robots face difficulty in operating and uncontrolled settings due to their separate mechanical construction operating in uncontrolled settings. Robotic components and biological organism tissues have different elastic modulus, which might lead to safety difficulties. The development of soft-bodied robots was prompted by the demand for robots that were both docile and safe. As a result of their continually changing bodies, soft-bodied robots have a greater degree of freedom than standard robots, allowing them to accomplish more complex tasks (Zhang et al. 2018).

Robotic systems of soft robotics and soft computer have typically relied on microcontrollers made of hard materials for most of their computing. But soft robots face obstacles in their fabrication process due to these hard materials, hence bio-based materials are used instead of hard materials. However, pneumatic networks and fundamental logic operations have been demonstrated in soft computers, which are still in the early stages of research. Along with this, the use of living cells as computers is a new field of biotechnology research (Angew 2019).

4.4.2 Bio-based Hydrogel in Soft Robotics

Hydrogels are formed when water and hydrophilic polymer chains are arranged in a network arrangement to form a gel. Because of the cross-linking of polymer chains, hydrogels can be transformed into three-dimensional elastic solids. The topology of the polymer network can be modified in order to sustain mechanical strains of up to 1,000 times its original value. owing to the high-volume proportion of water, hydrogels have a young's modulus in the range of 1–100 kPa, which is softer than that of other compliant materials owing to the high-volume proportion of water. Moreover, biological compatibility is determined by whether the hydrogel's polymer network is made up of non-toxic polymers or not. As a result, hydrogels can be employed in soft robotics to broaden the range of biological applications that can be explored (Lee et al. 2020) (Fig. 16).

They have a unique diffusion-based actuation system that allows their physical properties to be easily modified, and they can be sensorized utilizing a number of additives to get a range of results. Their high-water content, combined with their primarily biological composition, make it difficult for them to maintain stability and endurance over extended periods of time. However impedance layers that are parasitic to the hydrogel may make it difficult to make good contact between the hydrogel and the electrode, which may cause unwanted noise in the measuring equipment. As shape changes have no effect on the migration of mobile ions in the hydrogel network, the electrical resistivities of these hydrogels remain constant, rendering them sensitive to changes in both the length and area of the hydrogel network. As a

Fig. 16 Composition of polymer hydrogel (after Lee et al. 2020) (Hardman et al. 2022)

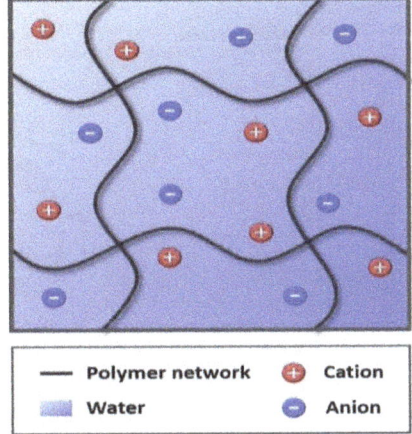

result of the inability to distinguish between the applied loads in different directions, their use to strain sensing is restricted (Zhang et al. 2018).

4.5   Neural Network

The brain is made up of billions of neurons that are linked together to form a complex neural network. In the neural network, communication occurs when bioelectrical signals known as action potentials (APs) are created and propagated across synapses between adjacent neurons, resulting in the transmission of information. A detailed examination of the neural activity is essential for understanding the brain functions and detecting brain illnesses, because the brain processes information through spatiotemporal activity patterns of the neural network (Fig. 17).

In neural network devices at the device interface, cell components must be supported by a scaffold or carrier. For better tissue communication and integration, a scaffold that can promote cell inclusion is preferred. The disintegration of the encapsulating scaffold must be synchronized with the growth and differentiation of cells within the scaffold. Fig. 1.16a illustrates how scar tissue formation can be prevented if a living interface mediates the difference between the "hard" device and the "soft" neural tissues. A low Young's-modulus hydrogel or soft conducting polymer is utilized to fabricate neural tissue-like electronics on the sciatic nerve using regular photolithography processes and a unique laser engraving method. Since then, many polymer-based neural interfaces, forms, and topologies have been created, all of which have shown improved long-term compliance with the device. There are still

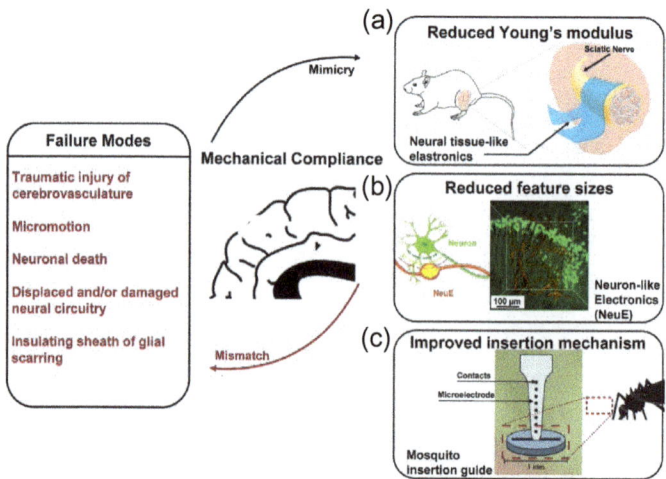

Fig. 17   Bioelectronic Neural Interfaces Inspired by the Mechanical Properties of Neural Tissue. (after woods et al. 2020)

orders of magnitude larger Young's moduli for the polymer materials utilized in these instances than Young's moduli of neural tissue, requiring ultrathin and macroporous electronics. This was first accomplished by woods et al. (2020), that a bioelectronic neural implant with Young's modulus similar to that of neural tissue could be made utilizing "elastronics," which comprises an elastic photoresist and a highly conductive soft hydrogel as the conductor (Woods et al. 2020). Figure 1.16b shows that the diameter and tip geometry of microwires can be varied to reduce the bending stiffness. The chronic immunological response to hard materials can be reduced using (Microthread electrodes) MTEs, which are polymer-coated 7-mm-diameter carbon fibre electrodes. In Fig. 1.16c, mesh electronics are injected into the brain using a syringe. A syringe filled with phosphate-buffered saline, together with the mesh electronics, can be injected into the target area of the brain via stereotaxic syringe injection. "Sewing machine-like" devices can be rapidly inserted into the specific brain regions while avoiding vital blood vessels to minimize the impact of the procedure. Neuronal interfaces can be implanted with better critical buckling force and precision by taking inspiration from mosquito skin-piercing fascicle mechanisms (Woods et al. 2020).

Moreover, living bioelectronics must encourage the establishment of synapses between the encapsulated cell population and the endogenous cell population to establish functional connections between the device and the target neurons to work properly. Furthermore, these encapsulated cells must sustain a meaningful connection with the implant electronics over an extended length of time.

4.5.1 Applications of Conducting Polymer and Hydrogel in Neural Networks

Bioelectronics can communicate with the nervous system directly by using the transducer signal created by electronic elements in their construction. In this case, metal microelectrode arrays (MEAs) are used to monitor the activity of several neurons simultaneously. Along with this, traditional metal electrodes are capable of probing and stimulating brain activity at the single-cell level due to their high temporal resolution. In recent decades, electrical recording via metal-based bioelectronics has made substantial contributions to our fundamental understanding of the neural activity, particularly in electrophysiology. Neurological illnesses such as Parkinson's disease are being studied extensively with the implantation of stimulating electrodes into the brain, which is becoming increasingly common nowadays (Fang et al. 2015).

(i) **Conducting polymer for neural interface**

In neural interfacing, the interface of bioelectronics requires the exchange of electronic flags because of the electrons of the electronic component. The use of conducting polymers rather than metals is recommended because they offer a variety of advantages over metals and inorganic semiconductors, including the ease of amalgamation and alteration, low-temperature and cost-effective handling, and compatibility with adaptable plastic substrates. Because of their soft nature and chemical resemblance to the biomolecule, bio-based conducting polymers such as poly(3,4-ethylenedioxythiophene) (PEDOT), poly(pyrrole) (PPy), and their derivatives are currently receiving a significant deal of attention for neurological recording and stimulation. Developing neural interfaces that integrate seamlessly with the brain while reducing the impact of brain penetration and implantation is an inherently difficult task. Buckling-free device implantation is used to overcome this issue, which comprises stiffeners and shape-memory polymers, which are inspired by the stimulus-responsive materials present in biology. Thin-film polymer probes coated with bioresorbable PEG or silk fibroin can be used to implant a flexible device (Shur et al. 2020).

(ii) **Mimicking morphology of Neurite**

Many developing technologies necessitate the development of biomaterials that look and behave like the brain tissue. To be effective, neural probes for brain–machine interfaces, micro physiological models of neurological diseases, scaffolds for neural tissue engineering, brain organoids, and brain proxies must have physical properties similar to those found in the brain tissue themselves. Implant rejection owing to the foreign-body response can be minimized in in vivo applications by ensuring that the mechanical fit between the implants and the surrounding brain tissue is as close as possible. Achieving accurate recapitulation of neurons and glial cells in native environments in-vitro is critical for the proper differentiation of the cells and their motility, function, and proliferation. Recapitulation is also important for studying cellular responses to chemical signals and new treatments in vitro (Fig 18).

Due to nanoscale building blocks of protein in the perineuronal net and the neural interstitial matrix, also known as the brain's extracellular matrix, the physical properties of the cerebral tissues differ depending on where they are located. Physical features such as nutrition, oxygen transport, and conductivity are crucial for the rational design of the brain tissue-mimicking hydrogels. Biomaterials that capture the diverse properties of brain tissue can be used in a variety of applications, including neural probes for brain–machine interfaces, brain organoids for studying neurogenesis or drug screening. They can also be used as injectable materials to replace current hard materials in the treatment of certain types of brain tumors, such as glioblastoma multiforme, among others (Fig. 19).

Fig. 18 Printable PEDOT:PSS coating on a neural interface. (after Shur et al. 2020)

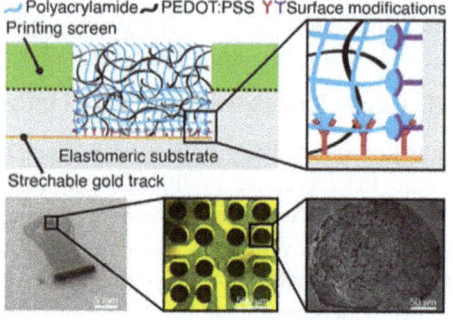

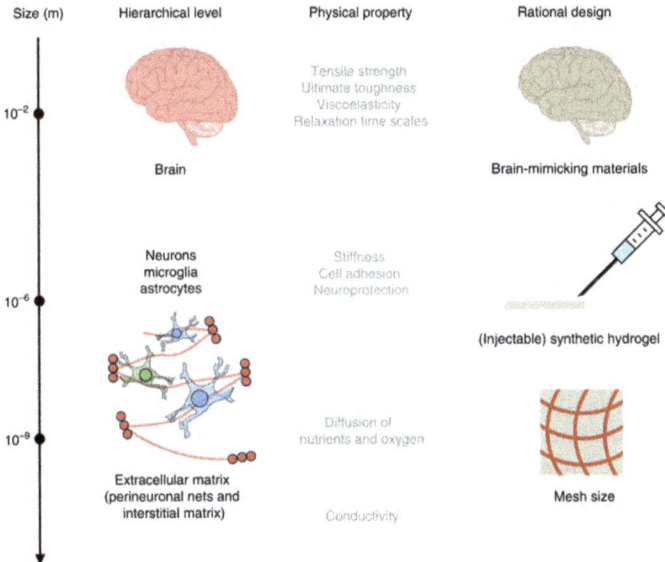

Fig. 19 Physical features of brain tissue that need to resemble at different scales (after Axpe et al. 2020)

(iii) **Polystyrene nanofiber**

Myelination is aided by polystyrene nanofibers with diameters that are equivalent to or more than 0.5 mm in diameter. However, normal axonal diameters fell within the range of their findings, meaning that the replicating neurite shape helps in the restoration of the endogenous neural tissue environment and the acceptance of the device by the tissue. A further benefit of imitating the structural characteristics of neurites is ongoing in chronic studies to improve the interface between the neural tissue and electrophysiological neural probes (Goding et al. 2018).

4.5.2 Challenges and Limitations in Bioelectronic–Neural Interface

The bioelectronics–neural interface is currently confronted with two primary obstacles. The first thing to note is that for bioelectronic signals to be of high quality, electrons and ions need to be coupled at metal/neural interfaces, where they can interact with each other to transmit and receive information. As a result of the resistance between the electrodes and electrolytes at the metal/neural interface, thermal noise, also known as Johnson noise, is generated at the electrode–neural interface. As a result, the long-term stability of metal/neural interfaces is complicated since the metal surfaces are positioned far away from soft and wet neural tissues. However, even though flexible and ultra-thin neural probes conform to the brain tissue and limit chronic immune responses, they necessitate an insertion approach that avoids mechanical deformation, a typical failure mode for these devices. Furthermore, engineered biomaterials and brain tissue are regularly found to be mechanically mismatched, and often they are compared on their stiffness property alone, and hence carefully investigated (Goding et al. 2018).

4.6 Biosensor

4.6.1 Classification of Biosensor

Biological or chemical processes are measured using biosensors, which translate produced signals into corresponding concentrations of the analyte in the reaction, which can then be compared to a standard curve. Biosensors are used in a wide variety of applications, including drug discovery, lab on a chip, biomedicine, food safety, diagnosis, environmental monitoring, security etc. A biosensor is made up of five components which are mentioned below:

(i) **Analyte**: an analyte is a material of interest that is intended to be detected by the instrument. Red blood cells, white blood cells, and haemoglobin are analytes in the tests devised for the complete blood count test, also known as a CBC blood test (Naresh and Lee 2021).

(ii) **Bioreceptor**: Bioreceptors are molecules specifically sensitive to the analyte they are detecting. Bioreceptors include a variety of enzymes, tissues, or cells that contain nucleic acids such as RNA or DNA, as well as antibodies. When the bioreceptors come into contact with the analyte, they generate signals in the form of light, heat, PH change, or mass change, depending on the situation. Bio-recognition is the word used to describe this process (Naresh and Lee 2021).

(iii) **Transducer**: An example of a transducer is one that transfers one form of energy into another. The transducer is a heart of a biological sensor; it is responsible for converting the energy created during the biorecognition process into detectable signals. Signalization is the term used to describe this process. When using most transducers, the amount of optical or electrical signals created by the transducer determines the number of analytes–bioreceptor interactions that occur (Naresh and Lee 2021; Bhalla et al. 2016).

(iv) **Electronics**: It is the electronic component of a biosensor that processes and encodes the transmitted signal to display it. Signal conditioning functions, such as amplification and digital signal conversion, are performed by complex electronic systems. The processed signals are subsequently displayed on the biosensor's display device, which analyses them (Naresh and Lee 2021).

(IV) **Display**: using a user interpretation system, such as the display technology on a device or a direct printer, numbers or curves that are understandable to the user are generated. This module frequently consists of the hardware and software that provide biosensor results that are easy to understand. A numeric,

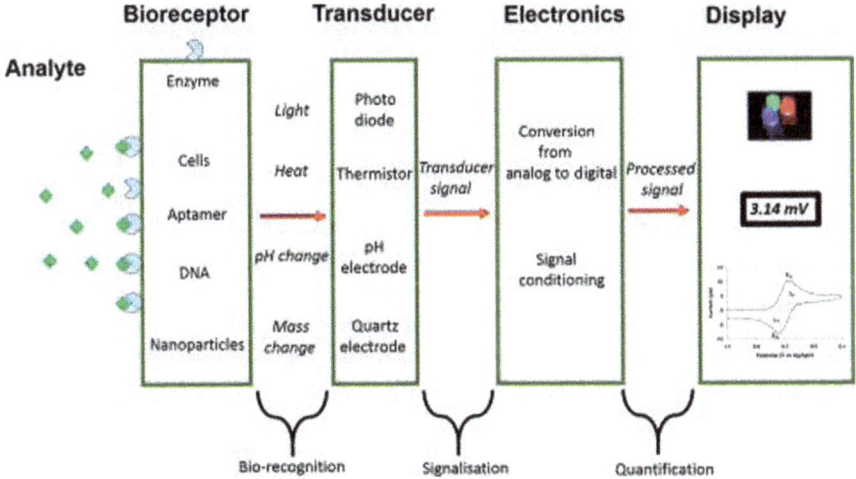

Fig. 20 Schematic of biosensor (after Bhalla et al. 2016)

visual, tabular, or picture-based output signal can be displayed on the display screen based on the end-user's requirements (Bhalla et al. 2016) (Fig. 20).

When an analyte interacts with a receptor, the transducer turns the frequency and rate of interaction into a quantifiable physical phenomenon. There are numerous ways in which the conversion might be interpreted, and biosensors are classified according to the sort of conversion technology that is employed. The following are the classifications of common conversion systems:

(i) **Optical techniques**

As a result of the interaction between the analyte and the receptor, changes in optical properties such as the intensity or frequency of absorption and emission, the resonance frequency of surface electrons, the refractive angle etc., might occur. Biosensors can record a wide range of spectrochemical characteristics utilizing a variety of different types of spectroscopies (absorption, fluorescence, phosphorescence, Raman, surface electron resonance, refraction, diffusion, etc.) (Bhalla et al. 2016).

(ii) **Electrochemical techniques**

Electrochemical detection is a conversion method used in biosensors to detect specific molecules. In conjunction with electrochemical approaches, optical detection methods, such as the very sensitive fluorescence optical approach, can be used to detect the presence of a substance. Electrochemical conversion can be particularly advantageous because many analytes do not have significant fluorescence and labelling them with a fluorescent marker is usually a time-consuming process. Due to the combination of the sensitivity of electrochemical measurements with the selectivity provided by the bioreceptor, the detection limit is comparable to that of fluorescence biosensors (Bhalla et al. 2016).

(iii) Mass-sensitive techniques

Another form of conversion used in biosensors is the measurement of tiny mass changes generated by the interaction of the analyte and the bioreceptor. In this approach, piezoelectric crystals are employed to generate electricity. These crystals vibrate at a certain frequency as a result of the introduction of an electrical signal. Due to chemical bonding, as the mass of the crystal increases, the oscillation frequency of the crystal changes. This change is electronically monitored and used to quantify the amount of mass added. Another class of mass-sensitive approaches employs the measurement of cantilever flexion caused by receptor–ligand contact for the purpose of analysis (Bhalla et al. 2016).

For more than two decades, researchers have been working on the development of cumulative biosensors that can detect multiple species at the same time. For each species tested, each array has a bioreceptor that can detect one of the several analytes present in the sample. The aggregate of these arrays uses the same excitation source and measuring process as the individual arrays for all of the species studied. A variety of other sensor components, including the sampling system, sensor element (including the converter), amplifier (including the amplifier), and other necessary pathways and circuits, are integrated into a single chip in assembled biochip systems (Bhalla et al. 2016).

4.6.2 Natural Polymers and Other Semiconducting materials used in Biosensors

(i) Chitosan

Chitosan's biocompatibility, stickiness, and excellent film-forming capacity, among other characteristics, have all contributed to the development of biosensors, which implies that it has a significant binding force for proteins and enzymes. Jain et al (Jain et al. 2019b). developed a biosensor interface by co-immobilizing adenosine deaminase (AChe) and choline oxidase (ChOx) onto palladium nanoparticles (Pdnano). On the surface of the gold electrode, the Pdnano has adsorbed on top of molybdenum disulfide (MoS_2) nanostructures that had been electrodeposited on the gold electrode (Jain et al. 2019b).

(ii) Cellulose

Cellulose is the primary structural component of plant cell walls and is now an extensively used biopolymer in a variety of applications. It is made up of polymer chains that are derived from glucose. Its excellent transparency, dimensional stability, and simplicity of modification distinguish it from other plastics on the market. Furthermore, because cellulose and its derivatives are biocompatible, they are appropriate substrates for the incorporation of physiologically active molecules into their structure and function. Because the typical carrier substrate for enzymes should be stable, inert, and resistant to mechanical changes, the cellulose matrix is also an excellent material for chemical covalent bonding, physical adsorption, and immobilization. It can be used to detect a variety of chemicals, including urea, lactic acid, glucose, genes, amino acids, cholesterol, and proteins, among others (Huang et al. 2021).

(iii) Metal–organic framework (MOF)

Metal–organic frameworks (MOFs), a novel type of porous hybrid material has attracted considerable research because of its intrinsic biodegradability and ability to integrate biocompatible components, making them attractive candidates for use in biological applications. MOF comprises metal ions and organic ligands that are tightly linked together by a strong coordination bond. MOFs have been used to detect DNA, RNA, enzymes, small molecules, and clinical diagnostics, among other applications. MOF has also been demonstrated to be an effective matrix for enzyme immobilization, owing to its superior properties of high surface area, big porosity, easy tenability of pore size, and changeable surfaces, among other things. MOF's high hierarchical surface area and its excellent porosity allow it to have a high loading capacity, while its good affinity inhibits enzyme leakage from the material (Mehrotra 2016).

(iv) Quantum dots

Quantum dots (QDs) are semiconducting nanocrystals that emit light when excited; they are another form of nanomaterial used in bioanalytical applications. Among the most extensively studied colloidal QDs, cadmium chalcogenides exhibit a broad absorption spectrum as well as a size-dependent restricted emission spectrum. This phenomenon is generated by the varying band gaps of the semiconductor material for different nanocrystal sizes (the smaller the particle, the lower the band gap), which results in different emission wavelengths from the electron–hole recombination exciton. Traditional optical transduction can be used to do efficient multiplexed analysis due to the availability of a wide range of emission wavelengths from QDs of varying sizes. To provide functional groups for bioreceptor immobilization and to anticipate potential toxicity concerns, QDs are now available with inert or biocompatible coatings, which can be applied to provide functional groups for bioreceptor immobilization and to anticipate potential toxicity difficulties. So long as the photophysical recombination event is not interfered with, nearly any biomolecule can be connected to these nanocrystals (Song et al. 2021).

(v) Carbon Nanotubes (CNTs)

Carbohydrate nanotubes (CNTs) are one-dimensional nanotubes extensively explored for applications in various fields, viz. biosensors, diagnosis, tissue engineering, cell tracking and labelling, drug administration, delivery of biomolecular therapeutics etc. Concentric graphite layers are stacked on top of fullerenic hemispheres to form the walls of carbon nanotubes (CNTs). Single-walled CNTs, double-walled CNTs, and multi-walled CNTs are hollow cylindrical tubes constructed of one, two, or more concentric graphite layers topped with fullerenic hemispheres. There are several different designs, excellent electrical and mechanical properties, high thermal conductivity, high chemical stability, superior electrocatalytic activity, low surface fouling, low over-voltage, and a high aspect ratio in this class of materials (surface-to-volume). Because of their high surface-to-volume ratio and novel electron transport characteristics, the electrical conductance of these nanostructures is greatly influenced by even minor surface disturbances, such as those associated with macromolecule binding, resulting in significant changes in electrical conductance (Song et al. 2021).

Biosensors and diagnostics based on carbon nanotubes (CNTS) have been utilized in healthcare, industrial, environmental monitoring, and food quality screening to detect analytes with high sensitivity and specificity. Through the use of electrochemical sensing, it has been possible to detect fructose, galactose, neurotransmitters, neurochemicals, amino acids, immunoglobulin, albumin, streptavidin, insulin, human chorionic gonadotropin, c-reactive protein, cancer biomarkers, cancer cells, bacteria, DNA, and a variety of other biomolecules. Multi-wall carbon nanotubes(MWCNTS) are used in all biosensing applications, and they are the most common. Such 1D nanomaterials, when used in nanosensor arrays, allow real-time and sensitive label-free bioelectronic detection and high redundancy and scalability.

4.6.3 Application of Biosensors in Biomedicine

The use of biosensors in the medical industry is rapidly expanding. Diagnosing diabetes requires precise glucose monitoring, which frequently calls for glucose biosensors. 85% of the global market for blood-glucose biosensors is accounted for by individuals. Also, the diagnosis of cancer and its therapy is critical in this as biosensors are being used to track cancer progression. Developing a biosensor technology could aid in the early detection of cancer and effective therapy delivery. Proteins released and expressed by cancerous cells can be measured using biosensors, which can detect the existence of tumors, as to whether they are benign or cancerous. Moreover, cancerous cells can also be determined by the results of the treatment. Detection of these diseases earlier could reduce the number of deaths associated with them. Fluorometric assays and enzyme-linked immunosorbent assays are some of the methods employed in this investigation to detect biomarkers. However, the methods outlined above are time-consuming and necessitate highly trained and professional employees. The outcome is the use of biosensors to monitor cardiac signs and give early detection (Song et al. 2021).

4.6.4 The Future Scope of Use of Nanoparticles in Biosensors

Nanoparticles used as biosensor components could pave the way for an entirely new class of biosensors in the future. Single-molecular sensors and high-throughput sensor arrays are now possible, thanks to the enhanced mechanical, electrochemical, optical, and magnetic capabilities of nanomaterials. To develop single-molecule multifunctional nanocomposites, nanofilms, and nanoelectrodes, researchers must figure out how to merge the structure and function of nanomaterials with biomolecular structures. This nanoscale composites' behavior on electrode surfaces is dictated by several fundamental principles that aren't well understood by present methodologies, including processing, characterization, interface difficulties, and the lack of readily available high-quality nanomaterials (Mehrotra 2016; Song et al. 2021).

5   Challenges and Limitations for Bio-based Materials in Bioelectronics

In bioelectronics, the development of bio-based conducting polymers and their subsequent processing has been a pivotal breakthrough. In addition to this, the bio-based polymer faces a number of difficulties in its use. PEDOT:PSS, PPy, polyaniline, and other bio-based conducting polymers lose conductivity due to oxidation, which removes the dopant from the material. The long-term electronic and biological stability of a material is essential for medical implants to have a long life span. When the temperature rises in the polymer, a fibrotic capsule is formed, which causes the overall impedance of the polymer to increase. On the other hand, several distinctive mechanisms, such as the surface modification or implant modification and loading with therapeutic drugs or surface gradient, might reduce the production of the fibrotic tissue. Proper sterilization of an implant can prevent inflammatory reactions from occurring when it comes into contact with the body's biological environment. A unique sterilization technique must be designed for each conducting polymer because there is no single sterilization process that works for all conducting polymers. PEDOT:PSS is sterilized using the autoclave process, whereas PPy is sterilized using the gamma method. While in the case of bio-implants, the implant's ability to adhere to the bodily tissue can be influenced by the addition of a functional polymer derivative. Arginine–glycine–aspartic acid (RGD), and other functional derivatives of carboxylic acids are all utilized to improve cell adherence.

Other materials such as hydrogel have been investigated in a limited way and research is in progress, particularly in wearable bioelectronics. The vast majority of wearable bioelectronics' research has concentrated on the development of hydrogel materials. The poor electrical conductivity (1 sm^{-1}) of conductive hydrogels and the hazardous raw materials utilized for manufacturing limit its usage.

Natural polymers such as polysaccharide, chitosan etc., are being used to prevent the minor elongation and breakage that silk nanofiber wires and fibres undergo when wet.

In bioceramics, researchers are more concerned with the adhesion of the bodily tissue and biocompatibility than about biodegradability. A bioresorbable ceramic that may obviate the need for a second treatment and other modulable materials that may reduce the implant's weight and increase the body contact is preferred.

Although we are all aware of the applications of bioelectronics, when it comes to the creation of bio-device interfaces, we still have a long way to go since we need to establish a proper technique of bioelectronics fabrication. The process of fabricating the device differs due to variations in the material synthesis and the way of construction of the device.

6 Future of Use of Bio-based Materials in Bioelectronics

High technological breakthroughs have made it possible to employ 3D printed implants consisting of bio-based conducting polymers. At the same time, bio-ink is being used as a medium for the interaction of the polymer with the bodily tissue in some cases. A critical factor in addressing the issue of dopant loss of polymer implants is the implant's ability to self-dope, which can be accomplished through the use of new sophisticated polymerization processes. Moreover, future developments in hydrogel bioelectronics must make use of the unique properties of hydrogels, which must be founded on a fundamental understanding of tissue–electrode interactions and design parameters, among other things. Hydrogel still has several issues to deal with, all of which can be resolved in the future and allow for further growth in this subject. There are three different approaches that can be used to improve the application of hydrogels in bioelectronics in the future. Improvements in working and biocompatibility can be made by creating hydrogels with specific characteristics; in this way, the functioning and biocompatibility will be more in line with the real tissues. The use of a cutting-edge hydrogel will ensure implant stability and prevent the implant from being removed from the body without permission. In order to simulate the self-healing function of the tissue, many self-healing hydrogels have been produced which are biodegradable. However, further research is needed in this area. The second method is to improve the adhesion of the hydrogel by robust integration, which is still a work in progress with many challenges to overcome. Due to robust integration, implantable hydrogel bioelectronic devices on deforming, fragile, and dynamic wet tissue such as the spinal cord, heart tissue, and muscle will be possible in the future by improving potential, electrical, and mechanical limits for a proper interface. The creation of an ionic conductive hydrogel using 3D printing and ink-jet printing is another component that is entirely reliant on future advancements in the sophisticated synthesis of the hydrogel for use in bioelectronics applications. In the field of bioceramics, the interaction of an artificial bone with an organic chemical in the body at the molecule level is the key to the future. By matching both the physical and chemical properties of the material to the metabolic requirements of the illness and bone tissue, it is possible to increase the feasibility of this approach. A boom in research is presently occurring in the fields of bone fixation, organ design, and 3D printable organ transplantation, all of which have a great deal of opportunity to improve with the help of advanced technology. Enhancement of bioactivity by gene activation, integration of sensing and bioactivity, and the development of smart bioceramics are future technologies.

References

Adeleye AT, Odoh CK, Enudi OC, Banjoko OO, Osiboye OO, Odediran ET, Louis H (2020) Sustainable synthesis and applications of polyhydroxyalkanoates (PHAs) from biomass. Process Biochem 96:174–193

Advances in soft bioelectronics for brain research and clinical neuroengineering - Sung-Hyuk S, Han SI, Joo H, Cha GD, Kim D, Choi SH, Hyeon T, Kim D. https://doi.org/10.1016/J.MATT. 2020.10.020

Alba NA, Sclabassi RJ, Sun M, Cui XT (2010) Novel hydrogel-based preparation-free EEG electrode. IEEE Trans Neural Syst Rehabil Eng 18(4):415–423. https://doi.org/10.1109/TNSRE. 2010.2048579 Epub 2010 Apr 26 PMID: 20423811

Angew (2019) Chem Int Ed. https://doi.org/10.1002/anie.201813402

Annu Rev Biomed Eng 2004. 6:41–75. https://doi.org/10.1146/annurev.bioeng.6.040803.140027

Axpe E, Orive G, Franze K, Appel EA (2020) Towards brain-tissue-like biomaterials. Nat Commun 11(1):3423. https://doi.org/10.1038/s41467-020-17245-x.PMID:32647269;PMCID: PMC7347841

Bambini F, Santarelli A, Putignano A, Procaccini M, Orsini G, Di Iorio D et al (2017) Use of supercharged cover screw as static magnetic field generator for bone healing, 2nd part: in vivo enhancement of bone regeneration in rabbits. J Biol Regul Homeost Agents 31:481–485

Bardhan S, Gupta S, Gorman ME, Haider A (2015) Biorenewable chemicals: feedstocks, technologies and the conflict with food production. Renew Sustain Energy Rev 51. https://doi.org/10. 1016/j.rser.2015.06.013

Bedian L, Villalba-Rodríguez AM, Hernández-Vargas G, Parra-Saldivar R, Iqba HMN Bio-based materials with novel characteristics for tissue engineering applications–a review. Int J Biol Macromol https://doi.org/10.1016/j.ijbiomac.2017.02.048.

Bellani S, Fazzi D, Bruno P, Giussani E, Canesi E, Lanzani G, Antognazza M (2014) Reversible P3HT/Oxygen charge transfer complex identification in thin films exposed to direct contact with water. J Phys Chem c 118:6291–6299. https://doi.org/10.1021/jp4119309

Beuerlein MJ, McKee MD (2010) Calcium sulfates: what is the evidence? J Orthop Trauma 24(Suppl 1):S46-51. https://doi.org/10.1097/BOT.0b013e3181cec48e PMID: 20182236

Bhalla N, Jolly P, Formisano N, Estrela P (2016) Introduction to biosensors. Essays Biochem 60(1):1–8. https://doi.org/10.1042/EBC20150001.PMID:27365030;PMCID:PMC4986445

Brown RM Jr, Czaja W, Young D, Jeschk M (2006) Microbial cellulose integration for wound healing. Provisional Patent 2914BRO

Carrodeguas RG, De Aza S (2011) α-Tricalcium phosphate: synthesis, properties and biomedical applications. Acta Biomater 7(10):3536–3546. https://doi.org/10.1016/j.actbio.2011.06.019 PMID: 21712105

Che Me R (2012) Natural based biocomposite material for prosthetic socket fabricatioN. ALAM CIPTA. Int J Sustain Tropical Design Res Pract 5:27–34

Chen GQ, Hajnal I, Wu H, Lv L, Ye J (2015) Engineering biosynthesis mechanisms for diversifying polyhydroxyalkanoates. Trends Biotechnol 33(10):565–574. https://doi.org/10.1016/j.tibtech.2015.07.007 PMID: 26409776

Chen GQ, Patel MK (2012) Plastics derived from biological sources: present and future: a technical and environmental review. Chem Rev 112(4):2082–2099

Chen S, Jiang Y, Xu Y, Fu J, He X, Su Y, Strnadel B (2019) Micro-environment regulation synthesis of calcium sulfate nanoparticles and its water removal application. Mater Res Express 6(10):1050b8. https://doi.org/10.1088/2053-1591/ab4070

Chua L (1971) Memristor-the missing circuit element. IEEE Trans Circuit Theory 18(5):507–519. https://doi.org/10.1109/TCT.1971.1083337

Conrad R (2021) Bio-based polymers. Kuraray specialty chemicals: elastomer division, www.elastomer.kuraray.com, 5 Mar. 2021, https://www.elastomer.kuraray.com/bio-based-polymers/

Czaja W, Krystynowicz A, Kawecki M, Wysota K, Sakiel S, Wroʹblewski P, Glik J, Nowak P, Bielecki S (2007) Cellulose: molecular and structural biology. in: Brown RM, Saxena IM (eds). Springer Dordrecht, The Netherlands

Eliaz N, Metoki N (2017) Calcium phosphate bioceramics: a review of their history, structure, properties, coating technologies and biomedical applications. Materials (Basel). 2017 Mar 2

Elliott G, Sawicki GS, Marecki A, Herr H (2013) The biomechanics and energetics of human running using an elastic knee exoskeleton. In: 2013 IEEE 13th international conference on rehabilitation robotics (ICORR), pp 1–6. https://doi.org/10.1109/ICORR.2013.6650418

Fang Y, Li X, Fang Y (2015) Organic bioelectronics for neural interfaces. J Mater Chem c 3. https://doi.org/10.1039/C5TC00569H

Fang J, Hunt K (2021) Mechanical design and control system development of a rehabilitation robotic system for walking with arm swing. Front Rehabilit Sci 2. https://doi.org/10.3389/fresc.2021.720182

Fischman G (1995) clare and Alexis. Materials and Applications, The American Society, Bioceramics

Fu T, Liu X, Gao H et al (2020) Bioinspired bio-voltage memristors. Nat Commun 11:1861. https://doi.org/10.1038/s41467-020-15759-y

Ghezzi D, Antognazza MR, Maccarone R, Bellani S, Lanzarini E, Martino N, Mete M, Pertile G, Bisti S, Lanzani G, Benfenati F (2013) A polymer optoelectronic interface restores light sensitivity in blind rat retinas. Nat Photonics 7(5):400–406. https://doi.org/10.1038/nphoton.2013.34.PMID:27158258;PMCID:PMC4855023

Goding J, Gilmour A, Aregueta robles U, Hasan E, Green R (2018) Living bioelectronics: strategies for developing an effective long-term implant with functional neural connections. Adv Funct Mater 28:1702969.https://doi.org/10.1002/adfm.201702969

Green RA, Baek S, Poole-Warren LA, Martens PJ (2010) Conducting polymer-hydrogels for medical electrode applications. Sci Technol Adv Mater 11(1):014107. https://doi.org/10.1088/1468-6996/11/1/014107.PMID:27877322;PMCID:PMC5090549

Gupta KK, Pal N, Mishra PK, Srivastava P, Mohanty S, Maiti P (2014) 5-Florouracil-loaded poly(lactic acid)-poly(caprolactone) hybrid scaffold: potential chemotherapeutic implant. J Biomed Mater Res A 102(8):2600–2612. https://doi.org/10.1002/jbm.a.34932 Epub 2013 Sep 3 PMID: 24038786

Hang LT (PDF) 6846793-Bioinstrumentation.Pdf I Le Thuy Hang - Academia.Edu." (PDF) 6846793-Bioinstrumentation.Pdf I Le Thuy Hang - Academia.Edu, www.academia.edu. Accessed 0 Jan 2002, https://www.academia.edu/30824186/6846793_Bioinstrumentation_pdf

Hardman D, George Thuruthel T, Iida F (2022) Self-healing ionic gelatin/glycerol hydrogels for strain sensing applications. NPG Asia Mater 14:11. https://doi.org/10.1038/s41427-022-00357-9

Hejazi R, Amiji M (2003) Chitosan-based gastrointestinal delivery systems. J Control Release 89(2):151–165. https://doi.org/10.1016/S0168-3659(03)00126-3

Heness G, Ben-Nissan B (2003) Innovative bioceramics. Mater Forum. 27

Houghton PJ, Howes MJ (2005) Natural products and derivatives affecting neurotransmission relevant to Alzheimer's and Parkinson's disease. Neurosignals 14(1–2):6–22. https://doi.org/10.1159/000085382 PMID: 15956811

Hsu H-J, Waris R, Ruslin M, Lin Y-H, Chen C-S, Ou K-L (2017) An innovative α-calcium sulfate hemihydrate bioceramic as a potential bone graft substitute. J Am Ceram Soc 101. https://doi.org/10.1111/jace.15181

Huang X, Zhu Y, Kianfar E (2021) Nano biosensors: properties, applications and electrochemical techniques. J Mater Res Technol 12,2021:1649–1672. ISSN 2238–7854. https://doi.org/10.1016/j.jmrt.2021.03.048

Inamdar N, Mourya VK, Tiwari A (2010) Carboxymethyl chitosan and its applications. Adv Mater Lett 1:11–33. https://doi.org/10.5185/amlett.2010.3108

Iwata T, Doi Y (1998) Morphology and enzymatic degradation of poly (L-lactic acid) single crystals. Macromolecules 31(8):2461–2467

Iwata T, Doi Y, Isono K, Yoshida Y (2001) Morphology and enzymatic degradation of solution-grown single crystals of poly (ethylene succinate). Macromolecules 34(21):7343–7348

Jaganathan SK et al (2014) Biomaterials in cardiovascular research: applications and clinical implications. BioMed Res Int 2014 (2014):459465. https://doi.org/10.1155/2014/459465

Jain U, Khanuja M, Gupta S, Harikumar A, Chauhan N (2019b) Pd nanoparticles and molybdenum disulfide (MoS2) integrated sensing platform for the detection of neuromodulator. Process Biochem 81:48–56. https://doi.org/10.1016/j.procbio.2019.03.019

Jain M, Zinjarde S, Gokhale D (2019a) Poly-Lactic acid (PLA): synthesis and biomedical applications. J Appl Microbiol 127. https://doi.org/10.1111/jam.14290

Jeong J, Kim JH, Shim JH, Hwang NS, Heo CY (2019) Bioactive calcium phosphate materials and applications in bone regeneration. Biomater Res 14(23):4. https://doi.org/10.1186/s40824-018-0149-3.PMID:30675377;PMCID:PMC6332599

Jirage AS, Baravkar VS, Kate VK, Payghan SA, Disouza JI (2013) Poly-βhydroxybutyrate: intriguing biopolymer in biomedical applications and pharma formulation trends. Int J Pharmaceut Biol Archives 4(6):1107–1118

Khan F, Tuason K, Sena A, Fagan JM (2015) Bioplastics: an alternative to petrochemical plastics in prosthetics. https://doi.org/10.7282/T3QV3PCK

Korri-Youssoufi K, Garnier F, Srivastava P, Godillot P, Yassar Y Laboratoire des Maté´riaux Mole´culaires, CNRS 2 rue Dunant, 94320 Thiais, France

Krombholz A, Ganster J (2013) Green week: fraunhofer shows bio-tiles and heat-resistant biopolymers - Press Release January 17, 2013." Fraunhofer-Gesellschaft. www.fraunhofer.de, 17 Jan. 2013. https://www.fraunhofer.de/en/press/research-news/2013/january/bio-tiles-and-heat-resistant-biopolymers.html

Köping-Höggård M, Vårum KM, Issa M, Danielsen S, Christensen BE, Stokke BT, Artursson P (2004) Improved chitosan-mediated gene delivery based on easily dissociated chitosan polyplexes of highly defined chitosan oligomers. Gene Ther 11(19):1441–1452. https://doi.org/10.1038/sj.gt.3302312. Kotlyar DS, Shum M, Hsieh J, Blonsk

Lee PY, Costumbrado J, Hsu CY, Kim YH (2012) Agarose gel electrophoresis for the separation of DNA fragments. J Visual Exp Jove 62:3923. https://doi.org/10.3791/3923

Lee Y, Song WJ, Sun JY (2020) Hydrogel soft robotics. Materials Today. Physics 15:100258

Legeros RZ (2008) Bioceramics and their clinical applications hydroxyapatite, 367–394.https://doi.org/10.1533/9781845694227.2.367

Li L, Ma Z, Xu P, Zhou B, Li Q, Ma J, He C, Feng Y, Liu C (2020) Flexible and alternant-layered cellulose nanofiber/graphene film with superior thermal conductivity and efficient electromagnetic interference shielding. Compos Part A Appl Sci Manuf 139():106134–. https://doi.org/10.1016/j.compositesa.2020.106134

Madsen EJ, Kohn DH (2021) Bioceramics. In: Biomedical engineering and design handbook, 3rd Edition, M. Kutz, Editor, McGraw-Hill, New York

Malmivuo J, Plonsey R (1995) Bioelectromagnetism. 1. Introduction

Malmivuo J (1999) Application of bioelectromagnetic methods in the detection of the electric sources of the brain and the heart

Mantione D, del Agua I, Sanchez-Sanchez A, Mecerreyes D (2017) Poly(3,4-ethylenedioxythiophene) (PEDOT) derivatives: innovative conductive polymers for bioelectronics. Polymers 9:354. https://doi.org/10.3390/polym9080354

Mao S, Zhang X, Sun B, Li B, Yu T, Chen Y, Zhao Y (2019) A bio-memristor with overwhelming capacitance effect. Electron Mater Lett. https://doi.org/10.1007/s13391-019-00150-x

Marasco PD, Hebert JS, Sensinger JW, Beckler DT, Thumser ZC, Shehata AW, Williams HE, Wilson KR (2021) Neurorobotic fusion of prosthetic touch, kinesthesia, and movement in bionic upper limbs promotes intrinsic brain behaviors. Sci Robot Sep 8;6(58):eabf3368. https://doi.org/10.1126/scirobotics.abf3368. Epub 2021 Sep 1. PMID: 34516746

Mazzolai B, Laschi C, Dario P, Mugnai S, Mancuso S (2010) The plant as a biomechatronic system. Plant Signal Behav 5:90–93. https://doi.org/10.4161/psb.5.2.10457

McPhee J (2018) Biomechatronic modelling will change the future. Research Features, research-features.com. Accessed 12 July 2018. https://researchfeatures.com/biomechatronic-modelling-will-change-the-future/

Mehrotra P (2016) Biosensors and their applications-a review. J Oral Biol Craniofacial Res 6(2):153–159. https://doi.org/10.1016/j.jobcr.2015.12.002

Mehrpouya M, Vahabi H, Barletta M, Laheurte P, Langlois V (2021) Additive manufacturing of polyhydroxyalkanoates (PHAs) biopolymers: materials, printing techniques, and applications. Mater Sci Eng C 127:112216. https://doi.org/10.1016/j.msec.2021.112216

Meng J, Xiao B, Zhang Y, Liu J, Xue H, Lei J et al (2013) Super-paramagnetic responsive nanofibrous scaffolds under static magnetic field enhance osteogenesis for bone repair in vivo. Sci Rep 3:2655. https://doi.org/10.1038/srep02655

Ming J, QinChai R, Ya-LiYuan 113806Biosensors and Bioelectronics, Volume 197,2022,113843,ISSN 0956–5663

Mohite BV, Patil SV (2014) A novel biomaterial: bacterial cellulose and its new era applications. Biotechnol Appl Biochem 61(2):101–110. https://doi.org/10.1002/bab.1148

Nagel V, Chu S, Forney J, Kosinski L, Viswanathan V (2017) Design and control of an assistive bionic joint for leg muscle rehabilitation. In: Proceedings of the ASME 2017 international mechanical engineering congress and exposition. Volume 3: Biomedical and biotechnology engineering. Tampa, Florida, USA. November 3–9, 2017. V003T04A046. ASME. https://doi.org/10.1115/IMECE2017-71143

Naidu D, Stopforth R, Bright G, Davrajh S (2012) A portable passive physiotherapeutic exoskeleton. Int J Adv Rob Syst 9:1–12. https://doi.org/10.5772/52065

Naresh V, Lee N (2021) A review on biosensors and recent development of nanostructured materials-enabled biosensors. Sensors 21:1109. https://doi.org/10.3390/s21041109

Niu Q et al (2020) Pulse-driven bio-triboelectric nanogenerator based on silk nanoribbons. Nano Energy 74:104837

P, Filip, Paszkiewicz, Samuel C. Wilson et al. Biomechatronics Lab. Biomechatronics Lab, www.biomechatronicslab.co.uk. May 2018. https://www.biomechatronicslab.co.uk/research/Project?ID=4

Park KM, Lee SY, Joung YK, Na JS, Lee MC, Park KD (2009) Thermosensitive chitosan-Pluronic hydrogel as an injectable cell delivery carrier for cartilage regeneration. Acta Biomater 5(6):1956–1965. https://doi.org/10.1016/j.actbio.2009.01.040 Epub 2009 Feb 4 PMID: 19261553

Peng J, Zhao J, Long Y, Xie Y, Nie J, Chen L (2019) Magnetic materials in promoting bone regeneration. Front Mater 6():268–. https://doi.org/10.3389/fmats.2019.00268

Petri DF (2015) Xanthan gum: a versatile biopolymer for biomedical and technological applications. J Appl Polymer Sci 132(23)

Qi, Yiming; Sun, Bai; Fu, Guoqiang; Li, Tengteng; Zhu, Shouhui; Zheng, Liang; Mao, Shuangsuo; Kan, Xiang; Lei, Ming; Chen, Yuanzheng (2019). A nonvolatile organic resistive switching memory based on lotus leaves. Chem Phys 516():168–174. https://doi.org/10.1016/j.chemphys.2018.09.008

Rafiqah SA, Khalina A, Harmaen AS, Tawakkal IA, Zaman K, Asim M, Lee CH (2021) A review on properties and application of bio-based poly (butylene succinate). Polymers 13(9):1436

Rajan KP, Thomas SP, Gopanna A, Chavali M. (2018) Polyhydroxybutyrate (PHB): a standout biopolymer for environmental sustainability. Handbook of Ecomaterials; Martínez, LMT, Kharissova, OV, Kharisov, BI, Eds, 1–23

Rebelo AMR, Liu Y, Liu C, Schafer, K-H, Saumer M, Yang G (2019) Carbon nanotube-reinforced poly(4-vinylaniline)/polyaniline bilayer-grafted bacterial cellulose for bioelectronic applications. Loughborough University. Journal contribution. https://hdl.handle.net/2134/37778

Riaz Rajoka MS, Zhao L, Mehwish HM, Wu Y, Mahmood S (2019) Chitosan and its derivatives: synthesis, biotechnological applications, and future challenges. Appl Microbiol Biotechnol 103(4):1557–1571

Rindge D (2008) Bioelectromagnetic therapy. Bioelectromagnetic Therapy, vol 09. No 04. www.acupuncturetoday.com. Accessed 1 Apr 2008, https://www.acupuncturetoday.com/mpacms/at/article.php?id=31705

Ruijin Y, Hongsheng L, Yizeng X, Deyu Y, Zaiquan S, Chunling Y (2016) Water soluble chitosan enhances bone fracture healing in rabbit model. Curr Signal Transduct Ther 11(1):28–32. https://doi.org/10.2174/1574362411666151231213944

Sadki S, Chevrot C (2003) Electropolymerization of 3,4-ethylenedioxythiophene, N-ethylcarbazole and their mixtures in aqueous micellar solution. Electrochim Acta 48:733–739. https://doi.org/10.1016/S0013-4686(02)00742-9

Salati MA, Khazai J, Tahmuri AM, Samadi A, Taghizadeh A, Taghizadeh M, Zarrintaj P, Ramsey JD, Habibzadeh S, Seidi F, Saeb MR, Mozafari M (2020) Agarose-based biomaterials: opportunities and challenges in cartilage tissue engineering. Polymers 12(5):1150. https://doi.org/10.3390/polym12051150

Shirakawa H, Louis EJ, MacDiarmid AG et al (1977) Synthesis of electrically conducting organic polymers: halogen derivatives of polyacetylene, (CH)x. J Chem Soc, Chem Commun 16:578–580

Shur M, Fallegger F, Pirondini E, Roux A, Bichat A, Barraud Q, Courtine G, Lacour SP (2020) Soft printable electrode coating for neural interfaces. ACS Appl Bio Mater 3(7):4388–4397

Siafaka PI, Barmbalexis P, Bikiaris DN (2016) Novel electrospun nanofibrous matrices prepared from poly(lactic acid)/poly(butylene adipate) blends for controlled release formulations of an anti-rheumatoid agent. Eur J Pharm Sci 10(88):12–25. https://doi.org/10.1016/j.ejps.2016.03.021 Epub 2016 Mar 30 PMID: 27039136

Singhal R, Takashima W, Kaneto K, Samanta S, Subramanian A, Malhotra B (2002) Langmuir-Blodgett films of poly(3-dodecyl thiophene) for application to glucose biosensor. Sensors and actuators B-chemical - SENSOR ACTUATOR B-CHEM. 86:42–48. https://doi.org/10.1016/S0925-4005(02)00145-4

Song M, Lin X, Peng Z, Xu S, Jin L, Zheng X, Luo H (2021) Materials and methods of biosensor interfaces with stability. Front Mater 7:583739. https://doi.org/10.3389/fmats.2020.583739

Sun H, Yang HL (2015) Calcium phosphate scaffolds combined with bone morphogenetic proteins or mesenchymal stem cells in bone tissue engineering. Chin Med J 128(8):1121–1127. https://doi.org/10.4103/0366-6999.1551214;10(4):334.doi:10.3390/ma10040334.PMID: 28772697;PMCID:PMC5506916

Sun B, Zhou G, Tao G, Zhou Y, Wu Y (2020) Biomemristors as the next generation bioelectronics. Nano Energy 75:104938. https://doi.org/10.1016/j.nanoen.2020.104938

Sun Q, Qian B, Uto K, Chen J, Liu X, Minari T Functional biomaterials towards flexible electronics and sensors, biosensors and bioelectronics. https://doi.org/10.1016/j.bios.2018.08.018

Szcześ A, Hołysz L, Chibowski E (2017) Synthesis of hydroxyapatite for biomedical applications. Adv Coll Interface Sci 249. https://doi.org/10.1016/j.cis.2017.04.007

Tahir Z, Alocilja E, Grooms D (2007) Indium tin oxide-polyaniline biosensor: fabrication and characterization. Sensors 7. https://doi.org/10.3390/s7071123

Thackray AC, Sammons RL, Macaskie LE, Yong P, Lugg H, Marquis PM (2004) Bacterial biosynthesis of a calcium phosphate bone-substitute material. J Mater Sci Mater Med 15(4):403–406. https://doi.org/10.1023/b:jmsm.0000021110.07796.6e PMID: 15332607

Thomas MV, Puleo DA (2009) Calcium sulfate: properties and clinical applications. J Biomed Mater Res B Appl Biomater 88(2):597–610. https://doi.org/10.1002/jbm.b.31269 PMID: 19025981

Valle G, Saliji A, Fogle E, Cimolato A, Petrini FM, Raspopovic S (2021) Mechanisms of neuro-robotic prosthesis operation in leg amputees. Sci Adv 7(17):eabd8354. https://doi.org/10.1126/sciadv.abd8354. PMID: 33883127; PMCID: PMC8059925

Vandamme EJ, De Baets S, Vanbaelen A, Joris K, De Wulf P (1998) Improved production of bacterial cellulose and its application potential. Polym Degrad Stab 59(1–3):93–99

Vieira E, Vieira T, Silva M, Santos M, Brito C, Bezerra R, Fialho A, Osajima J, Filho E (2018) Tuned hydroxyapatite materials for biomedical applications. https://doi.org/10.5772/intechopen.71622.

Wearable Bioelectronics: Enzyme-Based Body-Worn Electronic Devices-Jayoung Kim Itthipon, Jeerapan, Juliane R. Sempionatto, Abbas Barfidokht,Rupesh K. Mishra, Alan S. Campbell, Lee J.Hubble, and Joseph Wang.

Wei B, Liu J, Ouyang L, Kuo C-C, Martin D (2015) Significant enhancement of PEDOT thin film adhesion to inorganic solid substrates with EDOT-Acid. ACS Appl Mater Interfaces 7. https://doi.org/10.1021/acsami.5b03350

Wilson S, Vaidyanathan R (2017) Upper-limb prosthetic control using wearable multichannel mechanomyography. Int Con Rehabilit Robot (ICORR) 2017:1293–1298. https://doi.org/10.1109/ICORR.2017.8009427

Winkler T, Sass FA, Duda GN, Schmidt-Bleek K (2018) A review of biomaterials in bone defect healing, remaining shortcomings and future opportunities for bone tissue engineering: the unsolved challenge. Bone Joint Res. 7(3):232–243. https://doi.org/10.1302/2046-3758.73.BJR-2017-0270.R1.PMID:29922441;PMCID:PMC5987690

Winter JO, Cogan SF, Rizzo JF 3rd (2007) Neurotrophin-eluting hydrogel coatings for neural stimulating electrodes. J Biomed Mater Res B Appl Biomater 81(2):551–563. https://doi.org/10.1002/jbm.b.30696 PMID: 17041927

Woods GA, Rommelfanger NJ, Hong G (2020) Bioinspired materials for in vivo bioelectronic neural interfaces. Matter 3(4):1087–1113

Xu J, Guo BH (2010) Poly(Butylene Succinate) and its copolymers: research, development and industrialization. Biotechnol J 5:1149–1163. https://doi.org/10.1002/biot.201000136

Yamanaka S, Watanabe K, Kitamura N, Iguchi M, Mitsuhashi S, Nishi Y, Uryu MJ (1989) The structure and mechanical properties of sheets prepared from bacterial cellulose. Mater Sci 24:3141

Yang S, Zhang Y, Wang T, Sun W, Tong Z (2020) Ultrafast and programmable shape memory hydrogel of gelatin soaked in tannic acid solution. ACS Appl Mater Interfaces 12(41):46701–46709. https://doi.org/10.1021/acsami.0c13531 Epub 2020 Oct 1 PMID: 32960035

Yoshida T (2001) Synthesis of polysaccharides having specific biological activities. Prog Polym Sci 26(3):379–441

Yu, Zhang, Fu X, Duan D, Xu J, Gao X (2019) Preparation and characterization of agar, agarose, and agaropectin from the red alga Ahnfeltia plicata. J Oceanol Limnol 37. https://doi.org/10.1007/s00343-019-8129-6

Yuk H, Lu B, Zhao X (2018) Hydrogel bioelectronics. Chem Soc Rev 48. https://doi.org/10.1039/C8CS00595H

Zahari MAA, Lee SP, Kasim SR (2020) Synthesis of calcium sulphate as biomaterial. In: 3RD International postgraduate conference on materials, minerals & polymer (Mamip). https://doi.org/10.1063/5.0015693

Zhang D, Dubey VN, Yu W, Low KH (2018) Editorial: biomechatronics: harmonizing mechatronic systems with human beings. Front Neurosci 25(12):768. https://doi.org/10.3389/fnins.2018.00768.PMID:30459543;PMCID:PMC6232912

Zhang L, Lu JR, Waigh TA (2020) Electronics of peptide-and protein-based biomaterials. Adv Coll Interface Sci. https://doi.org/10.1016/j.cis.2020.102319

Zhang L-M, Wu C-X, Huang J-Y, Peng X-H, Chen P, Tang S-Q (2012) Synthesis and characterization of a degradable composite agarose/HA hydrogel. Carbohyd Polym 88:1445–1452. https://doi.org/10.1016/j.carbpol.2012.02.050

Zhang C, Wang W, Xi N, Wang Y, Liu L (2018) Development and future challenges of bio-syncretic robots. Engineering, ISSN 2095–8099, https://doi.org/10.1016/j.eng.2018.07.005.

de Sousa Costa LA, Inomata Campos M, Izabel Druzian J, de Oliveira AM, de Oliveira Junior EN (2014) Biosynthesis of xanthan gum from fermenting shrimp shell: yield and apparent viscosity. Int J Poly Sci

Biobased Materials and the Vast Domain of Environmental Pollution Control–A Critical Overview

Sukanchan Palit and Chaudhery Mustansar Hussain

1 Introduction

Environmental engineering and petroleum engineering are today in the middle of a great disaster and unending peril. The degradation of the environment and loss of fossil fuel resources are shocking the entire human civilization and are playing havoc with human society. The challenges of conversion of industrial wastes to hydrogen production need a deep rethought and introspection. The hydrogen economy should be integrated with a circular economy. Scientific resilience, the world of science and engineering challenges, and the need for biofuels and biobased materials will all lead a long, visionary, and understandable way to the true realization of renewable energy. Renewable energy is a marvel of energy engineering and science today. Hydrogen energy and the application of biobased materials is a much sought-after technology in today's avenues of human society. In this paper, the author deeply elucidates the need for bio-economy and the application of biofuels in the further emancipation of global science and engineering. Clean water and clean energy are today part and parcel of the circular economy. Reuse, reduce, and regenerate are the coin words of today's scientific endeavor in the circular economy. A new dawn in human civilization will surely emerge if academicians, scientists, researchers, and governments around the globe take affirmative steps in combating climate change and mitigating global warming. A new visionary era in the field of the hydrogen economy and renewable energy will surely open new paths and new avenues of scientific intellect

S. Palit (✉)
Department of Chemical Engineering, University of Petroleum and Energy Studies, Energy Acres, Post-Office-Bidholi Via Premnagar, Dehradun 248007, Uttarakhand, India
e-mail: sukanchan68@gmail.com

C. M. Hussain
Department of Chemistry and Environmental Sciences, New Jersey Institute of Technology, University Heights, Newark, NJ 07102, USA
e-mail: chaudhery.m.hussain@njit.edu

and scientific imagination in decades to come. There are today immense challenges in the applications of biobased materials in environmental protection. These are the salient features of this well-researched article. Apart from these domains, there is a deep scientific introspection in the field of biofuels.

2   The Aim and Objective of This Study

Biotechnology, bioengineering, and biological engineering are today in the midst of deep scientific comprehension and scientific ingenuity. Biobased materials and bio-nanocomposites are the smart materials and eco-materials of today's scientific research endeavor. The need for eco-friendly and smart materials and their future recommendations will veritably lead a long and visionary way in finding the truth and ingenuity of science and engineering in decades to come. Today global water issues and global climate change are spearheading science and engineering into a visionary era. The main aim and objective of this treatise are to elucidate upon the areas of bio-based materials and bionanocomposites. Today is the era of the fourth industrial revolution. Sustainable resource management and circular economy are today integrated with each other. Reuse, regenerate, and recycle are the visionary coin words of scientific revolution today. There is an immense application of biobased materials in environmental pollution control. The main vision of this paper is to delineate the vast and varied applications of biobased materials in water and wastewater treatment. A new chapter in the field of biological engineering will surely emerge if material science and composite science are effectively applied in environmental engineering science. Sustainability and sustainable resource management are the two other scientific zeniths of research pursuit today. Green or environmental sustainability is the visionary aisle of this research treatise. The authors of this article deeply stress the need for biobased materials and bionanocomposites in the sustainable future of the human race. Composite science, polymer science, material science, and nanotechnology should be integrated with each other for a greater emancipation of research and development initiatives in the global platform today. The authors in this treatise deeply stress the science of bio-based materials and their applications and scientific ingenuity in global economic development and sustainability. Mankind's immense scientific and engineering prowess will surely be bolstered if the applications of biological sciences and biotechnology are properly realized in decades to come.

3   The Need and Rationale of This Study

Humanity and science today stand in the middle of deep scientific contemplation and scientific and engineering profundity. The world of science and engineering needs to be revamped and reshaped as regards biological sciences, bio-engineering, composite science, and material science. Bionanocomposites are other areas of deep scientific introspection and have immense scientific and engineering challenges. In a similar vision, bio-based materials are the utmost need of science and civilization today. Thus the need and the rationale of this study. Sustainable development, sustainable resource management, circular economy, and environmental management are the veritable needs of the human race today. The vision of sustainable resource management needs to be overhauled as civilization treads forward. Scientific intellect, scientific perception, and futuristic vision of bio-based nanomaterials and bionanocomposites will be the torchbearers of the domain of composite science and material science. Man's immense scientific adjudication, mankind's vast knowledge prowess, and the futuristic vision of composite science and biological sciences will transform human intellect in the field of bio-based materials. The need and the rationale of this study should go beyond scientific imagination and scientific intellect. Thus, a newer scientific vision will evolve, which will lead human civilization toward vast scientific perception and scientific discerning.

4 Climate Change, Global Warming, and Sustainability

Climate change, global warming, and environmental and energy sustainability are the vexing issues of global research and development initiatives in the global landscape. Scientific triumph, deep scientific provenance, and futuristic vision will surely one day transform human society as regards climate change and global warming. United Nations Sustainable Development Goals deeply address global water scarcity, proper sanitation, and also climate change. Here comes the need for a concerted effort in combating climate change and global warming from scientists, researchers, policymakers, and governments across the world. Human scientific endurance is at its extreme as regards poverty alleviation and the development of the human habitat. Developing and disadvantaged nations around the world are in the middle of a serious scientific and engineering, which are heavy metal and arsenic groundwater contamination. In such a situation, Sustainable Development Goals are the utmost need of the hour and should be implemented in an effective way across all governmental regulations across the globe. Human population growth and rapid urbanization should be alleviated if mankind needs to be survived. Rapid urbanization and enhanced population are directly affecting the global climate change scenario. Thus human suffering and human struggle should be of greater concern across all levels of combat against climate change and global warming. Man's visionary scientific and engineering endeavor and the advancement of human civilization are today directly linked to each other. Thus, there is an immediate need for holistic global research and development initiatives in water science and technology and environmental remediation science. Social and economic sustainability in a similar vein is also the utmost necessity of the hour. Global human advancements totally depend on social, economic, environmental, and energy sustainability. Today in developing nations such as India and Bangladesh, water issues are ravaging human society and there is no feasible solution. Proper implementation of energy and environmental sustainability across all areas of governance will open newer vistas of research and development goals in decades to come.

5 Environmental Sustainability and the Visionary Future

Environmental sustainability and environmental protection are today aligned with each other. In a similar vein, social and economic sustainability should be applied to the scientific advancements of the human race.

The vast vision toward a sustainable future in the application of biobased materials or bio-nanocomposites needs to be reinvented and revamped with the course of scientific history and time. Sustainability, whether it is social, economic, energy, or environmental, needs to garner enough resources to make the world free from climate change and global warming. A sustainable global future involves free from water scarcity, free from improper sanitation, and good human habitat. In the proper realization and proper emancipation of sustainable development and environmental management, a strong and positive initiative is needed from engineers and scientists in the global scenario. Human perception, a deep interest in teaching and learning outcomes, and a strong research pedagogy will lead a long and visionary way to finding the truth and advancements in both environmental engineering and chemical process engineering. Bio-engineering and material science are today merged together in the research pursuit of bionanocomposites. The application of biobased materials in environmental pollution control and water remediation will surely open a new vista in the field of biological engineering and nanocomposites.

6 Recent Scientific Advancements in the Field of Biobased Materials and Their Applications

Biobased materials and their applications are today revolutionizing the field of biological sciences and biological engineering. There are today tremendous developments in the field of applications of biobased materials in environmental pollution control and water remediation. Mankind's immense knowledge prowess is in a state of disaster as regards industrial pollution control. In this section, the authors deeply elucidate significant achievements in the field of biobased materials applications. A newer revolutionary era in the field of biobased materials will emerge if scientists, engineers, and students across the globe concentrate and focus on drinking water treatment, industrial wastewater treatment, and groundwater remediation.

Hussain et al. (2006) deeply described with scientific and engineering far-sightedness polymer matrix nanocomposites, processing, manufacturing, and application in an overview. This review is designed to be a detailed and comprehensive resource for polymer nanocomposite research, including the fundamental structure and property relationship, manufacturing techniques, and vast applications of polymer nanocomposite materials. This review focuses on the scientific doctrines, scientific principles, and mechanisms in relation to the methods of processing and manufacturing with a discussion of commercial applications (Hussain et al. 2006). This review offers a comprehensive discussion on technology, modeling, characterization, processing, manufacturing, applications, and chemical process safety concerns. The authors discussed in detail polymeric nanocomposites, characterization techniques for nanocomposites, nanoplatelet-reinforced systems, and structure, properties, and applications of polymer nanocomposites. Carbon nanotube processing, dispersion, and orientation are the other hallmarks of this treatise. Current scientific and engineering challenges in the processing and manufacturing of nanocomposites are also described in detail in this paper (Hussain et al. 2006).

Yang et al. (2018) deeply discussed with vision and purpose applications of nanocomposite catalysts in biofuel production. Nanocomposite is used to describe a multiphase solid material where at least one of the phases has one, two, or three dimensions of less than 100 nm. Nanocomposites have been widely used in many diverse fields mostly in the industry of cosmetics, chemical catalysts, drug delivery, medical devices, optoelectronics, electronics, and magnetics (Yang et al. 2018). The authors discussed in minute detail bio-gasoline, alcohols and polyols, carbohydrates, lignocellulosic biomass, and lipids and lactones. Bio jet fuels from carbohydrates are the other areas of deep scientific introspection in this treatise. Renewable diesel fuel stands as a major pillar of this treatise. The vast applications of carbohydrates and carbohydrate-derived compounds and multifunctional catalysts are the other areas of research pursuit of this treatise (Yang et al. 2018).

Palaniappan (2017) discussed with cogent insight the applications of nanotechnology in biofuel production. The rapidly depleting energy resources are the greatest global challenge that the world is combating and mankind is forced towards exploring newer avenues in the research endeavor. The various nanomaterials production tools and the methods through which they are made functional and stable are the hallmarks of this study. Further to that, various literature studies are investigated in this paper. Increasing the price of crude oil and its refined products is a clear indication of the depletion of fossil fuel resources, namely coal, crude oil, and natural gas. In 1959, Feynman introduced the term "nanotechnology", which refers to the manufacture and the use of nanometer-scale materials. Since then, visionary research pursuit has been done around the globe in a wide variety of fields such as integrated circuits, food products, and energy conversion devices (Palaniappan 2017). The authors discussed in detail biogas and enhanced production using nanotechnology. This study provides and deeply explores an extensive overview of the new nanomaterials that have been investigated for their capabilities to improve the production of biofuels, be it biodiesel or biogas. A new chapter and a newer window of scientific wisdom will surely emerge as the science of biofuels and biogas overcome one visionary difficulty over another (Palaniappan 2017).

Humankind and civilization are in the middle of a vast environmental disaster that is water purification science and groundwater contamination of arsenic and heavy metals. In this entire treatise, the authors deeply stress and reiterate the needs for biotechnology, biological engineering, and material science in the further realization of research and development initiatives globally.

7 Recent and Significant Advances in the Field of Bionanocomposites

Bionanocomposites are the next generation smart and ecomaterials. Today, in the present-day human civilization, bionanocomposites have applications in diverse areas of human scientific endeavor. The authors in this treatise deeply describe and elucidate on certain significant advancements in the field of bionanocomposites.

Camargo et al. (2009) deeply discussed with scientific vision and scientific conscience nanocomposites, their structure, properties, and new application opportunities. Nanocomposites, a high-performance material exhibit unusual property combinations, unique design possibilities, and true vision. In this holistic review, the three types of matrix nanocomposites are presented in deep detail underling the need for these materials, their inherent processing methods, and some recent results on structure, properties, and potential applications (Camargo et al. 2009). Human scientific research pursuit is at its zenith as composite science and material sciences surpass one visionary boundary over another. Possible uses of natural materials such as clay-based minerals, chrysotile, and lignocellulosic fibers are deeply elucidated and highlighted. Being environmentally benign, the applications of nanocomposites offer unique and remarkable technology and business opportunities for several diverse sectors of the aerospace, automotive, electronic, advanced materials and biotechnology industries, and other chemical process industries. Nanocomposites are composites in which at least one of the phases shows dimensions in the nanometre range. A new visionary world in the domain of composite science and engineering is emerging as global technology-driven society treads forward (Camargo et al. 2009). The authors discussed in detail: (1) ceramic matrix nanocomposites, (2) metal matrix nanocomposites, and (3) polymer matrix nanocomposites. Processing of nanocomposites, the structure and properties, reinforcement in nanocomposite systems, and applications of nanocomposites are the other areas of scientific research pursuit of this well-researched treatise. A brief discussion on perspectives of nanocomposites is the other cornerstone of this paper (Camargo et al. 2009).

Shchipunov et al. (2012) discussed in detail bionanocomposites and green sustainable materials for the near future. Sustainability is today merged with diverse areas of science and engineering such as composite science and material science.

Bionanocomposites are a novel and unique class of nanosized materials. They veritably contain the constituent of biological origin and particles with at least one dimension in the range of 1–100 nm. There are fundamental differences with nanocomposites in the methods of preparation, properties, functionalities, biodegradability, biocompatibility, and applications. Bionanocomposite definition and classifications along with nanomaterials, biomaterials, and the methods of their preparations are deeply reviewed (Shchipunov 2012). Human scientific genre and engineering profundity in the field of composite science and material science will surely open new windows of innovation and instinct in research and development initiatives in decades to come. The authors discussed in minute detail definitions of composite materials, nanoparticles, and biopolymers. Chitosan-based bionanocomposites and their applications are the other areas of deep scientific introspection in this article. A deep scientific and engineering perspective in the field of bionanocomposites are the other hallmarks of this paper.

Okpala et al. (2014) discussed deeply the benefits and applications of nanocomposites. Materials and their developments are the fundamentals of today's human culture and society. The vast intensification in research of nanoscale materials in recent decades has occurred primarily due to their attractive potential and unique properties. This paper gives a broad definition of nanocomposites, nanofillers, the benefits of nanocomposites, the processing, milestones in nano development, the strength and limitations as well as vast and varied applications of nanocomposites (Okpala et al. 2014). The authors discussed in detail that carbon-nanotube-reinforced composites are the other cornerstones of this paper. The application of nanocomposites in oil and gas pipelines is the other unique feature of this treatise. Automobile industries, aircraft, electronics, food packaging, and environmental protection are the other areas of deep scientific comprehension of this treatise. Scientists and researchers across the world have made significant, resounding, and remarkable contributions in the field of nanocomposite science in the past two decades. Material sciences are today linked with biological sciences. A new era in the field of biomaterials is emerging as civilization, science, and engineering move forward in the right direction (Okpala et al. 2014).

Human scientific and engineering advancements in the field of bionanocomposites are groundbreaking and truly far-reaching. Biobased materials are the wonders of composite science today. A new era in composite science will surely usher in as scientific research and development initiatives move in the positive direction and towards a sustainable future.

8 Global Water Science and Technology Research and Sustainable Resource Management

Global water science and its research and development initiatives are today in the middle of deep scientific comprehension and vision. Sustainable resource management is the new coin word of today's fourth industrial revolution. In a similar vision, the circular economy is aligned with sustainable resource management. The idea of recycling, reusing, and reinventing are the goals of the circular economy today. Research initiatives in sustainable resource management are today in the process of a new beginning. Energy and environmental sustainability are today interlinked with the vast world of the circular economy today. Today the agony and plight of human civilization are unending as regards the provision of pure drinking water. Thus, the need for the application of green sustainability and sustainable resource management. A remarkable arena in the field of global science and engineering will surely evolve if there is a prime focus on the application of circular economy if every sphere of global research initiatives. Circular economy, green economy, and green nanotechnology are the scientific doctrines of human scientific endeavor today. Thus greater scientific emancipation in the field of water science and sustainable resource management is the utmost need of the hour. Human civilization's immense scientific stance, scientific provenance, and deep scientific discerning will widen the engineering perspective of sustainable resource management.

9 The Status of the Global Scenario in Biotechnology

Biofuels and biotechnology are marvels of science and technology in the global scenario. Today mankind is in the deep pursuit of alternate energy sources and renewable energy sources. Hence, the importance of biofuels. Biotechnology, biological sciences, and biological engineering are today the culmination of scientific vision and deep scientific provenance. Biotechnology is today the Holy Grail of biological engineering and environmental engineering today. The status of biotechnology research in the global scenario is exceedingly bright and groundbreaking. Water remediation and biotechnology and biological sciences are today akin to each other. A new field in science and engineering is the domain of biological engineering. In developed countries around the world, biological engineering is merged with chemical engineering and environmental engineering science. Thus the curriculum of biotechnology and biological engineering needs to be reinvented and re-envisioned as civilization treads forward. The authors in this entire treatise deeply elucidate the success of the application of biobased materials and bionanocomposites in the further emancipation of global science and engineering. Human science and engineering initiatives are today in the process of newer regeneration as regards environmental pollution control. A new era in the field of biotechnology will surely evolve if biotechnology and composite science move in the right direction of proper implementation of sustainable development goals.

10 Arsenic and Heavy Metal Contamination and Subsequent Environmental Remediation

Arsenic and heavy metal groundwater poisoning are today ravaging and destroying the entire scientific firmament. The world of challenges and scientific threats are immense as developing countries and many industrialized nations around the world grapple with this environmental disaster. Public health engineering and civil engineering in many under-developed countries around the world are in a similar vein in a state of immense catastrophe. In Bangladesh and the state of West Bengal, India, a vicious and monstrous calamity of arsenic groundwater contamination is veritably destroying the scientific landscape. Bioremediation, as well as nanoremediation, is the new-generation scientific innovation and technology. Inclusive economic growth, scientific resilience, and green sustainability are today aligned with each other. Today there are numerous arsenic treatment technologies which are:

- Iron coagulation,
- Alum coagulation,
- Coagulation/co-precipitation,
- Activated alumina adsorption,
- Iron-based sorbents,
- Reverse osmosis, and
- Nanofiltration.

Human race and engineering science are today in the middle of a huge crevice of vision, might, hope, and determination. Mankind's suffering as regards drinking water contamination and water scarcity are veritably in the process difficult threats, scientific and engineering triumphs. The utmost need of the hour is sustainable development and environmental management. Global water issues and water hiatus are today directly linked with environmental management. Over the course of time, man and mankind will unveil the immediate importance of the application of environmental management in arsenic and heavy metal groundwater remediation. A new scientific beginning in the field of circular economy, green economy, and hydrogen economy will start if humankind takes a deep interest in the application of sustainability, environmental management, and the proper implementation of United Nations Sustainable Development Goals.

Ghosh et al. (2019) discussed with vision, insight, and foresight technology alternatives for decontamination of arsenic-rich groundwater in a critical review. Arsenic contamination of groundwater is being reckoned as a global scientific and engineering problem as over 296 billion people residing in more than 100 countries have already reported to be affected by this environmental engineering issue. Developing countries like Bangladesh, India, Thailand, Taiwan, and Vietnam are totally ravaged due to this terrible event. The authors discussed oxidation, chemical oxidation, microbiological oxidation, coagulation-flocculation, adsorption, biological sorption, ion exchange, and membrane separation processes. Available treatment technologies for arsenic removal from groundwater are the hallmarks of this treatise. Treatment with bioorganisms is the other area of research pursuit. A critical review of arsenic decontamination techniques is the pillar of this paper (Ghosh et al. 2019).

Shannon et al. (2008) deeply discussed the vision and purpose of science and technology for water purification in the coming decades. One of the burning and vexing issues in the global scenario is the scarcity of pure drinking water and proper sanitation. Problems with water scarcity are going to be worse in the coming decades with water shortage occurring globally even in countries that are water rich. In this article, the authors deeply discussed disinfection, decontamination, reuse, and reclamation of drinking water. The work highlighted in this article plus the tremendous amount of research and development initiatives and efforts is veritably sowing the seeds of a revolution in water purification science. The article by Shannon et al. (2008) deeply and profoundly addresses these scientific and engineering problems (Shannon et al. 2008).

Mankind today needs to gear forward with newer challenges and a sustainable future. A strong scientific understanding and scientific discernment is truly needed in the path toward sustainable groundwater remediation tools. A new era in arsenic and heavy metal decontamination will surely widen our scientific views and engineering problems in water purification science (Abdelbasir and Shalan 2019; Mukherjee et al. 2015; Hashim et al. 2011; Saikia et al. 2019; Martinez et al. 2019; Ahmed 2018; Palit and Hussain 2020).

11 Global Scientific Strategies in the Field of Application of Biobased Materials

Global research and development initiatives and global strategies in the field of conversion of industrial wastes to hydrogen are today in the aisles of deep scientific and engineering divination. Renewable energy and hydrogen economy are the ultimate vision and the needs of human society today. Environmentally benign, low cost, and higher efficiency of the processes will be the torchbearers of a newer scientific genre and scientific imagination in the field of production of biohydrogen and biofuels. The ardor of science and engineering of biotechnology, biological sciences, and biological engineering will all lead to a largely visionary era in global engineering science. Population growth, vast economic growth, and rapid urbanization are today destroying the ecological and scientific landscape. Mankind's scientific and engineering expertise and knowledge prowess are at extreme peril. The strategies in the field of conversion of industrial wastes to hydrogen need to be thoroughly envisioned as man and mankind moves forward. A deep scientific thought and a deep scientific inventing and innovation are the utmost needs of the hour. Scientific strategies should target hydrogen economy and renewable energy resources. Thus the need for sustainable resource management and natural resource management which includes water and energy. A newer holistic domain in biological sciences and biological engineering will surely evolve if a circular economy is implemented in every sphere of scientific research pursuit (Ahuja 2019; Decourten et al. 2019; Shevah 2019; Bajpai et al. 2019).

12 Sustainable Resource Management, Integrated Water Resource Management, and the Progress of Human Society

Sustainable resource management is today veritably linked with water resource engineering and integrated water resource management. Human mankind today is in the middle of unending environmental disasters. These disasters include global warming, global water shortage, and degradation of ecological biodiversity. Degradation of the environment and ecological disbalance are burdens to human mankind today. Industrial water pollution and industrial wastewater management are creating an immense burden on the scientific domain today. The progress of the human race is at real peril as the environment gets tremendously degraded with the progress of science, technology, and mankind. Sustainable resource management and natural resource management are today connected by an unsevered cord. The provision of clean and pure drinking water is part and parcel of the natural resource. So here comes the importance of green or environmental sustainability (Metcalf and Souza 2019; Ahuja. 2019; Pandit and Gayatri 2020; Anukiruthika et al. 2020; Singh et al. 2020).

13 Biobased Materials, Water Science and Technology, and the Road to Scientific Wisdom

Water science and technology today need to be integrated with the vast domain of biofuels. Advancements in the field of biofuels are today not that developed yet in the path of new regeneration. Millions of global citizens are without clean drinking water. They are without proper sanitation, proper human habitat, and lack of education. In this entire article, the author reiterates the need for a hydrogen economy and eventually circular economy in the advancement of the human race. Human suffering is at its zenith today as regards water scarcity in many developing and disadvantaged countries around the world. Thus the need for scientific resilience, hope, and determination in research pursuit in water science and technology. The application of biofuels in providing clean energy will surely be the pathbreakers toward a newer era in the circular economy. Man and mankind will then usher in a new era in the science of biofuels, hydrogen economy, green economy, and finally circular economy. Potable and clean drinking water provision is surely a need of every nation around the world. The author of this entire article stresses the importance of the linkage between circular economy and water science and technology. A larger and wonderful era in the field of biofuels and renewable energy will emerge if human mankind and governments around the globe take strong and positive steps in enhancing hydrogen economy, circular economy, green economy, and green sustainability (Kaur et al. 2020; Gupta and Chatterjee 2017; Stollenwork 2003; Smedley 2003; Varma 2012; Sharma and Bhattacharya 2017).

14 Future Scientific Recommendations of This Study and the Visionary Future

Future scientific recommendations for the study of biobased materials and the application areas of biofuels are today in the avenues of deep scientific understanding and profundity. The circular economy is the coin word of today's scientific endeavor. Reduce, reuse, and recreate are the necessities of human society today. The visionary future in the field of biological sciences and biological engineering is today in the vistas of new learning and scientific intellect. The implementation of the United Nations Sustainable Development Goals, 2030, are the utmost need of the hour. Today human civilization is suffering immensely due to a lack of provision of basic human needs such as clean water and clean energy. Clean water and clean energy nexus will eventually pave the way toward a new visionary era in global science and engineering. Biotechnology and nanotechnology are marvels of science and engineering globally. They are veritably connected with human habitat and human well-being. Thus the author in this treatise stresses the interface of water–energy nexus with biofuels and hydrogen economy. A new dawn in the field of hydrogen energy and renewable energy will emerge if proper steps and engineering initiatives are taken with a clear vision forward.

15 Future in the Applications of Different Techniques in Biofuels

Biofuels applications are an integral part of global renewable energy initiatives. Global research and development initiatives and the deep scientific resilience, hope, determination, and grit will all lead to a newer generation of biological sciences and biotechnology. Human mankind is today in a detailed treatise in renewable energy and applications of biotechnology and biological sciences. Thus the need and rationale of this study. Future the applications of different techniques in biofuels should be targeted toward efficiency, environmentally safe, and low-cost technology. Human suffering in developing and disadvantaged nations around the world is immense as civilization moves forward. Thus the future of science and the fourth industrial revolution should target greater scientific emancipation on clean energy, clean water, circular economy, and scientific and engineering resilience. The process of converting industrial waste to bio-hydrogen will surely usher in a newer scientific hope, genre, promise, and determination in decades to come. The author deeply stresses these engineering issues with vision and scientific might. The future flow of scientific thoughts should also be targeted towards socio-economic sustainability, green sustainability, green economy, and blue or ocean economy. Renewable energy and the vast avenues of science and engineering will ultimately open newer windows of scientific rethinking and scientific ingenuity in the hydrogen economy in decades. It will be highly remarkable if a multidisciplinary team of scientists, engineers, researchers, governments, and policy-makers takes utmost and innovative steps in the future of the hydrogen economy.

16 Conclusion, Summary, and Environmental Perspectives

Human mankind and human destiny are in the middle of scientific comprehension, vision, and scientific divination. The environmental perspectives of the application of biofuels and biobased materials are in veritable needs of human society. Man and humankind's immense scientific verve and motivation are in the avenues of ever-growing disaster due to water shortage, loss of ecological biodiversity, and global climate change. A deep scientific thought and a well-researched scientific forbearance are the utmost needs of the moment. In this article, the author deeply reiterates the importance of hydrogen economy, renewable energy, and circular economy. Nano-energy is a remarkable area of scientific research pursuit in the global scenario. In a similar vein, biofuels and their applications are revolutionizing the entire gamut of renewable energy. Millions of people around the world are without clean drinking water and clean energy. Thus the need for deep rethought in the field of renewable energy, clean water, clean energy, and energy engineering science. Circular economy is today revolutionizing the global scientific scenario. Today is the path of the fourth industrial revolution. Advances in space technology, nuclear energy, nanotechnology, biotechnology, and genetic engineering will veritably change the face of human civilization and transform the fourth industrial revolution. Desalination and membrane science are immensely challenging areas of water science and technology globally. Engineering perspectives in chemical engineering, material science, and biotechnology will surely open new doors of scientific innovation in decades to come. The success of human civilization, the world of scientific validation, and the needs of humanity will change and transform the areas of renewable energy and hydrogen economy.

References

Ahmed S (2018) Bio-based nanomaterials for food packaging. Springer Publishing, Singapore,https://doi.org/10.1007/978-981-13-1909-9

Ahuja S (2019) Lessons learned from water disasters around the world, Chapter-16, Book- Evaluating water quality to prevent future disasters, Editor- Satinder Ahuja, Elsevier, Netherlands, pp 417–427. https://doi.org/10.1016/B978-0-12-815730-5.00016-8

Abdelbasir SM, Shalan AE (2019) An overview of nanomaterials for industrial wastewater treatment. Korean J Chem Eng 36(8):1209–1225. https://doi.org/10.1007/s11814-019-0306-y

Ahuja S (2019) Overview: evaluating water quality to prevent future disasters, Chapter-1, Book- Evaluating water quality to prevent future disasters, Editor- Satinder Ahuja, Elsevier, Netherlands, pp 1–12. https://doi.org/10.1016/B978-0-12-815730-5.00001-6

Anukiruthika T, Priyanka S, Moses JA, Anandharamakrishnan (2020) Characterization of green nanomaterials, Chapter-3, Book- Green Nanomaterials- processing, properties and applications, Editor- Shakeel Ahmed, Wazed Ali, Springer Nature Singapore Pte Ltd, pp 43–79. https://doi.org/10.1007/978-981-15-3560-4

Bajpai S, Alam N, Biswas P (2019) Present and potential water quality challenges in India, Chapter-4, Book-Evaluating water quality to prevent future disasters, Editor- Satinder Ahuja, Elsevier, Netherlands, pp 85–112. https://doi.org/10.1016/B978-0-12-815730-5.00004-1

Camargo PHC, Satyanarayana KG, Wypch F (2009) Nanocomposites: synthesis, structure, properties and new application opportunities. Mater Res 12(1):1–39

Decourten B, Romney A, Brander S (2019),The heat is on: complexities of aquatic endocrine disruption in a changing global climate, Chapter-2, Book- Evaluating water quality to prevent future disasters, Editor- Satinder Ahuja, Elsevier, Netherlands, pp 13–49. https://doi.org/10.1016/B978-0-12-815730-5.00002-8

Ghosh S, Debsarkar A, Dutta A (2019) Technology alternatives for decontamination of arsenic-rich groundwater-a critical review. Environ Technol Innov 13:277–303

Gupta DK, Chatterjee S (2017) Arsenic contamination in the environment: the issues and solutions. Springer International Publishing AG, Switzerland. https://doi.org/10.1007/978-3-319-543 56-7_7

Hashim MA, Mukhopadhyay S, Sahu JN, Sengupta B (2011) Remediation technologies for heavy metal contaminated groundwater. J Environ Manage 92(2011):2355–2388. https://doi.org/10.1016/j.jenvman.2011.06.009

Hussain F, Hojjati M, Okamoto M, Gorga RE (2006) Review article: polymer matrix nanocomposites, processing, manufacturing and applications: an overview. J Compos Mater 40(17):1511–1575

Kanchi S (2014) Nanotechnology for water treatment. J Environ Analyt Chem 1(2). https://doi.org/10.4172/jreac.1000e102

Kaur M, Mehta A, Bhardwaj KK, Gupta R (2020) Bionanomaterials from agricultural wastes, Chapter-10, Book- Green nanomaterials- processing, properties and applications, Editors-Shakeel Ahmed, Wazed Ali, Springer Nature Singapore Pte Ltd, pp 243–260.https://doi.org/10.1007/978-981-15-3560-4

Kuhlman T, Farrington J (2010) What is sustainability? Sustainability 2:3436–3448. https://doi.org/10.3390/su2113436

Martinez LMT, Kharissova OV, Kharisov BI (2019) Handbook of Ecomaterials. Springer International Publishing, Switzerland.https://doi.org/10.1007/978-3-319-68255-6

Metcalf JS, Souza NR (2019) Cynabacteria and their toxins, Chapter-5, Book- Evaluating water quality to prevent future disasters, Editor- Satinder Ahuja, Elsevier, Netherlands, pp 125–148. https://doi.org/10.1016/B978-0-12-815730-5.00006-5

Mukherjee S, Mukhopadhyay S, Hashim MA, Sengupta B (2015) Contemporary environmental issues of landfill leachate : assessment and remedies. Crit Rev Environ Sci Technol 45:472–590. Taylor and Francis Group, LLC.https://doi.org/10.1080/10643389.2013.876524

Okpala CC(2014) The benefits and applications of nanocomposites. Int J Adv Eng Technol V(IV):12–18

Palaniappan K (2017) An overview of applications of nanotechnology in biofuel production. World Appl Sci J 35(8):1305–1311

Palit S, Hussain CM (2020) Nanodevices applications and recent advancements in nanotechnology and the global pharmaceutical industry, Chapter, Book-Nanomaterials in diagnostic tools and devices, Editor-Suvardhan Kanchi, Deepali Sharma, Elsevier, Netherlands, pp 395–415. https://doi.org/10.1016/B978-0-12-817923-9.00014-6

Pandit P, Gayatri TN (2020) Introduction to green nanomaterials, Chapter-1, Book-Green nanomaterials- Processing, properties and applications, Editor- Shakeel Ahmed, Wazed Ali, Springer Nature Singapore Pte Ltd., 2020, pp 1–21. https://doi.org/10.1007/978-981-15-3560-4

Saikia J, Gogoi A, Baruah S (2019) Nanotechnology for water remediation, Chapter-7, Book- Environmental nanotechnology, Environmental chemistry for a sustainable world, Editors-N.Dasgupta et al., Springer Nature Switzerland.AG., 2019, pp 195–211. https://doi.org/10.1007/978-3-319-98708-8_7

Shannon MA, Bohn PW, Elimelech M, Georgiadis JG, Marinas BJ, Mayes AM (2008) Science and technology for water purification in the coming decades, Nature, Nature Publishing Group, London, U.K., pp 301–310, Vol 452j20 March 2008j. https://doi.org/10.1038/nature06599

Sharma S, Bhattacharya A (2017) Drinking water contamination and treatment techniques. Appl Water Sci 7(3):7:1043–1067. https://doi.org/10.1007/s13201-016-0455-7

Shchipunov Y (2012) Bionanocomposites: green sustainable materials for the future. Pure Appl Chem 84(12):2579–2607

Shevah Y (2019) Impact of persistent droughts on the quality of the Middle East water resources, Chapter-3, Book-Evaluating water quality to prevent future disasters, Editor- Satinder Ahuja, Elsevier, Netherlands, pp 51–84. https://doi.org/10.1016/B978-0-12-815730-5.00003-X

Singh P, Yadav SK, Kuddus M (2020) Green nanomaterials for wastewater treatment, Chapter-9, Book- Green nanomaterials- processing, properties and applications, Editors- Shakeel Ahmed, Wazed Ali, Springer Nature Singapore Pte Ltd., 2020, pp 227–242. https://doi.org/10.1007/978-981-15-3560-4

Smedley P (2003) Arsenic in groundwater- South and East Asia, Chapter-7, Book- Arsenic in groundwater: geochemistry and occurrence, Editots- Welch AH, Stollenwerk KG. Kluwer Academic Publishers, Boston, USA, pp 179–209. https://doi.org/10.1007/0-306-47956-7_7

Stollenwork KG (2003) Geochemical processes controlling transport of arsenic in groundwater: a review of adsorption, Chapter-3, Book- Arsenic in groundwater: Geochemistry and occurrence, Editors- Welch AH, Stollenwerk KG. Kluwer Academic Publishers, Boston, USA, pp 67–100. https://doi.org/10.1007/0-306-47956-7_3

Varma RS (2012) Greener approach to nanomaterials and their sustainable applications. Curr Opin Chem Eng 2012(1):123–128. https://doi.org/10.1016/j.coche.2011.12.002

Yang X, Tang K, Nasr A, Lin H (2018) The applications of nanocomposite catalysts in biofuel production, Chapter 12, Book- Multifunctional nanocomposites for energy and environmental applications, First Edition, Edited by Guo Z, Yuan C, Lu NL, Wiley-VCH Verlag GmbH &Co., pp 309–351

Recent Trends in Eco-Friendly Materials for Agrochemical Pollutants Removal: Polysaccharide-Based Nanocomposite Materials

Estefanía Baigorria, Laura M. Sanchez, Romina P. Ollier Primiano, and Vera A. Alvarez

In this chapter, recent progress in the investigation of polysaccharide-based nanocomposite materials for pesticide adsorption from wastewater is critically reviewed. The main properties and characteristics of the nanocomposites, as well as their adsorption performance toward several types of pesticides, are analyzed and discussed in terms of their efficiency and operability in sustainable agricultural practices.

1 Introduction

Water pollution due to different toxic substances, within them metals and organic compounds, is a thoughtful environmental problem. Furthermore, together with increasingly demanding regulations; water pollution has become a real source of concern but also a priority for several industrial sectors. Aromatic compounds and heavy metal ions, among other contaminants, are frequently found in the environment, and this finding can be mainly attributed to their extensive industrial uses. Within the common contaminants that can be found in wastewater, there are many associated with toxic or carcinogenic effects (Crini 2005).

The use of agrochemicals seems undeniably necessary to improve a wide range of crops, reducing pests of various types, among other advantages; whereas it enables significant increases in productivity. Together with these improvements, there exist

E. Baigorria · L. M. Sanchez · R. P. O. Primiano (✉) · V. A. Alvarez
Materiales Compuestos Termoplásticos (CoMP), Instituto de Investigaciones en Ciencia y Tecnología de Materiales (INTEMA), CONICET - Universidad Nacional de Mar del Plata (UNMdP), Av. Colón 10890, 7600 Mar del Plata, Argentina
e-mail: rominaollier@fi.mdp.edu.ar

E. Baigorria
Institute of Science and Technology, São Paulo State University (UNESP), Av. Três de Março, 511, Alto da Boa Vista, Sorocaba 51118087-180, Brazil

© The Author(s), under exclusive license to Springer Nature Singapore Pte Ltd. 2023
A. K. Mishra and C. M. Hussain (eds.), *Biobased Materials*,
https://doi.org/10.1007/978-981-19-6024-6_6

real risks of environmental pollution (Antonello Marocco et al. 2020). In addition to this, agrochemicals are related to some severe human beings' pathologies (Srivastava et al. 2009).

Since the agrochemicals present in both the soil and the surface water are substances with high persistence levels having mobility in the aqueous media (Derylo-Marczewska et al. 2010), and taking into account that they could be dangerous even at very low concentrations (including not detectable quantities or traces), they constitute a relevant kind of emerging contaminants. The growing concern about agrochemicals can be confirmed by the amount of published scientific works related to this topic (Crini 2005; Antonello Marocco et al. 2020; Srivastava et al. 2009; Jonoobi et al. 2015).

To avoid environmental contamination as well as human beings' health risks, strict legislation on toxic products and the development of several efficient technologies for the removal of agrochemicals from wastewater are undoubtedly necessary. In this sense, diverse processes and technologies are presently used for these purposes, as shown in Fig. 1. Among them, adsorption is one of the most important tools mainly due to its intrinsic simplicity, efficiency, and low cost (Smedt et al. 2015).

In recent years, nanocomposite adsorbents prepared by combining different nanoreinforcing materials and biopolymers have emerged as novel technological solutions for the sustainable preservation of the environment (Oliveira Sousa et al. 2021). Among biopolymers, polysaccharides, which are biological macromolecules obtained from plants, animals, and microbial-based resources, have been increasingly used for the development of several nanocomposite adsorbents for water remediation treatments. The presence of different functional groups that could be potentially useful as reaction and/or interaction sites for the removal of pollutants makes

Fig. 1 Processes and technologies that are presently used to remove agrochemicals

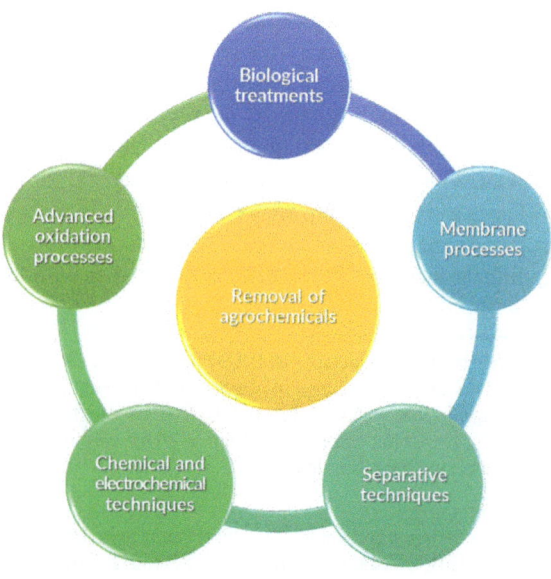

them suitable for this application. Chitosan, alginate, carrageenan, cellulose, and carboxymethyl cellulose are among the most reported polysaccharides.

The aim of this chapter was to report the recent progress in the investigation of polysaccharide-based nanocomposite materials for pesticide adsorption from wastewater. For this purpose, the main properties and characteristics of the nanocomposites, as well as their adsorption performances toward several types of pesticides were analyzed and discussed in terms of their efficiency and operability in sustainable agricultural practices.

2   Agrochemicals

2.1   Origin and Classification

Throughout the years, there have been several problems that decrease the productivity of agricultural crops. The presence of pests or problems in crop growth has affected agricultural activities worldwide for centuries. The existence of pests such as animals, microorganisms, and undergrowth dates back a long history. As early as 7000 BC, flies, locusts, and grasshoppers were known to invade crops. Farmers at that time carried out biological controls of these pests, intuitively selecting various methodologies, like planting seeds of resistant plants without having any scientific basis for such action. In China (1000 BC), they used mercury and sulfur to control crop pests (Ahmad et al. 2010; Baigorria and Fraceto 2021). The scientific study of pest control began only during the development of the Industrial Revolution (Ahmad et al. 2010; Baigorria and Fraceto 2021). Although the first reports indicate the use of products of natural origin for the control of various pests, the great industrial expansion led to the manufacture of various synthetic chemical products for this purpose. Since the nineteenth century, arsenic-based compounds began to be used as insecticides. Other pesticides used at that time were nicotine, tar oils, a mixture of copper sulfates and lime, and pyrethrum, among many others (Ahmad et al. 2010; Baigorria and Fraceto 2021). Around the end of that century and at the beginning of the twentieth century, the manufacture of various agrochemicals grew considerably as a plausible solution to increase the productivity of agricultural crops. Many of these chemicals were born as a conscious response to prevent the damage caused by various pests in the field of agriculture, while others helped in the promotion of crop growth (Handojo et al. 2020a).

Various chemicals, such as pesticides and fertilizers, are part of the so-called agrochemicals (Fig. 2) (Handojo et al. 2020a). Although the classification of these compounds is varied, those given by official international agencies or organizations, such as the World Health Organization (WHO), are the ones widely adopted by the scientific community.

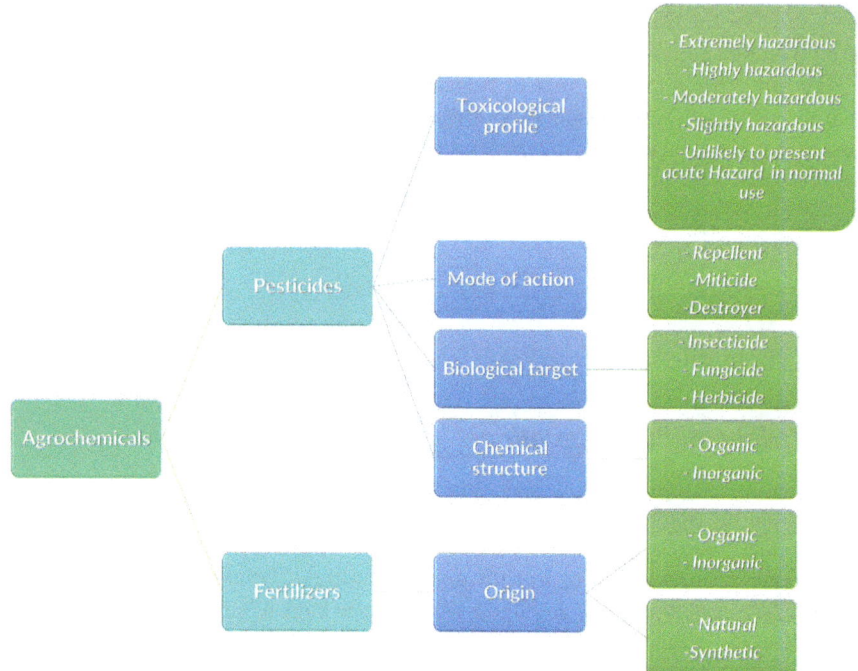

Fig. 2 Classification of agrochemicals

2.2 Pesticides

WHO defines pesticides as chemical substances that protect agricultural crops from pests and diseases such as insects, small animals, viruses, bacteria, fungi, and weeds (WHO 2020). The ways in which pesticides are classified are varied, ranging from the hazard they present, the mode of action on various pests (mitigator, repellent, or destroyer), to the biological target site and their chemical structure, as shown in Fig. 2. In this sense, the WHO classifies them according to their hazard/toxicity evaluated orally or dermally in laboratory animals, and it is expressed in terms of dermal median lethal dose (LD50) (WHO 2019). Thus, pesticides are divided into Class Ia (LD50oral < 5 mg/kg; LD50dermal < 50 mg/kg) or extremely hazardous, Class Ib or very hazardous (LD50oral 5–50 mg/kg; LD50dermal 50–200 mg/kg), Class II or moderately hazardous (LD50oral 5–2000 mg/kg; LD50dermal 200–2000 mg/kg; LD50dermal 200–2000 mg/kg), Class II or moderately hazardous (LD50oral 5–2000 mg/kg; LD50dermal 200–2000 mg/kg; LD50dermal 200–2000 mg/kg), Class III or slightly hazardous (LD50oral > 200 mg/kg; LD50dermal > 2000 mg/kg), and Class U or unlikely to present acute risk (LD50 ≥ 5000 mg/kg) (WHO 2020).

According to the biological targets of pesticides, it is possible to classify them as insecticides, fungicides, and herbicides (Baigorria and Fraceto 2021; Sanchez et al. 2020; Syafrudin et al. 2021; Ponnuchamy et al. 2021). Insecticides are those agrochemicals that control insects or small animals, present in large crops, orchards, and storage facilities, among others (Baigorria and Fraceto 2021; Sanchez et al. 2020; Syafrudin et al. 2021; Ponnuchamy et al. 2021). The Insecticide Resistance Action Committee (IRAC), on the other hand, classifies insecticides depending on their target mode of action and the physiological functions they affect (Sparks and Nauen 2015). The IRAC classifies insecticides as fast-acting to those that act on muscles or nerves, slow or moderately slow-acting to those that affect growth, and finally fast-acting to those that cause mitochondrial damage (Sparks and Nauen 2015). Fungicides, on the other hand, are those agrochemicals that act by preventing damage caused by fungi and act at the plant or seed level (Baigorria and Fraceto 2021). The Fungicide Resistance Action Committee subclassifies them considering the same criteria as IRAC (Funguicide 2019). Finally, those agrochemicals responsible for eliminating unwanted wild plants are called herbicides. Many authors consider growth-regulating agrochemicals as herbicides (Baigorria and Fraceto 2021; Sanchez et al. 2020; Syafrudin et al. 2021; Ponnuchamy et al. 2021). The Global Herbicide Resistance Action Committee (HRAC) presents a subclassification of herbicides depending on their specific or nonspecific target or site of action. According to their specific site of action, HRAC classifies them as Inhibitors of photosystems, Inhibitors of proteins, and inhibitors of enzymes. Considering the nonspecific site of action, HRAC classifies them as Enhanced Metabolic Detoxification, Altered Herbicide Distribution, and Multiple Resistance (HRAC 2020).

The chemical structure of pesticides is also an aspect to consider. The presence of various functional groups in the structures of pesticides is responsible for the sites of action or target organisms affected, the toxicity caused and the mechanisms of action developed. In this sense, they can be subclassified as inorganic or organic pesticides. Inorganic pesticides include mainly those based on heavy metals and sulfur. While among organic pesticides we find important variables such as triazines, phthalamides, benzoic derivatives, and carbamates, among others (Baigorria and Fraceto 2021; WHO 2020; Sanchez et al. 2020; Syafrudin et al. 2021; Ponnuchamy et al. 2021). Although pesticide classifications are varied, most of the criteria used consider similar aspects.

2.3 Fertilizers

Fertilizers are other members of the agrochemicals family. A fertilizer is any material of the organic or inorganic structure, of synthetic or natural origin, which is added to the soil with the aim of supplying various nutrients that contribute to plant growth (FAO 2020; Tripathi et al. 2020). Agricultural crops require for their growth and development various essential elements, most of which are usually available from the soil. However, due to intensive and continuous agricultural practices, a deficit of these nutrients is generated in the soil, which requires their replenishment. Fertilizers can be of natural or synthetic origin, whereas they could have both organic or inorganic chemical structures (Fig. 1). Inorganic fertilizers have the important advantage of being highly soluble in water, which increases their availability to plants. On the other hand, organic fertilizers are less available for plants due to their low water solubility (Tripathi et al. 2020). Elements considered as primary nutrients, such as nitrogen (N), potassium (K), and phosphorus (P), are those most required by plants and in great quantities (Tripathi et al. 2020). Although natural fertilizers exist, the synthetic field of these agrochemicals has opened up an unthinkable more efficient way to supply nutrients to crops. However, it must always be taken into account that the used fertilizers must be safe for the environment.

Most of these agrochemicals are also classified as emerging pollutants (EP). EPs have characteristics that vary dynamically over time and therefore their potential hazardousness also does it (Baigorria and Fraceto 2021; Ahamad et al. 2020). The growing existence of new agrochemicals with novel properties has caused an increase in the occurrence of EP and, consequently, greater environmental problems.

2.4 Environmental Implications, Remediation Treatments, and Regulations

Agricultural activities are an important link in the chain for generating improvements in the quality of life of society, as they contribute mainly to the generation of food. However, a quarter of the world's food crops are lost due to plant pests or diseases (FAO 2020). To counteract these losses, agrochemicals are used: in the past 3 decades, globally, fertilizers use has increased by 80% whereas pesticide use has increased 1.5 times (Baigorria and Fraceto 2021; Ponnuchamy et al. 2021; FAO 2020; Garba et al. 2021; Baigorria et al. 2021). Then, it is important to take into account that the excessive and inappropriate use of these agrochemicals generates ecological and environmental deteriorations, and some of them are irreversible. These undesired effects affect environmental matrices such as soil, air, and water, being the latter one of the most deteriorated.

Agrochemicals can be introduced into water bodies through various processes. Because they have different solubilities in water, agricultural runoff or drift can cause them to enter surface water bodies, while filtration processes favor their entry into groundwater (Baigorria and Fraceto 2021; Syafrudin et al. 2021). The incorrect application of agrochemicals combined with poor agricultural practices can cause direct spills of agrochemicals into water bodies. Besides, the improper disposal of the remains of these chemical products also generates damage in water basins (Baigorria and Fraceto 2021).

Several technologies are applied to remediate the agrochemical contamination of groundwater. These include biological, separative, and oxidative methods (Baigorria and Fraceto 2021; Ponnuchamy et al. 2021; Garba et al. 2021; Baigorria et al. 2021).

Among the methodologies involving biological processes are mainly bioremediation, phytoremediation, and biosorption. On the other hand, separative techniques involve coagulation, flocculation and precipitation methodologies, membrane technologies, and adsorption. As for oxidative techniques, we can find chemical oxidation, photocatalytic decomposition, and advanced oxidation processes (Baigorria and Fraceto 2021; Ponnuchamy et al. 2021; Garba et al. 2021; Baigorria et al. 2021). In many cases, the combination of several of these remediation techniques provides successful results. However, the combination of methodologies, such as hybrid technologies, implies higher economic costs. In addition, the generation of unwanted by-products after the removal treatments of these contaminants is another disadvantage. The correct choice of effective technologies for the remediation of these agrochemicals is of utmost importance. Adsorption techniques circumvent most of these aforementioned difficulties. Adsorption is considered to be the most appropriate methodology due to its low operational cost, lower energy expenditure, wide versatility, simple operational design, ease of recovery and reuse of the adsorbent materials used, elimination of small traces of contaminants, low generation of unwanted products and the wide variety of existing adsorbent materials (Baigorria and Fraceto 2021; Sanchez et al. 2020; Garba et al. 2021; Baigorria et al. 2021).

The correct application of technologies for the removal of agrochemicals must respect and comply with various national and international regulations that allow obtaining clean and safe water. This is due to the direct implications on public health caused by these problems. Several international regulations have been proposed by organizations such as the United States Environmental Protection Agency (US-EPA), the WHO, the European Union, and the Food and Agriculture Organization for the United Nations (FAO). These organizations have raised regulations at the international level for the use of agrochemicals (Baigorria and Fraceto 2021; Syafrudin et al. 2021). These norms and regulations, cover from the production of agrochemicals, their trade, and distribution, to their use, safety, and final disposal (Baigorria and Fraceto 2021; FAO 2020). However, compliance with these regulations depends on the countries where they are in force or adhered to, and their socioeconomic conditions, among many other aspects.

The implementation of these agricultural practices in a sustainable manner, applying national and international regulations and standards, is a solution to reduce environmental pollution. However, given the lack of consensus among the various countries, the application of methodologies for the removal of agrochemicals from environmental matrices, such as water, is of total relevance. The choice of ecologically friendly materials and technologies for this purpose, which complies with regulatory requirements, is a solution to this problem.

3 Nanocomposite Materials Based on Single Polysaccharides

Naturally occurring polymers are considered promising options for the development of novel water remediation devices. Among them, cellulose is the most abundant polymer on Earth and it also corresponds to the major constituent in waste biomass (Zheng et al. 2020; Rana et al. 2021). However, the high biopolymer abundance is not the only motivation to generate materials owing a higher added value: the presence of the –OH groups at C2, C3, and C6 positions of the glucose units offers multiple useful possibilities for diverse interactions (with both other materials components and pollutants) and also to generate other specific functionalities (Fig. 3). In this sense, composite cellulosic adsorbents able to remove a variety of agrochemicals have been generated by using natural and modified nanoclays, graphene and metal oxides, and metal sulfides (Narayanan et al. 2020, 2017; Khawaja et al. 2021; Gupta et al. 2017; Aris et al. 2020; Zhang et al. 2015; Komal et al. 2020). The nanoscale of the fillers is highly advantageous since they exhibit a larger surface area and higher interaction/activities than traditional materials (Jain et al. 2020; Handojo et al. 2020b).

Carboxymethyl cellulose (CMC) is a starting material in which the substitution of some cellulosic –OH groups by carboxymethyl ones, predominantly at the C2 position, is taking place (Narayanan et al. 2017). Firstly, the cellulose is activated

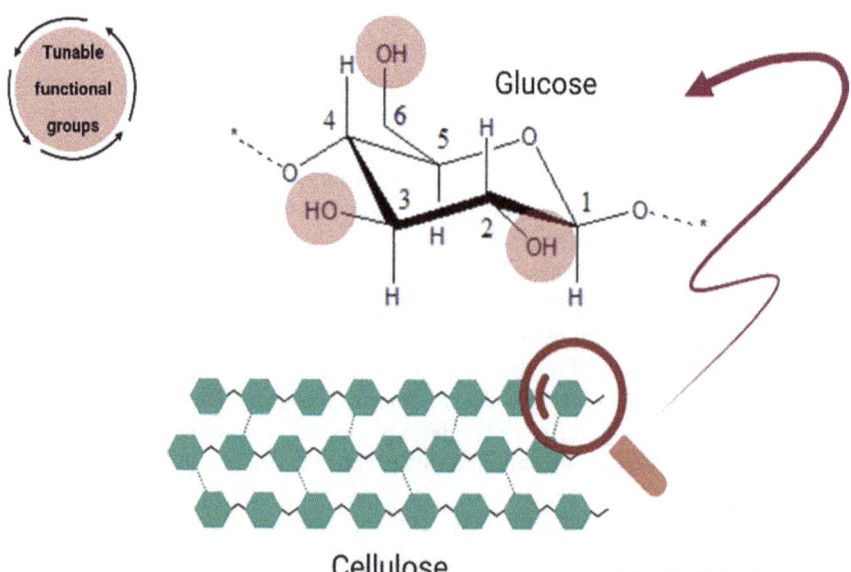

Fig. 3 Tunable functional groups of cellulose

by aqueous NaOH, and then the etherification reaction takes place. Khawaja and co-workers prepared magnetic graphene oxide (GO)–CMC nanocomposites containing a GO:CMC ratio of 1:100 (Khawaja et al. 2021). To this material, Fe magnetic nanoparticles were incorporated after being generated from $FeCl_3$ by its reduction and stabilization mediated by a green tea leaves extract. The adsorbents were tested as Atrazine removers from both model and real polluted water samples. It was found that an adsorbent dose of 100 mg for the treatment of a 1200 mg/L Atrazine aqueous solution exhibited a maximum adsorption capacity of 193.8 mg/g. The results indicated that the considered material is competitive with others previously reported.

Other widely employed cellulose derivative is cellulose acetate (CA), whose preparation implies the substitution of –OH groups by acetyl ones (Egot and Alguno 2018). Its intrinsic improved properties make it suitable for a variety of applications, including water remediation. Aris and collaborators prepared GO/CA nanocomposite materials through electrospinning (Aris et al. 2020). The specifically selected components, as well as the preparation technique, are oriented to control the materials morphology, and to also improve the adsorption capacity towards several polar agrochemicals: Ethoprophos, Chlorpyrifos, Methyl Parathion, and Sulfotepp. The authors have prepared and preliminary tested various formulations, including 1, 3, 5, 7, and 10 wt% of GO regarding CA contents, then found that the best performance was achieved by the 10 wt% of GO sample. Further pH-dependence, contact time, and adsorbent dosage influence studies were also conducted to determine the optimum treatment conditions. The highest activity for the removal of all agrochemicals was observed at pH 12 employing 5 mg of the 10 wt% of GO material for 15 min. It is important to mention that the researchers have found that the material is not suitable to be reused: very intense interactions might be taking place since a broad range of solvents were unable to restore the prepared sorbent.

GO was also used as starting material to generate graphene/cellulose composite materials after its reduction with hydrazine hydrate (Zhang et al. 2015). Zhang et al. have tested the adsorbent capability of this material towards six different triazine pesticides: Simeton, Cyprazine, Atrazine, Simazine, Ametryn, and Prometryn. Additionally, the authors have studied the intervening adsorption mechanism and the reuse possibilities. The desorption was performed in acetonitrile achieving more than 95% of pesticides recovery, then allowing to have adsorption percentages higher than 85% even after six removal cycles.

Commercial cellulose fibers and MnO_2 precursors were employed by Gupta and collaborators for nanocomposites preparation through a microwave-assisted hydrothermal method (Gupta et al. 2017). In order to optimize the experimental conditions for Toxaphene removal, Response Surface Methodology (RSM) was used the agrochemical concentration (2.0 and 20.0 mg/L), the pH (2 and 9), and the adsorbent dose (1 and 6 g/L) the selected input factors. The theoretical removal under the optimal conditions (Toxaphene 5 mg/L, pH 3.0, and adsorbent 5.0 g/L) was 96.8%, whereas the experimental value obtained was 96.5%.

Recently, Komar and coworkers have prepared, from sugarcane bagasse, nanocomposite materials having silanized cellulose nanofibers (SCNF) as polymeric matrix and CdS as the reinforcing component (Komal et al. 2020). The nanofibers were isolated from the agricultural waste via chemical and further mechanical treatment. After this, the silanization in accordance with other previous reports was conducted and the so-obtained SCNF was conveniently mixed with the CdS precursors. A heating step of 12 h at 180 °C in an autoclave was responsible for the nanoparticle synthesis. In this study, the researchers not only have completely characterized the nanocomposite material even explaining the stabilization mechanism of CdS by the polymeric matrix, but they also tested the removal efficiency of Chloropyrifos under a series of treatment conditions: pH, contact time, concentration of agrochemical and adsorbent dosage. Kinetic adsorption models as well as the materials regeneration and reuse capacities were also investigated. The CdS/SCNF nanocomposite was found to have very good performance as a reusable Chloropyrifos adsorbent.

A summary of the above-mentioned nanocomposite materials and the corresponding agrochemicals adsorbed is presented in Fig. 4.

Another abundant biopolymer, usually extracted from various brown marine algae of *Phaeophyceae* family, is alginate. This anionic polymer conforms to a whole family of linear copolymers containing blocks of α-L-guluronate (G) and β-D-mannuronate (M) residues being (1→4)-linked, and it is important to remark that alginates extracted from different sources vary in their G and M contents as well as their corresponding blocks length. One of the main reasons why alginate has become popular is related to the possibility of producing gels under the simple presence of di

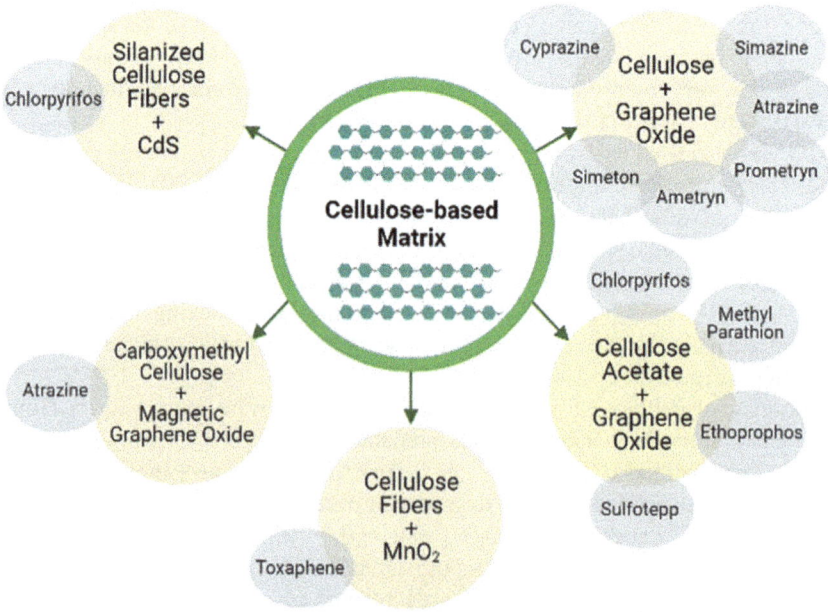

Created in BioRender.com

Fig. 4 Cellulose and cellulose derivatives nanocomposite materials for agrochemicals adsorption (Khawaja et al. 2021; Gupta et al. 2017; Aris et al. 2020; Zhang et al. 2015; Komal et al. 2020)

or trivalent cations, such as Ca^{2+} and Fe^{3+} ions (Skjåk-Bræk et al. 2012; Hasnain et al. 2020). Since alginate beads typically have poor mechanical properties, several fillers have been incorporated into them, such as natural or modified clays. Etcheverry and co-workers have prepared alginate beads containing different quantities of montmorillonite (0, 0.1, 1, and 4 wt%) with the aim to adsorb Paraquat herbicide (Etcheverry et al. 2017). Interestingly, the researchers found that the quantity of agrochemicals being adsorbed was linearly dependent on the clay content, and they checked that the removal activity has nothing to do with the polymeric matrix. However, as it serves as a useful container for easy removal after the applied treatment as well as it also offers the possibility to desorb and further reuse the beads, the preparation of this kind of gel system is still highly desired. Previously, two different pillared clays were encapsulated in Alginate beads by Lezehari et. al (Lezehari et al. 2010). The modified clays were aluminum-pillared montmorillonite and surfactant-modified-pillared montmorillonite, and it was found that both lead to an improvement in the beads removal capacities toward pentachlorophenol.

Mahmoud and collaborators presented other alginate beads containing nano-zero-valent iron (nZVI) potentially useful as chlorinated pesticides removers (Mahmoud et al. 2017). The nZVI particles were prepared from Fe^{+3} salts being reduced by $NaBH_4$, and then they were incorporated into the Alginate aqueous solution generated to be dropped into the Ca^{+2} cross-linking system. The effect of different operational parameters such as contact time, adsorbent dosage, pH, stirring rate, and initial pollutant concentration was deeply studied. Additionally, an artificial neural network (ANN) was employed to predict the ability to remove chlorinated pesticides.

From fungi cell walls and also from crustaceous exoskeletons, it is possible to isolate chitin and, after its deacetylation, the well-known chitosan (Jain et al. 2020). Chitosan has two different functional groups that can result useful from different points of view: on one side, depending on the working pH they can be responsible for electrostatic interactions and/or hydrogen bonding with different molecules and, on the other hand, they can be modified in order to have other new functionalities. This second-more-abundant polysaccharide in nature has been employed by Mahaninia and Wilson, among several other researchers, to prepare bare chitosan beads able to remove agrochemicals (Mahaninia and Wilson 2016). Contrarily to what was usually observed in alginate-based materials, the polymeric matrix has demonstrated a certain ability as an adsorbent by itself. Then, nanocomposite materials have also been prepared and tested looking forward to the adsorption of agrochemicals from polluted water bodies. For example, Swivedi et. al prepared chitosan beads crosslinked by NaOH and containing in situ prepared gold nanoparticles (Dwivedi et al. 2014). The chitosan's function is not only to act as a polymeric matrix but also play a stabilizing agent role during the particles preparation. This research has revealed that a small amount of the selected nanomaterial (0.5 wt.%) increases the sorption capacity of the chitosan beads toward methyl parathion more than twofold regarding the neat ones. Moreover, coagulation induced by pH was also used by Saifuddin and collaborators to prepare chitosan beads containing previously prepared silver nanoparticles (Lezehari et al. 2017). The authors conducted batch experiments and they find out that the adsorption capacity of the material was 0.5 mg of Atrazine per each gram of adsorbent after 65 min. After this, column adsorption studies and further desorption tests were also conducted by employing 0.1 M NaOH. Thus, the column was successfully regenerated and the materials loss of activity after five use cycles is near to 50%. In another recent work, adsorbent nanocomposite hydrogel beads composed of chitosan and a commercial organoclay were tested for the removal of carbendazim in batch tests (Baigorria and Fraceto 2022). The nano-reinforced beads achieved a removal capacity of 0.24 mg/g whereas the neat organoclay presented a removal efficiency of 0.69 mg/g. In view of these results, the authors explain that the operational facilities and the subsequent reuse of the composite hydrogels are important advantages compared to the tedious manipulation of neat organoclay in water decontamination treatments.

As it was previously done, a summary of the alginate- and chitosan-based nanocomposite materials mentioned in the last paragraphs as well as the corresponding agrochemicals adsorbed by them is presented in Fig. 5.

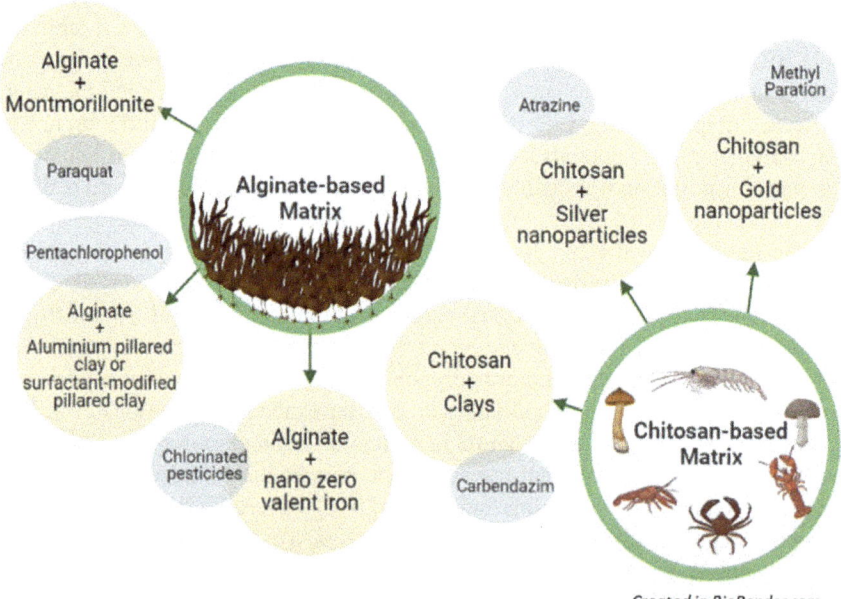

Created in BioRender.com

Fig. 5 Alginate- and Chitosan-based nanocomposite materials for agrochemicals adsorption (Etcheverry et al. 2017; Lezehari et al. 2010, 2017; Mahmoud et al. 2017; Dwivedi et al. 2014; Baigorria and Fraceto 2022)

4 Nanocomposite Materials Based on Polysaccharide Blends

At this point, it might be important to comment that the importance and usefulness of considering both natural polymers and nanomaterials as starting materials to generate bio-based eco-friendly nanocomposite adsorbents for agrochemical removal have already been exposed. However, sometimes the sole use of just one polymer matrix can lead to poor properties and further limited device performances. So, to better exploit the available resources, improve the properties of the matrix, and eventually achieve synergic effects, approaches in research have been headed towards the development of blending polysaccharide-based nanocomposite materials as novel adsorbent materials. Although much research has been conducted for the application of this kind of materials in the removal of a vast number of dyes and heavy metals, there is little available literature regarding the study of the removal of agrochemicals. Unfortunately, to date, none of the reported nanocomposites that have been proposed for this application are fully sustainable because the polymer matrices are composed of mixtures of a polysaccharide and a synthetic polymer. Indeed, no reports regarding the use of nanocomposites with mixed polysaccharide matrices or polysaccharides blended with other types of biopolymers have been found for any pesticide removal.

Barbosa et al. evaluated the adsorption of Paraquat herbicide on biocompatible and biodegradable polyacrylamide/CMC/zeolite nanocomposite hydrogels (Barbosa et al. 2018). This research evidenced the improved herbicide sorption performance by increasing the zeolite amount in the hydrogels. In another work, poly(methacrylic acid)-co-polyacrylamide-carboxymethylcellulose nanocomposite hydrogels containing different amounts of zeolite clinoptilolite were synthesized and sorption of Diquat in batch experiments was assed (Tanaka et al. 2021). A removal efficiency of 100% was achieved after 8 h. Although in the initial sorption process the pure hydrogel presented the highest removal of Diquat due to the highest chain elasticity, the zeolite-loaded nanocomposites were more effective in the final Diquat removal. Diquat sorption of 9.4 mg/g was obtained when 1.5% zeolite was present in the hydrogel nanocomposites. Therefore, the authors affirm that the application of these materials can help to avoid contamination by the excessive use of Diquat in agricultural practices.

Polyvinyl alcohol (PVA) is a hydrophilic, innocuous, and synthetic polymer, with different interesting properties such as good chemical stability, gel and film forming ability, biocompatibility, and mechanical strength. This useful polymer has been blended with different polysaccharides to prepare a great variety of nanocomposites with the aim of developing membranes and hydrogel beads for their application in water remediation treatments (Baigorria et al. 2020; Gholamali et al. 2020). Kaur and collaborators prepared chitosan–PVA nanocomposite hydrogels containing silver nanoparticles (AgNP) and tested them for the removal of Butachlor from an aqueous solution (Kaur et al. 2021). The nanocomposite hydrogels exhibited a Butachlor adsorption capacity of 96.87% at 30 °C and pH 3.0 at a concentration of 10 mg/L, which was 10% higher than neat chitosan-PVA matrix. At increasing

Butachlor concentrations, the removal efficiency of the materials declined. In addition, reusability tests of the nanocomposites evidenced that Butachlor removal decreased from 96.87% to 79.32% from the first to the fifth cycle.

Apart from being useful in the removal of pesticides by adsorption, nanocomposite materials based on polysaccharide blends could also be employed to encapsulate microorganisms with the aim of applying the resulting materials in bioremediation treatments. The immobilized microorganism technology has several advantages compared to the use of bare cells, mainly due to the fact that microbial cells would maintain good bioactivity and present better stability with longer service life and an easier extraction from the treated aqueous solution (Ursoiu et al. 2012). Penicillium sp. yz11-22N2 doped with nano Fe_3O_4 was immobilized within PVA-alginate gel beads and tested for Atrazine removal (Yu et al. 2018). The magnetic nanomaterial presented a removal efficiency of 91.2% at pH of 7, 28 °C, and 8.00 mg/L atrazine. The authors explained that active microorganisms played a major role in Atrazine removal, while a little atrazine was removed by adsorption. Moreover, nano Fe_3O_4 could promote Atrazine removal, and had a synergistic effect on active microorganisms.

5 Conclusions and Future Perspectives

Among biopolymers, polysaccharides are of great value in the present context of limited oil reserves and global warming associated with the use of synthetic polymers, due to their abundant availability, low cost, biocompatibility, eco-friendliness, and ease of modification. With the advent of nanotechnology, polysaccharide-based nanocomposite adsorbents synthesized by novel methods have carved a niche for themselves in the elimination of different pesticides from polluted water. Cellulose, chitosan, and alginate are the most employed polysaccharides for the development of such promising nanomaterials.

Although a lot of work has been done on the utilization of polysaccharide-based nanocomposites in water remediation treatments aiming to eliminate dyes, heavy metals, and drugs from contaminated water, more research is still needed regarding the study of their effectiveness in agrochemicals removal. Moreover, owing to the great potential and versatility of fully biobased nanocomposite materials prepared from polysaccharide blends, there is plenty of research work to be done to develop efficient and sustainable solutions for the removal of pesticides from water and wastewater. Thus, more research should be encouraged in this direction in the coming years.

Additionally, further research efforts should be focused on the fabrication and application of the above-mentioned adsorbent devices at full scale, which is always a key challenge that must be overcome when facing the real application of a product.

To sum up, we expect that this review will provide some inspiration for further research work, in the near future, about greener methods of fabrication and application of sustainable polysaccharide-based nanocomposite materials to eliminate harmful agrochemicals from polluted water bodies, and, thus, develop useful products that meet economic, environmental, and social needs.

Acknowledgements This work was supported by the Consejo Nacional de Investigaciones Científicas y Técnicas de la República Argentina (CONICET), the Agencia Nacional de Promoción de la Investigación, el Desarrollo Tecnológico y la Innovación (Agencia I+D+i) and the Universidad Nacional de Mar del Plata (UNMdP).

References

Ahamad A, Madhav S, Singh AK, Kumar A, Singh P (2020) in Sensors Water Pollut. Monit. Role Mater. Adv. Funct. Mater. Sensors., Pooja D, Kumar P, Singh P, Patil S (Eds), Springer, Singapore

Ahmad T, Rafatullah M, Ghazali A, Sulaiman O, Hashim R, Ahmad A (2010) J Environ Sci Heal Part C Environ Carcinog Ecotoxicol Rev 28:231–271

Antonello Marocco MP, Dell'Agli G, Sannino F, Esposito S, Bonelli B, Allia P, Tiberto P, Barrera G (2020) Process 8:1–26

Aris NIF, Rahman NA, Wahid MH, Yahaya N, Keyon ASA, Kamaruzaman S, Soc R (2020) Open Sci 7:1–19

Baigorria E, Galhardi JA, Fraceto LF (2021) J Clean Prod 295:126451

Baigorria E, Fraceto LF (2021) Adv Sustain Syst, 2100243

Baigorria E, Fraceto LF (2022) J Clean Prod 331:129867. Elsevier Ltd

Baigorria E, Cano LA, Sanchez LM, Alvarez VA, Ollier RP (2020) Environ Nanotechnology Monit Manag 14:100364

Barbosa D, Moura M, Aouada F (2018) Quim Nova 41:380–385

Crini G (2005) Prog Polym Sci 30:38–70

De Smedt C, Ferrer F, Leus K, Spanoghe P (2015) Adsorpt Sci Technol 33:457–485

Derylo-Marczewska A, Blachnio M, Marczewski AW, Swiatkowski A, Tarasiuk B (2010) J Therm Anal Calorim 101:785–794

Dwivedi C, Gupta A, Chaudhary A, Nandi CK (2014) RSC Adv 4:39830–39838. The Royal Society of Chemistry

Egot MP, Alguno AC (2018) Key Eng Mater 772:8–12

Etcheverry M, Cappa V, Trelles J, Zanini G (2017) J Environ Chem Eng 5:5868–5875

FAO Food and Agriculture Organization of the United Nations, Food Agric. Organ. United Nations (2020)

F. Funguicide Resistance Action Committee, Funguicide Resistance Action Committee, Ed. (España, 2019)

Garba ZN, Abdullahi AK, Haruna A, Gana SA (2021) Beni-Suef Univ J Basic. Appl Sci 10:1–19

Gholamali I, Asnaashariisfahani M, Alipour E (2020) Regen Eng Transl Med 6:138–153. Springer

Gupta VK, Fakhri A, Agarwal S, Sadeghi N (2017) Int J Biol Macromol 102:840–846. Elsevier B.V.

HRAC Herbicide Resistance Action Committee, Herbic. Resist. Action Comm.

Handojo LA, Ikhsan NA, Mukti R, Insarto A (2020a) in Agrochem. Detect. Treat. Remediat., 1st Editio, M. N. V. Prasad, Ed. (Butterworth-Heinemann

Handojo L, Ikhsan NA, Mukti RR, Indarto A (2020b) in Agrochem. Detect Treat Remediat (LTD, 2020b)

Hasnain MS, Jameel E, Mohanta B, Dhara AK, Alkahtani S, Nayak AK (2020) in Alginates Drug Deliv. INC

Jain M, Mudhoo A, Ramasamy DL, Najafi M, Usman M, Zhu R, Kumar G, Shobana S, Garg VK et al (2020) Environ Sci Pollut Res 27:34862–34905

Jonoobi M, Oladi R, Davoudpour Y (2015) Cellulose 22:935–969

Kaur R, Goyal D, Agnihotri S (2021) Carbohydr Polym 262:117906. Elsevier Ltd.

Khawaja H, Zahir E, Asghar MA, Rafique K, Asghar MA (2021) J Macromol Sci Part B Phys 60:1–21. Taylor & Francis

Komal, Gupta K, Kumar V, Tikoo KB, Kaushik A, Singhal S (2020) Chem Eng J 385:1–15. Elsevier

Lezehari M, Basly JP, Baudu M, Bouras O (2010) Colloids Surfaces A Physicochem Eng Asp 366:88–94

Lezehari M, Basly JP, Baudu M, Bouras O, Mahmoud AS, Saryel-deen RA, Mostafa MK, Peters RW, Tran VS et al (2017) Asian J Biochem 6:142–159. Elsevier Ltd

Mahaninia MH, Wilson LD (2016) Ind Eng Chem Res 55:11706–11715

Mahmoud AS, Saryel-deen RA, Mostafa MK, Peters RW (2017) in 2017 Annu. AIChE Meet. Minneapolis, MN

Narayanan N, Gupta S, Gajbhiye VT, Manjaiah KM (2017) Chemosphere 173:502–511

Narayanan N, Gupta S, Gajbhiye VT (2020) Int J Environ Sci Technol 17:4775–4786. Springer , Berlin, Heidelberg

Oliveira Sousa Neto V de, de Castro AJR, de Moura CP, Júnior GAM, Portela RR, Saraiva GD, do Nascimento RF (2021) Nanomater. Nanotechnol. Springer

Ponnuchamy M, Kapoor A, Senthil Kumar P, D.-V. N. Vo, Balakrishnan A, Mariam Jacob M, Sivaraman P, Environ. Chem Lett 19:2425–2463

Rana AK, Mishra YK, Gupta VK, Thakur VK (2021) Sci Total Environ 797:1–24. Elsevier B.V.

Sanchez LM, Ollier RP, Pereira AES, Fraceto LF, Alvarez VA (2020) Biopolym. Membr. Film, de Moraes MA, da Silva CF, Vieira R (Eds) Elsevier

Skjåk-Bræk G, Draget KI (2012) in Polym Sci A Compr Ref 10 Vol. Set 10. Elsevier B.V.

Sparks TC, Nauen R (2015) Pestic Biochem Physiol 121:122–128

Srivastava B, Jhelum V, Basu DD, Patanjali PK (2009) J Sci Ind Res (India) 68:839–850

Syafrudin M, Kristanti RA, Yuniarto A, Hadibarata T, Rhee J, Al-onazi WA, Algarni TS, Almarri AH, Al-Mohaimeed AM (2021) Int J Environ Res Public Health 18:468–473

Tanaka FC, Junior CRF, Fernandes RS, de Moura MR, Aouada FA (2021) J Polym Environ 29:3389–3400. Springer US.

Tripathi S, Srivastava P, Devi RS, Bhadouria R (2020) in Agrochem. Detect. Treat. Remediat., 1st Editio, M. N. V. Prasad, Ed. (Butterworth-Heinemann)

Ursoiu A, Paul C, Kurtán T, Péter F (2012) Molecules 17:13045–13061. Multidisciplinary Digital Publishing Institute

WHO World Health Organization, W. World Health Organization, Ed. (Chemical Safety and Healt Unit, 2020)

Yu J, He H, Yang WL, Yang C, Zeng G, Wu X (2018) Bioresour Technol 260:196–203. Elsevier

Zhang C, Zhang RZ, Ma YQ, Guan WB, Wu XL, Liu X, Li H, Du YL, Pan CP, Sustain ACS (2015) Chem Eng 3:396–405

Zheng F, Chen L, Zhang P, Zhou J, Lu X, Tian W (2020) Carbohydr. Polym 230:1–15. Elsevier

Protein-Based Biomaterials for Sustainable Remediation of Aquatic Environments

Pulak Pritam, Soumyaranjan Senapati, Shusree Prachi Palai, Jyotirmayee Giri, Manisha Dash, Bijayalaxmi Sahoo, Tapan Kumar Bastia, Prasanta Rath, and Alok Kumar Panda

1 Introduction

Water plays an important role in sustaining an ecosystem and also in developing a socioeconomic background of a particular geographical area. It is interesting to note that in today's world only about 25% of the human population has access to safe drinking water. Scarcity of clean and safe drinking water is a threat faced by humankind and is included as one of the 17 sustainable development goals by the UN for the year 2030. The growth in human population, environmental disasters, growing industrialization, and usage of artificial agricultural practices threaten the goal of safe drinking water. The major sources of water contamination are domestic sewage and wastewater, textile wastewater, landfills, mining activities, agricultural activities, etc. All these activities lead to the discharge of massive amounts of toxic compounds such as heavy metals, dyes, microplastics, and organic pollutants. Therefore, several water purification technologies such as distillation, filtration, and adsorption have been used to purify the water and make it suitable for drinking, but these methods are not sustainable. The need of the hour is developing a sustainable water purification technology that must be both environmental and techno-economical friendly. The removal of pollutants by adsorption is in general economical with minimum energy requirements.

Among all the adsorbing materials, proteins are considered to be the best sustainable adsorbing materials for water purification because they are mostly water-soluble and have good adhesive properties. In addition to this, specific sites in the proteins

P. Pritam · S. Senapati · S. P. Palai · J. Giri · M. Dash · B. Sahoo · T. K. Bastia · P. Rath · A. K. Panda (✉)
Environmental Science Laboratory, School of Applied Sciences, Kalinga Institute of Industrial Technology, Deemed to be University, Bhubaneswar 751024, India
e-mail: alok.pandafch@kiit.ac.in

P. Rath
e-mail: prathfch@kiit.ac.in

can be engineered to convert them into thin films or biomaterials with high surface area. Thirdly, proteins can be synthesized or extracted in large quantities at a very low cost by recombinant DNA technology or from agriculture, dairy, and food industries waste. Thus, converting protein-based biomaterials into good adsorbers for the treatment of water and aquatic environments is a sustainable and green solution.

In this chapter, we review the protein-based biomaterials formed from collagen, elastin, silk, and keratin and their applications in water and environmental remediation. First, we have described the structure, synthesis, and extraction process adopted for each of these above proteins. Then we discuss the biochemistry of each of these proteins in brief. Finally, the applications of each of these protein biomaterials in water and environmental remediation have been outlined.

2 Collagen

Have you ever wondered despite so many movable parts, e.g., bones, tendons, muscles, and organs how they get connected and stay within a finite place, i.e., our body? Well from ancient times to the modern era, people thought that there is some type of glue that held together our body. Later we know them as connective tissue, which is further composed of several kinds of protein and fat fibers. Among them "collagen" is the primary component of all connective tissues, including skin, tendon, ligament, and cartilage. The term "collagen" derives from the Greek words for "glue" and "to produce" and was first discovered in tissues that when boiled produce glue. The word "collagen" was coined in the nineteenth century to identify as a constituent of connective tissues that yields gelatin after boiling (Gorham 1991). So now we know that collagen is a major extracellular matrix (ECM) molecule that provides support for cell growth, self-assembles into cross-striated fibrils and is responsible for the mechanical resilience of connective tissues (Sorushanova et al. 2019).

2.1 Structure

Collagen is a secondary structure triple-helical protein, where the helices are composed of three polypeptide chains, each with the repeating units of tripeptide sequence Gly-X–Y, in which X and Y are usually Proline (Pro) and 4-hydroxyproline (4HyP), respectively (Rainey and Goh 2009). In the triple helix, -NH group of glycine and -CO group of proline (or 4-hydroxyproline) are boned by hydrogen bonds, where the Pro and 4HyP residues stabilize the triple helix structure through hydrophobic interactions. Here the core of the α helix is stabilized by hydrogen bonds between atoms of the peptide bond. The 4HyP residues in collagen, which are required to form strong noncovalent interactions within the collagen helix, are derived from

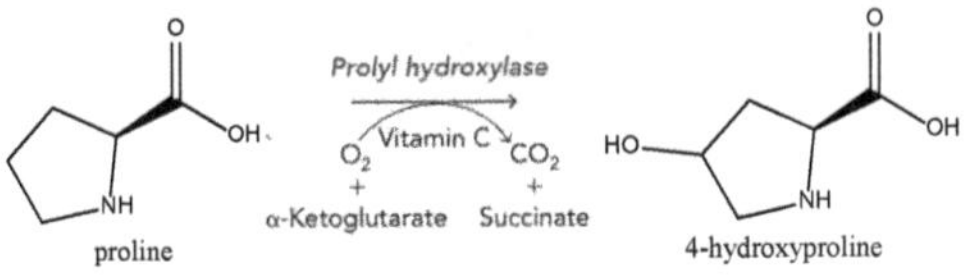

Fig. 1 Enzymatic conversion of proline to 4-hydroxy proline

proline   α-Ketoglutarate   Succinate   4-hydroxyproline

Prolyl hydroxylase

Vitamin C

O_2 CO_2

hydroxylation of proline by the enzyme prolyl hydroxylase, which uses (ascorbate) vitamin C as a cofactor (Fig. 1).

Collagen residues nearly (30%) consist of proline and hydroxyproline. After the amino acids are joined together, a specialized hydroxylating enzyme (prolyl hydroxylase) converts proline to hydroxyproline (Fig. 1). The triple helix is arranged so that every third residue on each chain is occupied by glycine inside the helix. The three individual collagen chains twist around each other in a superhelical pattern to form a rigid rod-like structure that differs from the α-helix. This molecule is called *tropocollagen,* which is 300 nm (3000 Å) long and 1.5 nm (15 Å) in diameter. The hydroxyproline and hydroxylysine residues form hydrogen bonds that hold the three strands together. The triple-stranded structure has a molecular weight of roughly 300,000, with each strand containing about 800 amino acid residues. Covalent connections between lysine and histidine residues link collagen intramolecularly and intermolecularly (cross-linked). This causes meat from older animals to be tougher than meat from younger animals because the quantity of cross-linking in tissue rises with age (Miller et al. 1980).

2.2 Sources

We can find collagen from both natural and synthetic sources. The natural collagens come from land animals (skin, tendon, bone) as well as from marine organisms (fish scale, fish skins). However, frequent use of animal collagens raises the possibility of getting allergenic diseases. Hence in later stages to prevent immunogenic problems synthetic collagen (viz. KOD) sources are invented. Apart from animal sources, we also obtain collagen fibers from several plants. But due to the high cost of production and low yield from plant sources, artificial collagens are preferred the most (Rodríguez et al. 2017; Browne, Zeugolis, and Pandit 2013; Felician et al. 2018; Ghomi et al. 2021).

2.3 Collagen Superfamily

The diversity of the collagen superfamily is determined by the presence of α-helices and their isoforms. This leads to the discovery of totally 28 vertebrate collagen types denoted as Roman letters (I–XXVIII) (Table 1) (Wu et al. 2009; Ricard-Blum 2011).

However, there is a novel epidermal collagen (type—XXIX) discovered which is very similar to the type—VI chain (Gara et al. 2008; Söderhäll et al. 2007).

Based on primary structure, the molecular weight, the size and shape of the terminal domains, and variation in the post-translation modifications, four collagen groups can be identified (Table 2) (Jenkins et al. 2005; Pace et al. 2003; Ramshaw,

Table 1 The collagen family

Collagen type	α-chain(s)
I	α1(I), α2(I)
II	α1(II)
III	α1(III)
IV	α1(IV), α2(IV), α3(IV), α4(IV), α5(IV), α6(IV)
V	α1(V), α2(V), α3(V), α4(V)*
VI	α1(VI), α2(VI), α3(VI), α4(VI)**, α5(VI)***, α6(V)
VII	α1(VII)
VIII	α1(VIII)
IX	α1(IX), α2(IX), α3(IX)
X	α1(X)
XI	α1(XI), α2(XI), α3(XI)
XII	α1(XII)
XIII	α1(XIII)
XIV	α1(XIV)
XV	α1(XV)
XVI	α1(XVI)
XVII	α1(XVII)
XVIII	α1 (XVIII)
XIX	α1(XIX)
XX	α1(XX)
XXI	α1(XXI)
XXII	α1(XXII)
XXIII	α1(XXIII)
XXIV	α1(XXIV)
XXV	α1(XXV)
XXVI	α1(XXVI)
XXVII	α1(XXVII)
XXVIII	α1(XXVIII)

*α4(V) chain is only synthesized by Schwann cells

** α4(VI) chain does not exist in humans

*** α5(VI) has been designated as α1(XXIX)

Table 2 Collagen groups

Groups	Description
Group 1	The majority of collagens that form filaments are found here (e.g., type I, II, III, V, XI, XXIV, and type XXVII). They are all 300 nm long and have triple helices with unbroken Gly–X–Y stretches. In the dermis, tendons, and other tissues, collagen strands are frequently a combination of type I, type III, and type V collagens. Heterotypic fibrils are distinct from homotypic fibrils, which are made up of only one type of collagen strand (e.g., collagen VII in anchoring fibrils of the dermo-epidermal junction)
Group 2	Homotypic fibril basement membrane collagens (e.g., type IV, VII, and XXVIII). Collagen type IV generates a fibrillar meshwork, whereas collagen type VII forms cross-striated fibrils with a distinct banding pattern as a result of antiparallel dimer interaction
Group 3	Short-chain collagens (e.g., type VI, VIII, X, and XXIX). They get their names from their triple-helical areas, which are usually only 100 to 150 nm long. Collagen types VI, VIII, and X form hexagonal lattices, whereas collagen type XXIX has a brief and unbroken triple-helical area surrounded by multiple von Willebrand factor A domains
Group 4	Multiple discontinuities exist in the triple-helical Gly–X–Y lengths of these collagens (e.g., type IX, XII, XIV, XVI, and types XIX to XXII). They contain fibril-associated collagens with triple helices that are disrupted (FACIT collagens). These collagens play specialized roles in numerous collagen fibrils, enhancing their functionality. As collagen type V, they also play a function in restricting lateral appositional growth (collagen type IX) and thereby controlling the diameter of collagen fibers in various tissues

Werkmeister, and Glattauer 1996; Rest, Garrone, and Herbage 1993; Bailey, Paul, and Knott 1998; Hulmes 2002; Sorushanova et al. 2019).

TYPE I COLLAGEN: Type I collagen molecules generally form cross-striated fibers of size larger than 400 Å in diameter. The properties of the fibers are determined by the degree of intra- and intermolecular cross-linking, orientation, and density of the fibrils. Their arrangement is in either, strong parallel alignment (e.g., in tendon and cornea), or they may cross each other (e.g., skin and organ capsules). The type I collagen fibers thus play the key role as a supporting element of high tensile strength and very limited elasticity in tissues that exhibit very little distensibility depending upon respective locations and functional demands (Gay and Miller 1983; Sorushanova et al. 2019). TYPE II COLLAGEN: These molecules exhibit far more restricted distribution than type I fibers in tissues. They are almost and exclusively found in cartilaginous structures in the body, where they are synthesized by chondroblasts. Therefore, the tissue matrix of cartilages is very rich in type II collagen, which represents up to 40% of the dry weight of cartilage (Gay and Miller 1983; Sorushanova et al. 2019). TYPE III COLLAGEN: These molecules are synthesized along with type I collagens by cells that synthesize predominantly type I collagen e.g., Skin fibroblasts both synthesize type I and type III collagen in about a 5:1 ratio, and the same cells can synthesize both types simultaneously.

2.4   Biosynthesis Overview

The collagen biosynthesis pathway (from gene transcription through secretion and aggregation of collagen monomers to functional fibrils) is a long and complicated process. The manufacture of collagen is initiated by mRNA transcription, which is encoded by tri helix chain combinations of distinct α-chain genes, depending on the type and isoform of collagen. The N-terminus of the collagen chain enters the endoplasmic reticulum (ER) lumen as pre-pro-collagen, which is then transformed into procollagen once the signal peptide is removed. The most notable feature of collagen biosynthesis is the fact that simultaneous beginning of synthesis and triple helix formation at the N-terminus and the C-terminus respectively. It requires the pro-α-chains to remain untied for the duration taken to complete translation of the α-chain, upon which three pro-α-chains align precisely at the C-terminus before triple helix formation initiates. Several chaperone proteins protect α-chains from getting snarled, including prolyl 4-hydroxylase (P4-H), protein disulfide isomerase (PDI), a homolog of heat shock protein 70 of the endoplasmic reticulum (BiP/Grp78), various peptidyl-prolyl cis–trans isomerases (PPIases) and heat shock protein 47 (hsp47) (Saha and Shamala 2011; Holmgren et al. 1999; 1998; Hudson et al. 2014; Mizuno et al. 2013; Pihlajaniemi, Myllylä, and Kivirikko 1991; Burjanadze 2000; Fratzl, n.d.; Engel and Prockop 1991; Koide and Nagata 2005; Raghunath, Bruckner, and Steinmann 1994; Sorushanova et al. 2019) (Table 3).

2.5   Biomaterial from Collagen and Their Applications

Biomaterials prepared from collagen are used extensively in many environmental bioremediation processes. Desimone and co-workers have shown that collagen-based hydrogels have the ability to adsorb Remazol black B dye (Tuttolomondo et al. 2015). In the same study, they have shown that in an alkaline solution stronger retention of the dye is achieved. They have suggested that the bonding between the protein and the dye is covalent in nature and took place by Michael's reaction. In another study, Kalirajan et al. have shown that collagen-based hybrid scaffolds incorporated titanium oxide displayed both absorption and adsorption behavior. These hybrid scaffolds helped in the removal of oil and dye molecules from the contaminated water (Kalirajan et al. 2019). In this study, they have seen that the hybrid scaffold can remove both the cationic and anionic dyes from the contaminated water. Nagaraj et al. have shown that collagen-titanium oxide nanoparticles photolytically degraded Rhodamine B under visible light (Nagaraj et al. 2021). Collagen–polyurethane–chitosan hydrogels help in removing lead ions from water (Irene et al. 2020).

Table 3 Summary of collagen biosynthesis [Reprinted from Ramshaw, J. A. M., Werkmeister, J. A. & Glattauer, V. Collagen-based Biomaterials. *Biotechnol. Genet. Eng. Rev.* 13, 335–382 (Copyright 1996), with permission from Elsevier]

Location and enzymes	Biosynthesis process ↓	Events and enzymatic actions
Cellular matrix	Collagen α chain genes	Transcription
Cellular matrix	Pro-collagen pre mRNA	mRNA processing, methylation and poly-adenylation
Cellular matrix	Pro α chain mRNAs production	Translation
Endoplasmic reticulum (1) Signal peptidase (2) Prolyl 4-hydroxylase (3) Prolyl 3-hydroxylase (4) Lysyl hydroxylase (5) Hydroxy-lysyl galactosyl-transferase (6) Hydroxy-lysyl glucosyl-transferase (7) Oligosaccharyltransferasease (8) Protein disulfide isomerase (9) Prolyl-peptidyl cis/trans isomerase (10) Bip (11) Hsp47	Pro α chains (collag'en precursor)	Secondary modifications (1) Removal of signal peptide (2) 4-hydroxylation of proline (3) 3-hydroxylation of proline (4) Hydroxylation of lysin (5) O-glycosylation of hydroxylysine (6) O-glycosylation of galactosyl hydroxylysine (7) N-glycosylation of Asn-X-Scr/Thr (8) Native disulfide bond formation (9) Interconversion of the cis and transform of proline peptide bonds (10) Oligomeric assembly (11) Stabilization of procollagen triple helix and prevent the premature aggregation of procollagen
Golgi apparatuses (1) Exo-glycosidase and transferase Cellular matrix (2) Procollagen N-proteinase (3) Procollagen C-proteinase (4) C-proteinase enhancer (5) lysyl oxidase	Triple helical pro-collagen molecules	(1)Processing of N-linked oligosaccharides Secretion (2) Removal of N-propeptides (3) Removal of C-propeptides (4) Modulate C-propeptides processing Aggregation (5) Cross-link initiation Fibril formation
ECM (Extra Cellular Matrix)	Collagen in tissue	

3 Keratin

Keratin belongs to an insoluble, high-sulfur content, and filament-forming protein family called scleroproteins. This is the most abundant biopolymer present in vertebrates as lining to certain cell membranes, hair, hooves, feathers, etc. Due to the advancement of techniques and desirable intrinsic properties like biocompatibility, response to hydration, stiffness, strength, and as a readily available and renewable material, it has been used as a raw material in fiber-reinforced polymer production.

3.1 Sources of Keratin

By the presence of net sulfur content, keratins are divided into two groups. Soft or epithelial keratins have lower sulfur content (~1%) and hard or trichocyte keratins (~5%). The epithelial keratins are found in the stratum of the skin having low cysteine content (<3%) and help to stabilize the cells in epithelia whereas trichocyte keratins are found in wool, hairs, feathers, nails, horns and hoofs of mammals, reptiles and birds, containing a relatively high level of cysteine (4–17%) and helps in structural scaffolding. (Rajabi et al. 2020).

3.2 Structure of Keratin

Keratins are of two types, i.e., α type and β type. α-keratin is a coiled heterodimer structure of 7 nm diameter intermediate filament. β-keratin is of 3 nm diameter intermediate filament in a plated sheet. (Wang et al. 2016). Generally, mammalian keratins are made up of α-keratin, whereas in the case of Aves and Reptiles keratin is mostly of β-keratin type (Fig. 2).

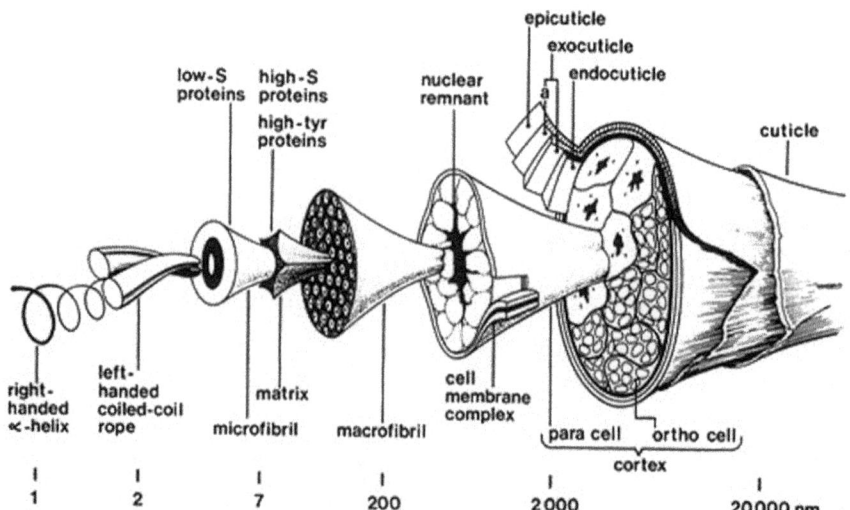

Fig. 2 Schematic Diagram of the wool fiber drawn by Bruce Fraser, Tom Mac Rae, and colleagues. (https://csiropedia.csiro.au/wool-fibre-structure/)

Biochemistry of keratins shows that α-keratin intermediate filaments are present as a mixture of low-sulfur proteins, while the matrix consists of high-sulfur and high-glycine–tyrosine proteins. But in the case of β-keratin, the filament and matrix are incorporated into one single protein. Hence α-keratin molecular mass (40–68 kDa) is larger than that of β-keratin (10–22 kDa). Molecular analysis of keratins shows that all types of keratins possess a central domain and two terminal domains (Feroz et al. 2020; Zhang and Fan 2021) (Figs. 3 and 4).

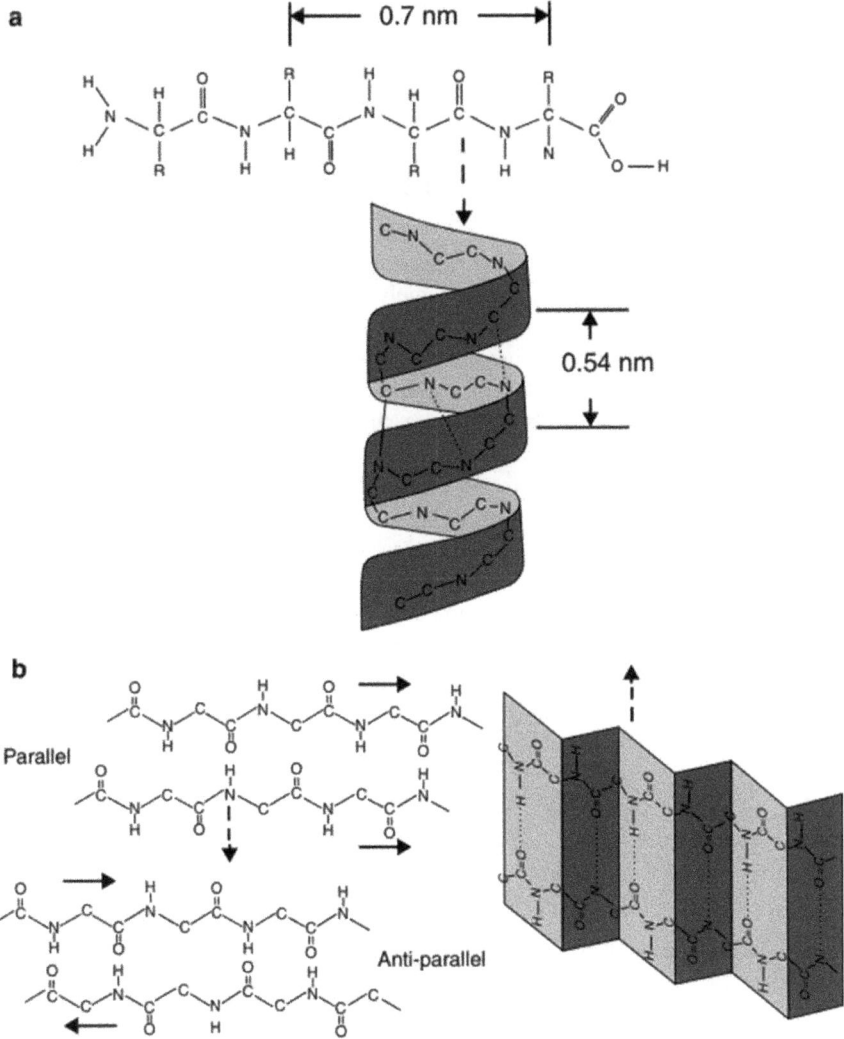

Fig. 3 Structure of keratins. **a** α-keratin helical structure, **b** β-keratin crystalline sheet structure (Murr 2021)

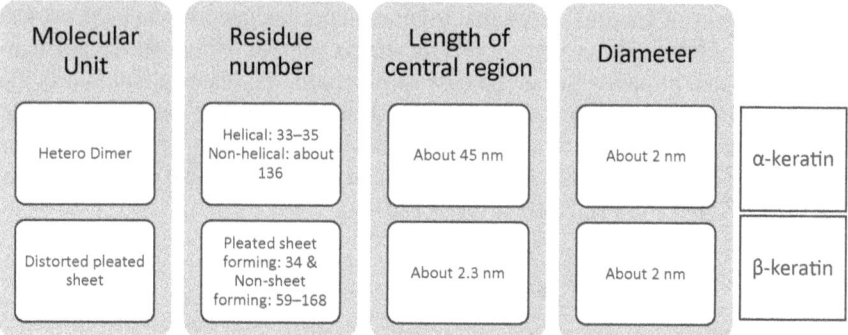

Fig. 4 Comparison of α- and β-keratin

3.3 Biomaterial of Keratin and Its Application

Keratin-based polymers are used in various stages of our everyday life. Being exhibit great mechanical durability and are extremely biocompatible, they are transformed into 3D scaffolds, sponges, films, and hydrogels for various environmental applications. Copello and co-workers have shown that keratin-titanium oxide nanoparticles have the ability to absorb trimethoprim from wastewater and degrade it (Villanueva et al. 2020). In the same study, they have shown that adsorption and the degradation capability of the nanocomposites is highly dependent on the temperature. Keratin-derived biopolymers helped in the removal of toxic trace elements from the synthetic water. In the same study, they have shown that modification of the biopolymers enhances their adsorption ability in the removal of contaminants from wastewater (Zahara et al. 2021). Song et al. have shown that mechanically robust keratin film formed by mesoscopic molecular networking helps in the removal of dyes from wastewater. They have also shown that these composite films have higher regeneration and recycling performance (Song et al. 2020).

4 Elastin

4.1 Source and Isolation

Novel biomaterials with innovative and advanced characteristics have discovered various uses in biomedicine and can perform a significant role in areas including such controlled drug release or tissue regeneration, as well as synthetic biology in general. Biomedical applications for delivery of drugs and tissue regeneration can be chemical or natural, and compostable or not, depending on the requirement. Because their qualities are defined by the chemical and physical functionalities and series of their constitutive monomers, protein-based biopolymers prompted by natural sources of proteins in the extracellular matrix are strong contenders as key elements of effective drug delivery devices and augmented alternatives in regenerative medicine (Santos et al. 2020). Elastin is an extracellular matrix protein that accounts for approximately 30% of the artery's dry weight. In soft tissue, elastin is the most abundant multicellular elastic protein. The soluble precursor tropoelastin is synthesized and secreted before elastin generation can initiate. Elastin-based biomaterials have the ability to rebuild elastic tissues by increasing local elasticity and facilitating adequate cellular conversations and triggering. It confers resistance and flexibility to a broad variety of human body tissues and organs. The usage of biomaterials derived from extracellular protein polymers is beneficial because they naturally acquire desired traits for tissue regeneration, for example, associating cellular movement, with cell signaling, and bio-degradability wherever applicable. Stem cells are unspecialized cells that have the ability to distinguish cells from a variety of tissue lineages. They are critical in promoting biological growth and are extremely elaborate in tissue repairing (Ozsvar et al. 2015).

The use of such elastin-based biomaterials for biomedical applications has grown steadily in recent years, with several biopolymers increasingly being used in the production of hydrogels, scaffolds, and drug delivery systems. Because of their biological activities, elastin protein and its derivatives are of specific interest. It gives organs and tissues like skin, blood vessels, lungs, bladder, ligaments, cartilage stretch, flexibility, and strength. Elastin-like polypeptides are also synthetic polymers that mimic the configuration and composition of elastin. Because of their structural properties, elastin and its derivatives are considered low in carcinogenicity and immunogenicity. As a result, elastin and elastin-like substances have a huge upside for biological and medical applications such as drug delivery, wound repair, and tissue regeneration. The mechanical survey of elastic fibers at the micro and macro levels affirms their impressive elastic and adaptable features (Wise et al. 2009).

Because of its deformability, self-assembly, long-term cohesion, and biological activity, elastin is growing in popularity as a biomaterial. Remarkably, the mechanical and physical properties of biomedical products and drug delivery systems are modeled and managed to improve using the elasticity and biological activity of elastin in combined effect with natural or synthetic polymers (Skopinska-Wisniewska et al. 2009).

4.2 Biochemistry of Elastin

Elastin is a protein with around 800 amino acid residues. Tropoelastin is highly hydrophobic, water-soluble, and non-glycosylated. Tropoelastin is an exceedingly elastic protein that can be stretched to around eight times its resting length. The tropoelastin molecule is made up of two kinds of domains that are encrypted by different exons: hydrophobic domains rich in nonpolar amino acids and hydrophilic domains. Desmosine and iso-desmosine are the two most common innate elastin cross-links, with each one concerning four lysine residues that are cross-linked by lysyl oxidase (Costa, Silva, and Boccaccini 2018).

Elastin is embedded by a single-copy gene and is simulated as a soluble precursor protein by various cells, including fibroblasts, smooth muscle cells, and endothelial cells, and settles down to give tropoelastin upon secretion. The dermal fibroblast cell is the principal source of tropoelastin in the skin. Tropoelastin is made up of interchanging hydrophobic and hydrophilic contexts. Tropoelastin is extensively covalently linked in the matrix after it is synthesized, resulting in insoluble elastin fibers. The crosslinking method is divided into three stages: tropoelastin coacervation, followed by lysyl oxidase family oxidation of lysine side chains, and subsequent crosslinking of juxtaposed tropoelastin molecules. Coacervation recruits lysyl oxidase to oxidatively deaminate tropoelastin lysine residues, permitting crosslinking of the protein, culminating in insoluble elastin fibers. The capacity to execute elastin to a tissue is thus extremely beneficial and can be accomplished using elastin biomaterials. Elastin is biocompatible and nonimmunogenic as a native ECM protein, making it an excellent replacement for surgical sutures (Wen et al. 2020).

4.3 Application of Elastin-Based Biomaterials

Firmness in the structure of tropoelastin, elasticity, and bioactivity, coupled with its ability to self-assemble, make it an extremely coveted contender for biomaterial fabrication. In recent times, many research groups have shown that elastin-based biomaterials can be used for the removal of heavy metals from wastewater (Kostal et al. 2005). In the same study, they have shown that the introduction of six histidine tags into the elastin molecule enhances the heavy metal removal efficiency of elastin. Nose and co-workers have shown that the introduction of metal-binding sequence to elastin exhibits the ability to bind to cadmium and zinc metal ions, which is a useful strategy to scavenge heavy metals from wastewater.

5 Silk

5.1 Source and Isolation

The Silk Road, which ran from China to India, Persia, and Europe, has historically linked many civilizations and fostered economic development. Silk was the primary and most important fiber in that era. The most recognized silk is made from the cocoons of mulberry silkworm larvae Bombyx mori cultured in isolation which is called sericulture. Silk is primarily produced by insect larvae during the whole metamorphosis, but certain bugs, for example, web spinners and rasping crickets, yield silk all through their lives.

Lepidoptera larvae convert the silk-protein polymers which are developed into fibers, with silkworms, spiders, scorpions, and flies. Upon biosynthesis in epithelial cells, silk proteins are commonly generated within special-purpose glands, accompanied by discharging further into the lumen of such glands. Silk composition, framework, and characteristics differ tremendously depending on the origin. The domesticated silkworm, Bombyx mori, and spider silks like Nephila clavipes are the most thoroughly studied. Some additionally evolved spiders can synthesize multiple kinds of silk (Altman et al. 2003).

Silkworm silks, which have invariant features and are more plentiful, have achieved widespread acknowledgment and implementation. Silkworm silk is divided into two types based on the worm's food source: mulberry silk and non-mulberry silk (Kundu et al. 2014a, b).

Lepidopteran creatures of the Bombiciidae and Saturniidae families produce economically valuable silks. Antheraea mylitta, a silk-producing tropical non-mulberry Tasar silkworm, is found all across India in a variety of geographic climates. Almost all of the silk protein acquired from cocoons is fibroin, which accounts for 80% of total silk proteins, while sericin accounts for only 20%. The fibroin of the non-mulberry silkworm P. ricini has been found to differ from mulberry silk fibroin in both biochemical and autoimmune aspects (Dash et al. 2006).

Silkworm silk proteins are comprised primarily of core fibroin protein and the glycoprotein sericin, which binds fibroin. The use of fibroin in biomedical applications is made possible by its distinctive mechanical behavior, good biocompatibility, and good biocompatibility. Based on the two main proteins that make up silks–fibroin and sericin. The sericin glue conglutinates two fibroin brins in a silk cocoon, giving rise to silk fibers. The amorphous adhesive protein sericin can be extracted using a thermochemical technique called degumming. Following extraction, the fibroin protein is the most popularly utilized noticed silk protein, with a broad array of applications. Surprisingly, the regeneration process and extent of sericin removal have a large effect on the properties of the regenerated fibroin, along with its shelf-life (Kundu et al. 2014a, b).

Silk proteins were extracted from cocoons using an ethylenediamine solution. At pH 5.5, fibroin and sericin were separated by the precipitation process. Sericin displayed distinct bands on gel electrophoresis, while fibroin did not. The gel filtration technique is used to separate the components of fibroin and sericin. Purification of the smallest component in the sericin fraction was done by chromatography and gel electrophoresis displayed a single band. The composition of amino acid was confirmed to have a molecular wt. of 24,000 (Tokutake 1980).

Silk is made up of two kinds of proteins: silk fibroin and sericin. Sericin accounts for approximately 20–30% of overall cocoon weight. It is distinguished by its high serine content and 18 amino acid compositions, which includes essential amino acids. The method of isolation influences the solubility, molecular weight, and gelling properties of sericin. It has many uses in pharmaceuticals and cosmetics, including wound curing, bioadhesive moisturizing, and anti-aging (Padamwar et al. 2005).

5.2   Characterization

Analysis of single-crystalline materials has been done by XRD, but it has constraints in unveiling the structure of polycrystalline silk fibroin. Initial attempts obtained an overall representation of its molecular entity, sorely missing atomic resolution specifics. However, if by recrystallizing fibroin it can turn to a state of a single crystal, the results may not be reflective of the fibrous state. In addition to XRD, mathematical and molecular models, and NMR spectroscopy have been investigated as substitute procedures for investigating fibroin crystal structure. Even though amino acid residues, glycine and alanine, have long been recognized as the primary compositions involved in the creation of the crystallites of silk fibroin, the presence of larger residues such as serine and tyrosine in the sheet crystallite was only recently confirmed by NMR. However, there is some degree of local disorder in the structure, which is frequently identified as amorphous by XRD techniques (Ryu et al. 2019).

5.3   Biochemistry of Silk

The silk released into the atmosphere by the silkworm is made up of two main proteins: sericin and fibroin, with fibroin serving as the structural center of the silk and sericin serving as the sticky substance that surrounds it. Fibroin is formed by the amino acids Gly–Ser–Gly–Ala–Gly–Ala and is composed of beta-pleated sheets. Domesticated silkworm strains have been reared to sustain in a diverse array of tropical and temperate areas in order to expand the rural inhabitants' products. Local silkworm varieties that nourish on natural plants, on the other hand, could provide a more self-sustaining and eco-friendly relevant product in many of these same areas. Small-scale wild silk production in remote areas of India allows farmers to create income-generating assets, markets, and skills. Cultivating wild silk allows farmers to enter a niche market that is less viable than the domesticated silk market but has the potential to be just as profitable if not more profitable. Wild silk is a niche market component, so its cost is not strongly linked to input costs, which can nullify conservation measures (Asakura et al. 2018).

5.4   Applications on Silk-Based Biomaterials

Silk fibers have greater ecological integrity than globular proteins because of broad hydrogen bonding, the hydrophobic behavior of protein, and substantial crystallinity. Most solvents, which include water, dilute acid, and alkali, are immiscible in silks. Thorough scrutiny of spider dragline, silk proteins revealed facts on the organization of the very small sheet crystals in the fibers. The several beneficial inherent assets of silk fibroin, such as its shimmering look, smoothness in the texture, good biodegradability, functionalization, and so on, have been demonstrated in a variety of applications. Most notably, applications benefit from its enhanced mechanical properties, as well as the material's low cost and abundant supply. Silk fibroin is now a valued textile as well as an appealing biomaterial in a variety of pharmaceutical grounds, primarily drug supply, optics, sensing, diagnostics, and so on. Li et al. and Freddie et al. have shown that regenerated silk fibroins blended with cellulose enhanced alkali ion's adsorption affinity (Li et al. 2008; Freddi et al. 1995). Baek et al. reported the adsorption of copper ions by blending silk fibroin with wool keratose (Li et al. 2008). Similarly, Campagnola et al. used silk blended orange-peel-based powder for the removal of methylene blue dye (Campagnolo et al. 2019).

6 Conclusions

In this chapter, the biomaterials from four proteins, i.e., collagen, keratin, elastin, and silk, and their applications in water remediation have been discussed. Before arriving at the application of the biomaterials based on these four proteins each protein source, isolation and biochemistry have been discussed extensively. Lastly, each of the protein-based biomaterials application in water remediation has been discussed.

References

Altman GH, Diaz F, Jakuba C, Calabro T, Horan RL, Chen J, Lu H, Richmond J, Kaplan DL (2003) Silk-based biomaterials. Biomaterials 24(3):401–16.https://doi.org/10.1016/s0142-9612(02)003 53-8

Asakura T, Kametani S, Suzuki Y (2018) Silk. Encyclopedia Polymer Sci Technol. Wiley.https://doi.org/10.1002/0471440264.pst339.pub2

Avila R, Isabela M, Barroso LGR, Sánchez ML(2017) Collagen: a review on its sources and potential cosmetic applications. J Cosmet Dermatol 17(1):20–26. https://doi.org/10.1111/jocd.12450

Bailey AJ, Paul RG, Knott L (1998) Mechanisms of maturation and ageing of collagen. Mech Ageing Dev 106(1–2):1–56. https://doi.org/10.1016/s0047-6374(98)00119-5

Browne S, Zeugolis DI, Pandit A (2013) Collagen: finding a solution for the source. Tissue Eng Part A 19(13–14):1491–1494. https://doi.org/10.1089/ten.TEA.2012.0721

Burjanadze TV (2000) New analysis of the phylogenetic change of collagen Thermostability. Biopolymers 53(6):523–528. https://doi.org/10.1002/(sici)1097-0282(200005)53:6%3c523::aid-bip8%3e3.0.co;2-7

Campagnolo L, Morselli D, Magrì D, Scarpellini A, Demirci C, Colombo M, Athanassiou A, Fragouli D (2019) Silk Fibroin/orange peel foam: an efficient biocomposite for water remediation. Adv Sustain Syst 3(1):1800097. https://doi.org/10.1002/adsu.201800097

Costa F, Silva R, Boccaccini AR (2018) Fibrous protein-based biomaterials (Silk, Keratin, Elastin, and Resilin Proteins) for tissue regeneration and repair. Elsevier, Peptides and proteins as biomaterials for tissue regeneration and repair. https://doi.org/10.1016/b978-0-08-100803-4.000 07-3

Dash R, Mukherjee S, Kundu SC (2006) Isolation, purification and characterization of silk protein Sericin from cocoon peduncles of tropical Tasar silkworm, Antheraea Mylitta. Int J Biol Macromol 38(3–5):255–258. https://doi.org/10.1016/j.ijbiomac.2006.03.001

Engel J, Prockop DJ (1991) The Zipper-like folding of collagen triple helices and the effects of mutations that disrupt the zipper. Annu Rev Biophys Biophys Chem 20(1):137–152. https://doi.org/10.1146/annurev.bb.20.060191.001033

Felician FF, Xia C, Qi W, Xu H (2018) Collagen from marine biological sources and medical applications. Chem Biodiversity 15 (5):e1700557. https://doi.org/10.1002/cbdv.201700557

Feroz S, Muhammad N, Ranayake J, Dias G (2020) Keratin—based materials for biomedical applications. Bioactive Mater 5(3):496–509. https://doi.org/10.1016/J.BIOACTMAT.2020.04.007

Fratzl P (n.d.) Collagen: structure and mechanics, an introduction. Collagen. Springer, US. https://doi.org/10.1007/978-0-387-73906-9_1

Freddi G, Romanò M, Massafra MR, Tsukada M (1995) Silk fibroin/cellulose blend films: preparation, structure, and physical properties. J Appl Polym Sci 56(12):1537–1545. https://doi.org/10.1002/app.1995.070561203

Gara SK, Grumati P, Urciuolo A, Bonaldo P, Kobbe B, Koch M, Paulsson M, Wagener R (2008) Three novel collagen VI chains with high homology to the A3 chain. J Biol Chem 283(16):10658–10670. https://doi.org/10.1074/jbc.m709540200

Gay S, Miller EJ (1983) What is collagen, what is not. Ultrastruct Pathol 4(4):365–377. https://doi.org/10.3109/01913128309140589

Gorham SD (1991). Collagen. Biomaterials. Palgrave Macmillan, UK. https://doi.org/10.1007/978-1-349-11167-1_2

Holmgren SK, Bretscher LE, Taylor KM, Raines RT (1999) A Hyperstable collagen Mimic. Chem Biol 6(2):63–70. https://doi.org/10.1016/s1074-5521(99)80003-9

Holmgren SK, Taylor KM, Bretscher LE, Raines RT (1998) Code for collagen's stability deciphered. Nature 392(6677):666–667. https://doi.org/10.1038/33573

Hudson DM, Werther R, Weis MaryAnn, Wu, Jiann-Jiu, Eyre DR (2014) Evolutionary origins of C-terminal (GPP)n 3-hydroxyproline formation in vertebrate tendon collagen. PLoS ONE 9(4):e93467–e93467. https://doi.org/10.1371/journal.pone.0093467

Hulmes DJS (2002) Building collagen molecules, fibrils, and Suprafibrillar structures. J Struct Biol 137(1–2):2–10. https://doi.org/10.1006/jsbi.2002.4450

Irene D, Reyes-Ruiz C, Castañeda-Calzoncit CE, Claudio Rizo JA, Flores-Guía TE, Cabrera-Munguía DA, Cano-Salazar LF, Becerra-Rodriguez JJ (2020) Evaluation of collagen-polyurethane-chitosan hydrogels for lead ions removal from water. Mediterranean J Basic Appl Sci 4(2):93–104. https://doi.org/10.46382/MJBAS.2020.4209

Jenkins E, Moss JB, Pace JM, Bridgewater LC (2005) The new collagen gene COL27A1 contains SOX9-responsive enhancer elements. Matrix Biology J Int Soc Matrix Biology 24(3):177–184. https://doi.org/10.1016/j.matbio.2005.02.004

Kalirajan C, Hameed P, Subbiah N, Palanisamy T (2019) A facile approach to fabricate dual purpose hybrid materials for tissue engineering and water remediation. Sci Rep 9(1):1040. https://doi.org/10.1038/s41598-018-37758-2

Koide T, Nagata K (2005) Collagen biosynthesis. Topics Current Chem. Springer, Berlin, Heidelberg. https://doi.org/10.1007/b103820

Kostal J, Prabhukumar G, Loi Lao U, Chen A, Matsumoto M, Mulchandani A, Chen W (2005) Customizable biopolymers for heavy metal remediation. J Nanopart Res 7(4–5):517–23.https://doi.org/10.1007/s11051-005-5132-y

Kundu B, Kurland NE, Bano S, Patra C, Engel FB, Yadavalli VK, Kundu SC (2014a) Silk proteins for biomedical applications: bioengineering perspectives. Prog Polym Sci 39(2):251–267. https://doi.org/10.1016/j.progpolymsci.2013.09.002

Kundu B, Kurland NE, Yadavalli VK, Kundu SC (2014b) Isolation and processing of silk proteins for biomedical applications. Int J Biol Macromol 70:70–77. https://doi.org/10.1016/j.ijbiomac.2014.06.022

Li X-G, Wu L-Y, Huang M-R, Shao H-L, Hu X-C (2008) Conformational transition and liquid crystalline state of regenerated silk fibroin in water. Biopolymers 89(6):497–505.https://doi.org/10.1002/bip.20905

Miller MH, Némethy G, Scheraga HA (1980) Calculation of the structures of collagen models. Role of Interchain interactions in determining the triple-helical coiled-coil conformation. 2. Poly(Glycyl-Prolyl-Hydroxyprolyl). Macromolecules 13(3):470–478. https://doi.org/10.1021/ma60075a003

Mizuno K, Boudko S, Engel J, Bächinger HP (2013) Vascular Ehlers-Danlos syndrome mutations in Type III collagen differently stall the triple helical folding. J Biol Chem 288(26):19166–19176. https://doi.org/10.1074/jbc.M113.462002

Murr LE (2021) Structures and properties of keratin-based and related biological materials BT—handbook of materials structures, properties, processing and performance. In: Murr LE (ed). Springer International Publishing, Cham, pp 1–24

Nagaraj S, Cheirmadurai K, Thanikaivelan P (2021) Visible-light active collagen-TiO2 Nanobio-sponge for water remediation: a sustainable approach. Cleaner Mater 1https://doi.org/10.1016/j.clema.2021.100011

Ozsvar J, Mithieux SM, Wang R, Weiss AS (2015) Elastin-based biomaterials and mesenchymal stem cells. Biomater Sci 3(6):800–809. https://doi.org/10.1039/C5BM00038F

Pace JM, Corrado M, Missero C, Byers PH (2003) Identification, characterization and expression analysis of a new Fibrillar collagen gene, COL27A1. Matrix Biol 22(1):3–14. https://doi.org/10.1016/s0945-053x(03)00007-6

Padamwar MN, Pawar AP, Daithankar AV, Mahadik KR (2005) Silk Sericin as a moisturizer: an in vivo study. J Cosmet Dermatol 4(4):250–257. https://doi.org/10.1111/j.1473-2165.2005.00200.x

Pihlajaniemi T, Myllylä R, Kivirikko KI (1991) Prolyl 4-hydroxylase and its role in collagen synthesis. J Hepatol 13:S2-7. https://doi.org/10.1016/0168-8278(91)90002-s

Raghunath M, Bruckner P, Steinmann B (1994) Delayed triple helix formation of mutant collagen from patient with osteogenesis imperfecta. J Mol Biol 236(3):940–949. https://doi.org/10.1006/jmbi.1994.1199

Rainey JK, Cynthia Goh M (2009) A Statistically derived parameterization for the collagen triple-helix. Protein Sci 11(11):2748–2754. https://doi.org/10.1110/ps.0218502

Rajabi M, Ali A, McConnell M, Cabral J (2020) Keratinous materials: structures and functions in biomedical applications. Mater Sci Eng C 110. https://doi.org/10.1016/j.msec.2019.110612

Ramshaw JAM, Werkmeister JA, Glattauer V (1996) Collagen-based biomaterials. Biotechnol Genet Eng Rev 13(1):335–382. https://doi.org/10.1080/02648725.1996.10647934

Rezvani Ghomi E, Nourbakhsh N, Kenari MA, Zare M, Ramakrishna S (2021) Collagen-based biomaterials for biomedical applications. J Biomed Mater Res. Part B Appl Biomater 109(12): 1986–99. https://doi.org/10.1002/JBM.B.34881

Ricard-Blum S (2011) The collagen family. Cold Spring Harb Perspect Biol 3(1):1–19. https://doi.org/10.1101/cshperspect.a004978

Ryu M, Honda R, Cernescu A, Vailionis A, Balčytis A, Vongsvivut J, Li J-L et al (2019) Nanoscale optical and structural characterisation of silk. Beilstein J Nanotechnol 10(April):922–929. https://doi.org/10.3762/bjnano.10.93

Saha I, Shamala N (2011) Investigating Diproline segments in proteins: occurrences, conformation and classification. Biopolymers 97(1):54–64. https://doi.org/10.1002/bip.21703

Santos M, Serrano-Dúcar S, González-Valdivieso J, Vallejo R, Girotti A, Cuadrado P, Arias FJ (2020) Genetically engineered elastin-based biomaterials for biomedical applications. Curr Med Chem 26(40):7117–7146. https://doi.org/10.2174/0929867325666180508094637

Skopinska-Wisniewska J, Sionkowska A, Kaminska A, Kaznica A, Jachimiak R, Drewa T (2009) Surface characterization of collagen/elastin based biomaterials for tissue regeneration. Appl Surf Sci 255(19):8286–8292. https://doi.org/10.1016/j.apsusc.2009.05.127

Söderhäll C, Marenholz I, Kerscher T, Rüschendorf F, Esparza-Gordillo J, Worm M, Gruber C et al (2007) Variants in a novel epidermal collagen gene (COL29A1) are associated with atopic dermatitis. PLoS Biol 5(9):e242–e242. https://doi.org/10.1371/journal.pbio.0050242

Song K, Qian X, Zhu X, Li X, Hong X (2020) Fabrication of mechanical robust keratin film by mesoscopic molecular network reconstruction and its performance for dye removal. J Colloid Interface Sci 579(November):28–36. https://doi.org/10.1016/j.jcis.2020.06.026

Sorushanova A, Delgado LM, Wu Z, Shologu N, Kshirsagar A, Raghunath R, Mullen AM et al (2019) The collagen Suprafamily: from biosynthesis to advanced biomaterial development. Adv Mater 31(1). https://doi.org/10.1002/adma.201801651

Tokutake S (1980) Isolation of the smallest component of silk protein. Biochem J 187(2):413–417. https://doi.org/10.1042/bj1870413

Tuttolomondo MV, Galdopórpora JM, Trichet L, Voisin H, Coradin T, Desimone MF (2015) Dye–collagen interactions. Mechanism, kinetic and thermodynamic analysis. RSC Adv 5(71):57395–57405. https://doi.org/10.1039/C5RA08611F

van der Rest M, Garrone R, Herbage D (1993) Collagen: a family of proteins with many facets. Elsevier, Extracellular Matrix. https://doi.org/10.1016/s1569-2558(08)60198-8

Villanueva ME, Puca M, Bravo JP, Bafico J, Orto VCD, Copello GJ (2020) Dual adsorbent-photocatalytic keratin–TiO $_2$ nanocomposite for trimethoprim removal from wastewater. New J Chem 44(26):10964–10972. https://doi.org/10.1039/D0NJ02784G

Wang B, Yang W, McKittrick J, Meyers MA (2016) Keratin: structure, mechanical properties, occurrence in biological organisms, and efforts at Bioinspiration. Prog Mater Sci 76(March):229–318. https://doi.org/10.1016/J.PMATSCI.2015.06.001

Wen Q, Mithieux SM, Weiss AS (2020) Elastin biomaterials in dermal repair. Trends Biotechnol 38(3):280–291. https://doi.org/10.1016/j.tibtech.2019.08.005

Wise SG, Mithieux SM, Weiss AS (2009) Engineered Tropoelastin and elastin-based biomaterials. Adv Protein Chem Struct Biol. Elsevier. https://doi.org/10.1016/s1876-1623(08)78001-5

Wu J-J, Weis MA, Kim LS, Carter BG, Eyre DR (2009) Differences in chain usage and cross-linking specificities of cartilage Type V/XI collagen isoforms with age and tissue. J Biol Chem 284(9):5539–5545. https://doi.org/10.1074/jbc.M806369200

Zahara I, Muhammad Arshad M, Naeth A, Siddique T, Ullah A (2021) Feather keratin derived sorbents for the treatment of wastewater produced during energy generation processes. Chemosphere 273(June):128545. https://doi.org/10.1016/j.chemosphere.2020.128545

Zhang W, Fan Y (2021) Structure of Keratin. Methods Molec Biol (Clifton, N.J.) 2347:41–53. https://doi.org/10.1007/978-1-0716-1574-4_5

Green Sustainability and Arsenic Groundwater Remediation in Developing countries—A Far-Reaching Review

Sukanchan Palit and Chaudhery Mustansar Hussain

1 Introduction

The world of the science of sustainable development, environmental management and environmental engineering science today is in the middle of scientific vision and divination. Today civilization, science and engineering are in a similar vein in the midst of deep scientific and engineering introspection. Global water issues and water scarcity are plaguing the vast scientific and technological landscape today. The provision of basic amenities of human society such as drinking water and proper sanitation is in the middle of a deep disaster. Heavy metal and arsenic groundwater and drinking water contamination are veritably destroying the vast global scientific horizon. Human mankind is thus in the avenue and vista of real danger. In this chapter, the authors deeply portray the need for the application of the science of sustainability in groundwater remediation. Sustainable development whether it is environmental, energy, social or economic is the veritable need of human society today. The visionary words of Dr. Gro Harlem Brundtland, former Prime Minister of Norway on the science of sustainability, need to be re-envisioned as civilization, engineering and science moves ahead. Today human civilization and the planet are at crossroads. Green sustainability and water and wastewater treatment are two opposite sides of the visionary coin. Man's vision, humankind's immense scientific prowess and the futuristic vision of environmental pollution control will all be the torch-bearers towards a newer scientific era in the field of drinking water and industrial

S. Palit (✉)
Department of Chemical Engineering, University of Petroleum and Energy Studies, Energy Acres, Post-Office-Bidholi Via Premnagar, Dehradun, Uttarakhand 248007, India
e-mail: sukanchan68@gmail.com

C. M. Hussain
Department of Chemistry and Environmental Sciences, New Jersey Institute of Technology, University Heights, Newark, NJ 07102, USA
e-mail: chaudhery.m.hussain@njit.edu

wastewater treatment. Arsenic drinking water and groundwater contamination are devastating many developing, developed and emerging economics around the world. Bangladesh and the state of West Bengal, India, are being confronted with this vexing and monstrous water scarcity issue. Thus there is an immediate need for scientific vision, scientific resilience and deep scientific provenance in the field of drinking water and groundwater remediation around the world. A deep scientific introspection in the field of environmental or green sustainability will surely open newer doors of scientific innovation and scientific and engineering ingenuity in environmental and water remediation in decades to come.

2 The Vision of the Study

Chemical process engineering, chemical technology, biochemical engineering and environmental engineering science are on the path of newer scientific rejuvenation. The application of the science of sustainability and sustainable development in human progress and advancement are today aligned with the science of environmental protection and environmental engineering science. Rapid industrial growth and intense mass manufacturing are today degrading the global environment and resulting in loss of ecological biodiversity. Heavy metal and arsenic groundwater contamination are immensely burning and enigmatic issues of human scientific progress today. Human scientific and academic rigor are thus in a state of distress. In South Asia mainly India and Bangladesh, arsenic and heavy metal groundwater poisoning are causing havoc to the human population. Thus human race stands in the middle of deep scientific introspection. Immense health-related issues are also creating immense scientific difficulties in the progress of human civilization. The scientific ingenuity and the scientific fortitude of the science of environmental remediation and industrial wastewater treatment thus need to be reinvigorated with the passage of scientific history and time. The deep glimpse of the application of green or environmental sustainability in the field of arsenic and heavy metal groundwater remediation is elucidated in deep details in this chapter.

3 The Need and the Rationale of This Study

Technology and engineering science of arsenic and heavy metal remediation are today in the avenues of newer regeneration. In many developing and developed nations around the world, lack of clean drinking water and proper sanitation is a huge burden to human civilization and its progress. Here comes the need and the rationale of this treatise. The authors with deep scientific introspection delineate the recent advances in green sustainability and techniques of arsenic groundwater remediation tools. Today newer innovations and newer discoveries in the field of arsenic and heavy metal groundwater and drinking water are changing the vast scientific firmament. Arsenic and heavy metal groundwater contamination have serious health effects and can cause serious illnesses. Thus the need for a detailed treatise in the field of environmental sustainability and newer innovations in the field of groundwater remediation techniques. Developing countries, developed countries and emerging economies around the world are today plagued due to arsenic and heavy metal groundwater poisoning. Bangladesh and India are today ravaged by the ever-growing crisis of arsenic groundwater contamination. Tremendous health effects such as cancer are today destroying the scientific scenario. Mankind's immense scientific and engineering prowess and vision are thus in a state of disaster.

4   Green Sustainability and the Vast Vision for the Future

Green sustainability or environmental sustainability is the utmost need of the hour in the scientific progress of human civilization today. Global water issues and global climate change are urging the scientific domain to innovate new technologies in the true emancipation of green or environmental sustainability. Global climate change, global warming and loss of ecological biodiversity are threatening the planet earth. The vast vision for the future in the field of green sustainability and heavy metal groundwater remediation needs to be revisited and revamped as civilization progresses forward. There are immense need for environmental and energy sustainability in human society today. So the need for a detailed scientific treatise in the field of green sustainability and water and wastewater treatment. Today there is an immense need for sustainable development in human society. The vast vision for the future should be targeted towards that direction. Research directions and research acumen in the field of engineering and technology should be directed towards further emancipation of water and industrial wastewater treatment. Sustainability whether it is environmental, energy, social or economic is the ultimate vision of tomorrow's human scientific progress. The vision, the purpose and the ardor of science and engineering in the global scenario should be targeted towards sustainability and environmental management. Integrated water resource management and wastewater management should be the pillars of human scientific endeavor today. The authors in this article deeply elucidate the cause of the crisis and catastrophe of arsenic groundwater contamination in South Asia mainly Bangladesh and India. Scientific destiny, deep scientific provenance and scientific subtleties will eventually open new doors of innovation in the field of green sustainability, green engineering and industrial and groundwater treatment. A new visionary era in the field of sustainable development and environmental management will surely emerge if science and technology of environmental protection and water and wastewater treatment move in the right scientific directions. Environmental sustainability and its vast vision for the future will surely be emboldened if scientists, engineers, researchers, students, the civil society and nations around the world takes concerted efforts in eradicating the burning issues of arsenic and heavy metal groundwater poisoning. A deep and a profound scientific contemplation in the field of drinking water and industrial wastewater are thus the need of the hour.

5 Arsenic and Heavy Metal Groundwater Remediation in South Asia and Other Developing Countries

Arsenic and heavy metal groundwater remediation in developing countries, developed countries and emerging economies are on the path of newer scientific rejuvenation. Immense research endeavor in the areas of groundwater remediation is today futile. Thus the need for a concerted and coordinated effort from researchers, engineers, scientists, governments and the civil society. In developing nations around the world, the provision of clean drinking water and proper sanitation is highly neglected. Also environmental, energy and socio-economic sustainability are in a state of disaster. Different scientific innovations in the area of arsenic groundwater and drinking water remediation in developing countries such as India and Bangladesh are absolutely futile. Arsenic groundwater contamination in India and Bangladesh is the world's largest environmental disaster. Environmental management and sustainable development are today two opposite sides of the visionary coin. Without sustainable development, the human civilization cannot move forward. Water remediation issues are today connected to environmental sustainability by an unsevered umbilical cord. Reuse, recycle and reduce are the coin words of scientific vision and scientific fortitude in the field of environmental engineering science and sustainable resource management. The status of the human planet and humankind is deeply at the crossroads. Arsenic and heavy metal groundwater contamination are world's largest scientific and technological catastrophes. A deep scientific introspection and a deep scientific vision in the scientific domain of water purification will surely go a long and visionary way in the true realization of environmental engineering science and global environmental sustainability. Today, chemical engineering, biological sciences, biotechnology and geological sciences are the interdisciplinary areas of scientific research pursuit which need to be readdressed and restructured in confronting arsenic and heavy metal groundwater contamination. The contribution of areas of applied sciences such as physics, chemistry, biology and material sciences in water purification and environmental remediation will surely open new windows of innovation, scientific opportunities and immense scientific and engineering potential. Arsenic groundwater remediation and subsequent water purification in developed and developing nations around the world and its vast scientific and academic rigor will also unravel the new world of research and development initiatives in global science and technology.

6   Various Arsenic Remediation Technologies in the Global Scenario

Arsenic remediation technologies need to be re-envisioned and revamped as human civilization moves forward. So drinking water remediation and industrial wastewater treatment are the needs of the hour. Arsenic and heavy metal groundwater and drinking water contamination are monstrous burdens to human civilization particularly India and Bangladesh. Developed and developing countries around the world are highly concerned with this largest global environmental disaster. Thus scientific innovations and research and development initiatives around the world are targeted towards the greater realization of water and wastewater treatment techniques. Palit et al. (2019) elucidate in detail modern scientific and engineering tools of arsenic remediation of groundwater and drinking water. The remedial measures for the removal of arsenic in groundwater are classified into domestic measures and industrial measures. Domestic measures involve (1) rapid small-scale column test, (2) use of natural geological material, (3) desalination, (4) oxidation, (5) co-precipitation, and (6) subterranean arsenic removal technology. Industrial measures involve (1) ion exchange, (2) hybrid technology based on ion exchange and electrodialysis, (3) use of nanoparticles, and (4) magnetite-graphene hybrids (Palit et al. 2019). A new visionary epoch in the field of water remediation is today emerging as man and mankind moves towards new scientific boundaries. Arsenic remediation technologies in the global scenario thus need to be reshaped and restructured as mankind and scientific research pursuit crosses one frontier over another (Palit et al. 2019).

Zamora-Ledezma et al. (2021) deeply elucidate and explain profoundly heavy metal water pollution and a fresh outlook about hazards, novel and conventional remediation methods. Human mankind, science and technology are today in the vistas of immense scientific vision, forbearance and scientific truthfulness (Zamora-Ledezma et al. 2021). Rapid industrialization and urbanization are today changing drastically the vast scientific and engineering landscape globally. Water pollution is one of the global scientific challenges that society must address in the twenty-first century vastly aiming to improve water quality and drastically reduce human and ecosystem health aspects. Scientific perspectives, deep scientific and engineering progeny will today widen environmental engineering and sustainable development paradigm (Zamora-Ledezma et al. 2021). In this work the author deeply discusses some of the recent and relevant scientific findings related to the release of heavy metals, the vital domain of environmental health concerns and the materials and technologies available for removal (Zamora-Ledezma et al. 2021). Today a remarkable era in the field of environmental engineering is slowly emerging. Scientific profundity and deep scientific discernment are the utmost needs of the hour in the research pursuit in water remediation today. The advantages, disadvantages and the drawbacks of conventional and non-conventional heavy metal removal methods are critically discussed and deliberated given immense importance to those related to adsorption, nanostructured membranes and plant-mediated remediation tools. Some of the commercial products currently used in the proliferation of the science of environmental and water remediation are discussed in minute details. Scientific fervor and engineering vision of heavy metals removal from industrial wastewater and its scientific and knowledge prowess will veritably open new windows of innovation and instinct in the field of holistic world of environmental pollution control (Zamora-Ledezma et al. 2021). Wastewater treatment techniques can be classified into primary, secondary and tertiary treatments. Primary treatment focuses on the removal of organic matter and suspended solids from wastewater through physical and chemical processes. Microfiltration, ultrafiltration, chemical filtration, coagulation and flocculation are the important primary technologies when involved with a high concentration of heavy metals. Secondary treatments (anaerobic or aerobic) are based on natural living organisms capable of converting organic and inorganic contaminants into simpler and safer substances, which result in higher removal efficiency. Tertiary treatment technologies involve chemical oxidation, electrochemical precipitation, crystallization, distillation and photocatalysis, adsorption, membrane technologies, and ion exchange technologies. In this stage of the treatment procedure, wastewater is converted into safe and potable water (Zamora-Ledezma et al. 2021). The authors in this article discussed with scientific and technological far-sightedness sources, effects and regulations on heavy metal pollution, detection which includes spectroscopic detection, electrochemical methods of detection, and optical methods of detection, removal of heavy metals from water, factors affecting heavy metal removals such as the effect of pH, the effect of temperature, the effect of ionic strength, and the effect of natural organic matter. Conventional treatments discussed are chemical precipitation, coagulation/flocculation, ion exchange, membrane technologies and electrochemical technologies. Non-conventional treatments discussed are adsorption, microbial fuel cells, and the vast domain of nanotechnology Zamora-Ledezma et al. (2021).

Saikia et al. (2019) deeply discussed and elaborated in details nanotechnology for water remediation. Scarcity of fresh drinking water has escalated to be one of the global environmental engineering and public health engineering issues. Traditional water and wastewater treatment technologies are today not adequate enough to produce safe water due to the increasing demand of water coupled with stringent environmental health guidelines and emerging recalcitrant pollutants (Saikia et al. 2019). A new scientific genre and a new engineering vision and profundity are today ushering in the scientific landscape. The advent of nanotechnology has given immense scope and vast opportunities for the removal of heavy metal, microorganisms and organic pollutants from wastewater and has veritably emerged to be a dynamic branch of science and engineering (Saikia et al. 2019). This is due to their unique physiochemical and biological properties compared to the bulk materials. Nanomaterials and engineered nanomaterials are the ingenious materials and smart materials of tomorrow. Deeply exploiting these properties of high specific surface area and surface activity has resulted in the excessive use and study of nanomaterials in wastewater remediation (Saikia et al. 2019). The use of various nanomaterials, including carbon-based nanomaterials, metal and metal oxides nanoparticles were deeply focused on, and their mode of action towards wastewater remediation was discussed. Water is the most essential substance for all forms of life on earth. Nearly

71% of the earth's surface is covered with water, but only 1% is veritably useful for drinking. Over 1.1 billion people lack supply of drinking water and nearly 1.8 billion children die every year from diarrhea primarily due to water and wastewater contamination (Saikia et al. 2019). The traditional methods of water purification have serious drawbacks such as high energy requirement, incomplete pollutant removal and generation of toxic sludge. Filtration can be used to remove and mitigate the contaminants but that produces a highly concentrated sludge in one phase which is toxic and extremely difficult to dispose. Thus there is an immediate requirement for more efficient and less power consuming for treating municipal wastewater (Saikia et al. 2019). Thus the need of vision, verve and scientific motivation in the field of water and wastewater remediation. Nanowires, nanotubes, films, particles, quantum dots, and colloids are different methodologies of nanomaterials which have at least one dimension less than 100 nm. At this nanoscale, materials often exhibit different physical, chemical and biological properties compared to their bulk counterparts (Saikia et al. 2019). The authors discussed in details adsorption onto nanomaterials, photocatalytic water treatment using nanoparticles, disinfection of wastewater using nanoparticles, nanomembrane in wastewater treatment, and the limitation of nanoparticles used in wastewater treatment. Exploiting and redirecting scientific research in nanomaterials and wastewater treatment have proven to be highly promising. Different treatment technologies such as adsorption, photocatalysis, disinfection and membrane separation have been successfully utilized to get maximum efficacies and maximum scientific and engineering output. Nanotechnology is one of the revolutionary domains of science and technology today. Today environmental remediation and nanotechnology are two opposite sides of the visionary coin. Research and development forays in the field of nanomaterials and engineered nanomaterials will surely usher in a new era in the field of environmental pollution control in decades to come Saikia et al. (2019).

Science and technology of water and wastewater remediation are today in the avenues of much scientific comprehension and vision. A detailed scientific treatise will open newer windows of environmental protection science in years to come. Soil and groundwater remediation are the challenges of science and engineering today. In the similar premise, the application of nanotechnology in arsenic and heavy metal remediation of drinking water and groundwater will usher in a newer epoch in the field of engineering sciences and applied science. The author in this entire treatise addresses these water and wastewater issues.

7 Recent Scientific Advancements in the Field of Green or Environmental Sustainability

Green or environmental sustainability is the scientific imperatives of human mankind today. Sustainable development such as energy or environmental is also today the need of the human civilization today. In this section, the authors deeply elucidate recent advancements in the field of green or environmental sustainability.

The United States Environmental Protection Agency Report (2011) deeply eluci-dated with cogent insight sustainability and US EPA (Environmental Protection Agency). Recognizing the importance of sustainability to its work, the United States Environmental Protection Agency has been rigorously examining applications in a variety of areas in order to better incorporate sustainability into decision making in the agency. Sustainable development and environmental management are today opposite sides of the visionary coin. The agency has also undertaken several sustain-able development initiatives and can claim success in developing the process of sustainability (United States Environmental Protection Agency Report 2011). This agency in the United States takes strong initiatives in implementing sustainability in the human society. Today, the vision of science and engineering in the global scenario are aligned with the application of both environmental sustainability and environmental management (United States Environmental Protection Agency Report 2011). The domains of environmental remediation, water resource management and wastewater management and control are of immense importance in the progress of civilization, science and technology. This report deeply elucidates (1) history of sustainability, (2) a sustainability framework for Environmental Protection Agency, (3) sustainability assessment and management and process, tools and indicators, (4) the relation between risk assessment and risk management, (5) the changing of the culture in US Environmental Protection Agency and (6) benefits of sustainability approach in US Environmental Protection Agency (Zamora-Ledezma et al. 2021). The growing identification of sustainability and environmental management as both a process and a goal to ensure long-term human well-being and human progress that does not threaten the continued availability of critical natural resources is based on four definite and converging drivers. The first is the recognition that current approaches aimed at decreasing existing risks, however immensely successful, are not capable of avoiding the complex problems in the United States, such as the widening gaps between rich and poor, depletion of natural resources, social inequality, loss of ecological biodiversity, climate change and disruption of nutrient and food cycles. Second sophisticated tools are immensely available to address the complex and chal-lenging issues that go beyond the current risk management domain. Third, sustain-ability is being used as a common approach to address broader social, environmental and economic issues by international bodies in which the United States is an active and a strong participant. Finally, the potential economic value of sustainability to the United States is recognized to not merely decrease environmental risks but also to optimize the social and economic benefits of environmental protection and remedi-ation (United States Environmental Protection Agency Report 2011). A new genre and a new beginning in human civilization and human scientific progress will surely emerge if the scientific domain, the civil society and governments around the world moves in the right direction in the effective implementation of environmental and energy sustainability. US Environmental Protection Agency today is moving in that right and positive direction. Each agency of the United States government has distinct responsibilities for various social, environmental and economic aspects of sustainable development and environmental management. This report deeply addresses and envisions these scientific and technological issues of sustainable development. Sustainability is based on a simple and long-recognized factual foundation: everything that humans require for their survival and well-being depends, directly or indirectly, on the natural environment and the ecology (United States Environmental Protection Agency Report 2011). The definition of sustainability was envisioned by Dr. Gro Harlem Brundtland, former Prime Minister of Norway in the year 1987. Sustainability is the ability to exist constantly and with happiness. According to "Our Common Future", sustainable development is defined as development that "meets the needs of the present without compromising the ability of future generations to meet their needs". The three dimensions of sustainable development are environmental, social and economic. Thus a deep scientific understanding and scientific discernment is the absolute need of the hour. This report deeply enumerates and explains these issues (United States Environmental Protection Agency Report 2011).

Environmental Protection Agency, Ireland Colgan and Donlon (2010) deeply elucidates with scientific vision and lucid insight environment, heritage and local government. A new scientific and social beginning are the pillars of sustainability whether it is environmental, social or economic. This report addresses science and sustainability and research-based knowledge for environmental protection (Saikia et al. 2019). Mankind, science and technology are today at the crossroads of an uncertain future as regards sustainable development and environmental management. This report deeply discusses (1) environmental research and policy, (2) support for environmental policies such as climate change, water quality, water and resource management, air quality, industrial pollution control, environment and health, and socio-economics, (3) support for knowledge economy and 4)future directions and futuristic priorities (Colgan and Donlon 2010). A remarkable scientific and technological vision in the field of sustainability will usher in a new era in human mankind and human scientific progress. The pivots of this well-researched treatise are (1) policy development and implementation, (2) green innovation and (3) research capacity in the visionary realization of environmental protection and environmental pollution control (Colgan and Donlon 2010). The four main environmental challenges facing and confronting Ireland in recent years are:

- Limiting and strongly adapting to climate change.
- Reversing environmental degradation.
- Re-envisioning environmental considerations across all sectors of economy.
- Complying with environmental regulations.

The report also underpins the role that science, engineering, research and development activities play in responding to environmental protection challenges. The policy making process involves analysis, formulation and implementation (Colgan and Donlon 2010). The other areas of deep scientific introspection are horizon scanning, fundamental research, experiments and modeling, data analysis and assessment, reports and research briefs, policy options, method development, modeling and guidance and assessment and feedback. Man's immense scientific vision and scientific perseverance, mankind's engineering vision and the futuristic vision of environmental engineering science will today go a long and visionary way in the true emancipation of global research and development initiatives. This report is a watershed text in the field of environmental protection in Ireland (Saikia et al. 2019). A new day and a new beginning in the field of environmental remediation, water remediation and industrial pollution control will surely usher in as human mankind moves forward towards a greater scientific vision (Colgan and Donlon 2010).

Goni et al. (2015) deeply discussed with scientific vision and scientific conscience environmental sustainability and its research growth and trends. Today environmental sustainability and environmental protection are two opposite sides of the visionary coin. The number of research in sustainability development is highly increasing along with the importance of the concept of sustainability (Goni et al. 2015). The vision of human mankind, the vast knowledge prowess of science and engineering and the futuristic vision of environmental remediation and water pollution control will eventually lead a long and visionary way in the true realization of today's environmental engineering, chemical engineering and the vast world of nanotechnology. Research and development forays in environmental sustainability are the immediate needs of the hour as humankind moves in the right direction. A profound scientific introspection is also the utmost need of the hour (Goni et al. 2015). The vision of this paper is to outline the research trend, redefine the literature categorization, and the vast research focuses of environmental sustainability engineering research from the deep perspective of historical evaluation. The research in sustainability whether it is energy, environmental, social or economic is growing rapidly and two research focuses appear at the highest count—water science and technology research and industrial pollution control and prevention (Goni et al. 2015). The vision of sustainable resource management, its purpose and scientific potential are also the immediate needs of the hour as humankind, science and technology moves towards one scientific paradigm over another. This well-researched treatise contributes vastly to the direction of future research in the discipline of sustainability. World Commission on Environment and Development of the United Nations defined sustainable development as "visionary development that meets the needs of the present without compromising the ability of future generations to meet their own needs" (Goni et al. 2015). The vision and purpose of "sustainable development" still needs to be readdressed and re-envisioned as human civilization treads towards a newer scientific and engineering paradigm. Today sustainability is a huge colossus with a definite vision and purpose of its own. The definition of sustainable development relates deeply to three interlocking goals: environmental, economic and social. Academicians in every field of science and technology are today contributing their empirical research to better understand their issues such as materials and technologies, green engineering and green technology, pollution control and prevention, energy harvesting and management, and water research (Goni et al. 2015). The growth of research focus on green engineering/manufacture is very slow in today's scientific and engineering scenario. Therefore there is a need for academic researchers to focus their study more on green engineering/green technology and manufacture. The impact of green engineering on sustainable development is the need of the hour and the deep scientific vision of this treatise (Goni et al. 2015).

Annual report and on the environment in Japan (2016) deeply elucidated with scientific vision and scientific far-sightedness global warming countermeasures. This report delineates with vision, purpose and vast scientific ingenuity (1) an international framework to address global warming, (2) Japan's global warming countermeasures, (3) green finance, (4) reconstruction after the Great East Japan earthquake, (5) biodiversity and ecosystem services, (6) low carbon society and (7) futuristic vision of global biodiversity. The year 2015 was a historic one for the global environment. In October,2015, the United Nations General Assembly adopted 2030 Agenda for Sustainable Development, a vastly universal set of goals and targets for achieving sustainable development in the global scenario. In December, the 21st Conference of the Parties to the United Nations Framework Convention on Climate Change (COP21) adopted the Paris Agreement the first legally binding agreement since the Kyoto Protocol eighteen years earlier (Report and on the Environment in Japan 2016). Today the vision, aim and objective of climate change mitigation and environmental pollution control mitigation are in the avenues of newer scientific regeneration. The ingenuity and scientific forbearance of the field of global warming elimination are also in the path of newer scientific rejuvenation. The main theme of the 2016 Annual Report on the environment is "a new stage for global warming mitigation and countermeasures". It explains the scientific truth and the actions Japan is taking to combat and confront global warming in the vast context of international discussion regarding measures on global warming. A newer scientific imagination, scientific comprehension and a newer scientific destiny are on the rise as science, civilization and mankind treads forward. In this report, the vast scientific and technological commitments of Japan as a nation in the field of Global Climate Change mitigation are lucidly delineated. Japan's National Plan for adaptation to the Impacts of Climate Change are vastly delineated with sound scientific and engineering fundamentals (Report and on the Environment in Japan 2016). It indicates basic adaptive measures for government ministries, sectors and domains to take in seven sectors-Agriculture, Forests/Forestry, Fisheries, Water environment/Water resources, Natural ecosystems, Natural disasters/coastal areas, human health, industrial activity, and the life of citizens and urban life. The vision and purpose of scientific research pursuit in the field of environmental engineering and environmental remediation will surely be emboldened in the global scenario if concerted and coordinated effort are taken in the research and development strides (Report and on the Environment in Japan, 2016).

Man's immense scientific verve and vision, mankind's vast scientific and engineering prowess and the scientific truth of human civilization will veritably ensure a new scientific and technological order in environmental protection science and chemical process engineering. Sustainable development, environmental management and the futuristic vision of integrated water resource management and wastewater management will unveil and uncover newer areas of research and development initiative. The author in this article deeply elucidates these intricate scientific and engineering research questions and the academic rigor.

8 Recent Scientific Advances in the Field of Arsenic Groundwater Remediation

Arsenic groundwater remediation, water and wastewater treatment are the utmost needs of human civilization and human scientific progress today. In the similar vein, drinking water treatment is also the utmost need of the hour. In this section, the authors deeply pronounce some recent advancements in the field of arsenic groundwater remediation. Today mankind's and civilization's immense gift to humanity is nanotechnology, nanomaterials and engineered nanomaterials. The world of science and technology stands immensely mesmerized at the application of nanotechnology to diverse areas of science and technology. There are today immense and sound applications of nanotechnology in groundwater and drinking water remediation. In this section, the author deeply pronounces the scientific subtleties and the vast scientific profundity in the application of nanotechnology and nanomaterials in groundwater remediation. A new epoch will surely ensue in the field of environmental protection and water remediation if sound and effective research and development initiatives are pursued by scientists, engineers, civil society and governments across the world. Ecological vulnerability and loss of ecological diversity are the pallbearers towards a greater scientific understanding and a greater scientific revelation in the field of environmental remediation and groundwater decontamination of heavy metals such as arsenic. The author deeply elucidates and pronounces these environmental engineering issues in minute details.

Hashim et al. (2011) deeply discussed with vast and versatile scientific vision and scientific foresight remediation technologies for heavy metal contaminated groundwater. The contamination of groundwater by heavy metal, originating either from natural soil sources or from anthropogenic and manmade source is a matter of immense concern to human health and global public health engineering scenario (Goni et al. 2015). Remediation of contaminated groundwater is a matter of immense priority since billions of citizens around the world use it for drinking water purpose (Goni et al. 2015). In this well researched article, thirty-five approaches for groundwater treatment have been reviewed and classified under three large categories which are chemical, biological/biochemical/biosorption and physico chemical treatment processes. Selection of a suitable technology and a suitable procedure for contaminant reduction at a particular site is one of the challenging job due to extremely complex soil chemistry and aquifer characteristics and there are no thumb rules. Keeping the sustainability issues and environmental ethics in mind, the technologies and scientific innovations encompassing natural chemistry, bioremediation, and biosorption are highly recommended to be adopted in the appropriate cases. In many places, two or more techniques can work synergistically for better scientific directions (Hashim et al. 2011). Research and development directions in the field of groundwater remediation and water and wastewater treatment are thus in the vistas of deep scientific vision and engineering ingenuity. In this treatise, the author deeply stresses on sources, chemical property and speciation of heavy metals in groundwater, technologies for the treatment of heavy metal contaminated groundwater, chemical treatment technologies, in-situ treatment by using reductants, reduction by using iron-based technologies, soil washing and biological, biochemical and biosorptive treatment technologies (Hashim et al. 2011). Enhanced biorestoration, and biosorption of heavy metals are the other areas of deep scientific endeavour. Permeable reactive barrier is an innovation of environmental engineering science and chemical process engineering. The United States Environmental Protection Agency (1989) deeply elucidated with vision and cogent insight and defined Permeable Reactive Barrier an "emplacement of reactive media in the sub-surface designed to intercept a contaminated plume, provide a flow path through the reactive media and transform the contaminants into environmentally accepted forms". A new visionary concept in the field of groundwater remediation and drinking water treatment is emerging as biological treatment and physico-chemical treatment gains immense importance. The authors deeply discussed the application of Permeable reactive barrier in the wide domain of physico-treatment technologies in heavy metal groundwater remediation (Hashim et al. 2011). The concept behind Permeable Reactive Barrier is that a permanent, semi-permanent or replaceable reactive media is placed in the sub-surface across the flow path of a plume of contaminated groundwater which move through it under its natural gradient, thereby creating a passive treatment system. This is a novel area of environmental engineering treatment and water remediation procedure. Permeable reactor barriers can be classified as (1) sorption-zeolites, humic materials and oxides, (2) chemical reaction-zero-valent metals-(Fe, etc.), minerals, and (3) biological treatment-oxygen and nitrate releasing compounds, organic materials. Electrokinetic remediation of soil are the other pivots of this well researched scientific endeavour (Hashim et al. 2011). In a critical discussion of environmental protection science and water remediation science, the authors deeply discussed chemical treatment technologies, biological, biochemical and biosorptive treatment technologies, and physico-chemical treatment technologies. Scientific vision and deep scientific ingenuity in the field of water remediation, industrial wastewater and drinking water treatment are greatly enhanced as science and engineering in the global firmament reached new visionary heights (Hashim et al. 2011). Groundwater treatment technologies have with deep scientific vision have come a long way since its inception in the last decade of the twentieth century. Geologists, chemical engineers, environmental engineers and nanotechnologists are today contributing immensely in the larger emancipation of the science of water and wastewater treatment. Much research pursuit has been done on numerous technologies ranging from simple ex-situ physical treatment technologies to complex in-situ microbiological and biological treatment tools. The authors profoundly discuss these important scientific and engineering research problems of heavy metal groundwater remediation (Hashim et al. 2011).

A new world in the domain of environmental engineering science and environmental protection science is slowly emerging today as science and engineering crosses one visionary boundary over another. The vast engineering vision and scientific prowess in the field of water and wastewater treatment will today usher in a domain of newer futuristic vision and futuristic flow of scientific thoughts. Nanotechnology and application of nanomaterials and engineered nanomaterials in diverse areas of science and engineering are the revolutionary domains of science and technology. The author in this treatise deeply addresses these scientific and engineering issues with a greater emancipation of science and technology in the global firmament.

Hassan (2018) discussed with scientific farsightedness poisoning and risk assessment of arsenic in groundwater. Groundwater is the main source of safe drinking water in many nations around the world but much of that groundwater have been found to be contaminated with arsenic and heavy metals. It is highly ironic that so many tubewells have been installed for drinking water that are safe from water-borne diseases but highly contaminated with arsenic and other metals (Hassan 2018). It is estimated that more than 300 million people in 70 countries worldwide are at deep risk of groundwater arsenic poisoning. Mankind's immense scientific vision and vast knowledge prowess are at risk due to this marauding environmental engineering catastrophe. Apart from Bangladesh and the neighbouring state of West Bengal, India, which between them have the largest problem, there have been serious warnings from Argentina, Chile, Taiwan, Vietnam, China, Pakistan, Thailand and south western part of the United States of America. The situation of arsenic poisoning in Bangladesh are addressed deeply in this book. Groundwater arsenic toxicity presents a extensively a new dimension of hazard in Bangladesh apart from the disastrous and marauding calamities such as floods, cyclones, tidal surges, famine and infectious diseases (Hassan 2018). This book deeply discusses, (1) arsenic poisoning through ages, (2) the global scenario of groundwater arsenic catastrophe, (3) spatial mapping, spatial planning and public participation of groundwater arsenic discontinuity, (4) chronic arsenic exposure to drinking water, (5) epidemiological and spatial assessment of risk from groundwater arsenic exposure, (6) arsenic induced health and social hazard and the subsequent survival strategies, and (7) arsenic poisoning in Bangladesh and legal issues of responsibility (Hassan 2018). This book can contribute modules on public health engineering, environmental studies, water quality, Geographical Information System and applications, human rights and legal issues. Arsenic is a serious carcinogen and only a small quantity can result a serious health hazard. The impacts of arsenic exposure on human health at different dose levels are deeply discussed in this well researched treatise. Risk assessment is an important scientific and engineering issue in the global scenario. Explaining and exploring risk characterization in terms of probabilistic environmental health risk and spatial risk zoning of arsenic toxicity are the vision and purpose of this book (Hassan 2018).

Palit et al. (2019) elucidated with cogent insight and scientific and technological far-sightedness arsenic contamination in South Asian regions and the difficulties, challenges and vision for the future. Arsenic and heavy metal groundwater contamination are vexing and burning issues in many developing and developed nations around the world. The South Asian countries of Afghanistan, Bangladesh, India, Nepal, Pakistan and Sri Lanka face the severest degree of arsenic contamination in groundwater and drinking water containing high levels arsenous acid, arsenic acid, and their derivatives that lead to severe health issues (Palit et al. 2019). Chronic health issues are also the result of this world's largest environmental disaster. The health problems due to arsenic contamination are arsenicosis, cancer, neuropathy, and respiratory disorders. Country wise status of arsenic contamination in South Asia is grave, disastrous and highly thought-provoking (Palit et al. 2019). It is a raging fear amongst the scientific community in South Asia, developing and developed nations around the world. Difficulties and challenges associated with arsenic contamination in South Asia are in the similar vein highly grave and scientifically mind-boggling. The points of concern are lack of awareness, absence of sophisticated technology and accessibility of sound environmental engineering tools. Domestic measures and domestic environmental engineering tools are (1) rapid small column test, (2) use of natural geological material, (3) desalination, (4) oxidation, (5) coprecipitation, and (6) Subterranean arsenic removal technology. Industrial measures are (1) ion exchange, (2) hybrid technology based on ion exchange and electrodialysis, (3) use of nanoparticles, (4) carbon nanotubes, (5) titanium-based nanoparticles, (6) zirconium oxide nanoparticles and (7) magnetite—graphene hybrids (Palit et al. 2019). The deleterious effects inflicted by high arsenic concentrations in groundwater and drinking water are continuously exceeding the prescribed limits of 10 ppb set by World Health Organization, sometimes increasing beyond 10 ppb. Thus the need of sustainable and efficient integrated water resource management and also in the similar vein wastewater management system. Human scientific progress and scientific vision in the field of drinking water and groundwater treatment are the utmost needs of the hour (Palit et al. 2019). As per recent reports and global research and development strides, in Bangladesh alone, 80 million people are suffering due to unavoidable dependence on arsenic contaminated water and cases of arsenic poisoning and associated cancer aliments are on the rise. The vision, the purpose and the scientific subtleties of human race are the ultimate needs of the hour and will lead a long way in the true realization of environmental engineering science and chemical engineering science (Palit et al. 2019).

Abdelbasir and Shalan (2019) deeply discussed with scientific vision and scientific perseverance nanomaterials for water and industrial wastewater treatment. Mankind, science and technology today stands in the middle of deep scientific forbearance and scientific divination. Industrial wastewater is today an universal issue in the global forefront. Numerous organic pollutants, heavy metals and recalcitrant compounds are today present in the water and wastewater at extreme concentrations. Efficient purification and separation techniques are highly needed to remove these pollutants from drinking water and industrial wastewater (Abdelbasir and Shalan 2019). In this positive direction and sound research and development initiatives, wastewater treatment is one of the many applications of nanomaterials and engineered nanomaterials. This review article deeply highlights the use of nanomaterials for the removal of different polluting materials from industrial wastewater with a deep focus on metal and metal oxide nanomaterials, carbon-based nanomaterials and nanofiber/nanocomposite membranes. The areas of emergent nanomaterials in efficient removal of recalcitrant compounds are the other areas of deep scientific introspection of this treatise (Abdelbasir and Shalan 2019). Wastewater is the water containing substances that have an unfavorable effect on the quality of water making it useless to drink. Provision of drinking water and proper sanitation are highly neglected in many developing and developed nations around the world. These are the vision of United Nations Sustainable Development Goals (Hashim et al. 2011). Wastewater is produced from numerous sources as in residential areas, industrial areas, commercial areas, agricultural lands etc. Rapid industrialization, environmental degradation and loss of ecological biodiversity are today urging the scientific domain to move towards newer innovation and newer scientific instinct (Hashim et al. 2011). Common constituents of wastewater are inorganic substances like solutes, heavy metals, metal ions, ammonia along with gases, complex organic compounds such as excreta, protein, natural organic matter, plants materials, food, nitrate, and several other contaminants in ground water, drinking water, surface water and industrial wastewater. A newer scientific and technological order is in the process of scientific vision and regeneration as civilization and humankind moves forward. The science and engineering of environmental remediation thus needs to be emboldened and reassessed as mankind confronts environmental engineering issues (Hashim et al. 2011). Many biological, physical and chemical treatment tools are today exploited to handle industrial wastewater and drinking water. The common materials and the subsequent treatment techniques like activated carbon, oxidation, reverse osmosis membranes, and activated sludge are today not sufficient to get rid of complex polluted water that contains pharmaceuticals, surfactants, various industrial chemicals and other recalcitrant compounds. Many nanomaterials like carbon nanotubes, nanomembranes, zeolites and dendrimers are highly aiding in the improvement of more competent treatment pathways and procedures amongst the other methodologies and techniques (Abdelbasir and Shalan 2019). Nanotechnology avenues can be utilized to address the many complications of water quality to support environmental engineering stability with the help of industrial wastewater treatment systems and chemical engineering systems. In this well-researched treatise, the author deeply discusses sources and compositions of industrial wastewater, industrial wastewater treatment processes, and the application of nanomaterials for industrial wastewater treatment and also groundwater treatment (Abdelbasir and Shalan 2019). The areas of deep scientific contemplation are metal and metal oxide nanoparticles, silver nanoparticles, iron nanoparticles, and iron oxide nanoparticles. Carbon-based nanoadsorbents and graphene-based nanomaterials and their vast and visionary applications are the other pillars of this well-researched treatise. Case studies of electroplating industry, dairy wastewater treatment and membranes and membrane processes are the other areas of deep scientific introspection. Retaining and reuse of nanomaterials are the other pivots of this scientific research pursuit. Regenerating and reusing nanoadsorbent materials from an aqueous solution is highly difficult and may create environmental engineering problems, as the adsorbed compounds need to develop and innovate sample pretreatment and separation tools (Hashim et al. 2011). Chemical process engineering and environmental engineering systems are two opposite sides of the visionary coin. Man's immense scientific verve and motivation, the difficulties and barriers of environmental engineering science will surely pave the way towards a newer engineering and scientific order in water and environmental remediation. The treatment of industrial wastewater, the world of challenges in desalination science and the futuristic vision of drinking and groundwater treatment will eventually open new doors of innovation in the field of environmental engineering. The authors validates these issues in this treatise (Abdelbasir and Shalan 2019).

Science, technology and engineering are today transcending and surpassing vast and versatile frontiers. Arsenic and heavy metal drinking water and groundwater contamination are in veritable disastrous state and are the major challenges of civilization and science today. In this well researched treatise, the authors reiterates with vision and scientific prowess the need of water purification and water science and technology in the true emancipation of global scientific endeavour (Hassan 2018; Gupta et al. 2017; Chatterjee et al. 2017; Chakraborti et al. 2018; Mishra et al. 2016; Mazumder et al. 1998; Palit and Hussain 2018; Palit 2018).

9 Arsenic Groundwater Remediation, the March of Science and the Visionary Future

Today, the status of arsenic and heavy metals groundwater remediation is highly grave. Billions of people are dying due to the evergrowing concerns of drinking water poisoning. Scientific validation, scientific verve and deep scientific motivation are the absolute needs of the hour. This crisis is unending and a deep scientific truth and scientific introspection are the ultimate needs of the hour. In this section the authors with cogent insight deal with the primary issues of arsenic and heavy metal contamination and subsequent remediation in water scarce regions around the world.

Gupta et al. (2017) deeply discussed with deep scientific far-sightedness arsenic contamination from historical aspects to the present. Arsenic (As) is well known as the "king of poisons". The word "arsenic" results and envisions an appalling reaction for its devastating mutagenic, carcinogenic and teratogenic effects. Human civilization and green chemistry and green engineering are the pillars of any scientific pursuit in water remediation today (Gupta et al. 2017). The author in this article profoundly depicts historical uses of arsenic, different sources of arsenic in the environment, some epidemiological studies of arsenic exposure from water and food sources, toxic effects, metabolic pathways and remediation measures for arsenic decontamination. Arsenic gains entry into the human body through various means, and much chronic measures even at low levels are enough for the onset of visible toxic symptoms due to contamination. As per the elemental scientific abundance, arsenic ranks twentieth in nature, fourteenth in sea water and twelfth in human body (Gupta et al. 2017). The environmental health effects are simply appalling as regards drinking water contamination. Under natural conditions, groundwater may contain different concentrations and wide ranges of arsenic mainly as a result of resilient and recalcitrant effect of water–rock interfaces. The Asian region is veritably much more affected by Arsenic as compared to other regions of the world. Global aquifers of Bangladesh and West Bengal, India represent the most severe Arsenic contamination and related health problems in the citizens residing in those areas. Remediation technologies are ion exchange, ultrafiltration, excavation, chemical precipitation, reverse osmosis, adsorption, solidification and stabilization and electrocoagulation. Huge and evergrowing concerns over Arsenic contamination in groundwater is a challenging task and a vexing issue. Thus the needs of effective, low cost and ecology-friendly technologies will surely open new doors of innovation and scientific vision in years to come (Gupta et al. 2017).

Chatterjee et al. (2017) discussed and deliberated with scientific and engineering vision prospects of combating arsenic and the physico-chemical aspects. The authors discussed in details lime softening, coagulation/filtration, sorption, ion exchange, arsenic adsorption using fixed bed, arsenic removal by layered double hydroxide, nanomaterials in arsenic removal, reverse osmosis and nanofiltration. Arsenic is a Class A human carcinogen and its exposure results in serious health effects including cancer. Thus the immediate urge for research and development in the field of arsenic and heavy metal groundwater remediation (Chatterjee et al. 2017). The author compared common arsenic removal technologies such as Fe coagulation, Alum coagulation, co-precipitation, activated alumina adsorption, iron-based adsorbents, reverse osmosis and nanofiltration. A wide area of water remediation and environmental integrity is slowly evoing. Subterranean arsenic removal technology stands as a major pillar of all arsenic removal technologies. It is effective, environmental friendly and immensely low cost. Arsenic is a Group A human carcinogen. Intake and ingestion of arsenic contaminated water is affecting 137 million people around the world. For the removal of arsenic, physico-chemical techniques will evolve into a newer scientific and engineering order. The author deeply addresses these engineering vision and purpose (Chatterjee et al. 2017).

Mankind and man's vision needs to be reorganized as regards water and wastewater treatment. In this entire treatise, the authors pronounces and reiterates the immediate needs of pollutant removal technologies. These will lead to furtherance of science and technology globally.

10 The Scientific Sagacity and Scientific Ingenuity of Heavy Metal and Arsenic Groundwater Remediation and the March of Engineering Science

Today is the world of fourth industrial revolution, circular economy and green economy. Protection of environment is the prime objective of human civilization and human scientific progress today. Engineering and technology's advancements today depends on environmental remediation, water remediation, green sustainability and circular economy. Regenerate, recreate, recycle and reuse are the coin-words of today's industrialization and global urbanization. Arsenic and heavy metal groundwater and drinking water contamination are creating immense scientific challenges and difficulties. Some of the important challenges, difficulties and vision are elucidated in details in this chapter.

Chakraborti et al. (2018) discussed in minute details arsenic occurrence in groundwater. Geogenic groundwater arsenic contamination and poisoning affects over 296 individuals' lives in more than 100 countries worldwide. Arsenic occurs in five valence states, among which the inorganic forms arsenite, the most toxic, and arsenate are the veritably most abundant species in groundwater. In the last years of twentieth century, groundwater from only five countries including Taiwan, China and India (West Bengal), Bangladesh and Thailand were reported as severely contaminated sites with several arsenicosis patients. The authors discussed in minute details arsenic in the food chain, dermal effects and prevalence of diabetes (Chakraborti et al. 2018). The vast domain of environmental protection science, environmental engineering and public health engineering are in a true scientific need of revamping and reorganization. Cardiovascular effects, neurological effects and cancer effects due to groundwater poisoning are detailed in this paper. Arsenic and genotoxic effects are the other cornerstones of this paper. Human civilization's deep scientific stance and engineering needs as regards arsenic poisoning are the ultimate pillars of this paper (Chakraborti et al. 2018).

Mishra et al. (2016) deeply discussed current status of groundwater arsenic contamination in India and recent advancements in removal techniques from drinking water. The authors discussed in details arsenic contamination in North and Central India, arsenic contamination in East India, and emerging treatment technologies for arsenic removal. The ultimate solution of the use of surface and rain water are also detailed in this paper. The treatment tools discussed are (1) ceramic microfiltration membrane unit, (2) DRDO arsenic removal filter, and (3) low cost laterite based arsenic filter. The march of science and engineering will surely be emboldened and envisioned if low-cost and environmentally friendly techniques are developed in the future (Mishra et al. 2016).

Mazumder et al. (1998) discussed with cogent insight arsenic levels in drinking water and the prevalence of skin lesions in West Bengal, India. A broad cross-sectional survey was deeply conducted between April 1995 and March 1996 to investigate and explore arsenic-associated skin lesions of keratosis and hyperpigmentation in West Bengal, India and to determine their relationship to arsenic water levels. The authors did a deep study of the study area and population, interview and medical examination, water sampling and arsenic measurement. In summary, it can be found out that clear exposure–response relationship between the prevalence of skin lesions and both arsenic water levels and dose per body weight, with males showing greater prevalence of both keratosis and hyperpigmentation. There is a clear evidence that the risks were somewhat greater for those who are malnourished. A deep scientific understanding and scientific and research resilience in the field of drinking water treatment will ameliorate this ever-growing crisis (Mazumder et al. 1998).

An unmitigated human crisis and human suffering is prevalent in many developing nations around the globe- the arsenic drinking water contamination. Man's as well as mankind's vision stands defeated as human suffering in different manners affects global population. The author reiterates these scientific and engineering issues with vision and mights.

11 Future Scientific Recommendations of This Study, Future Flow of Scientific Thoughts and the Vision Behind it

Future scientific recommendations and future flow of scientific thoughts in the field of green sustainability and water remediation should be targeted towards greater scientific emancipation in the field of desalination and novel separation processes. Today water remediation and groundwater contamination are burning issues in the vast scientific landscape. The vast vision for the future in the field of green sustainability and environmental pollution control needs to be revisited and re-envisioned as science and civilization moves forward (Hassan 2018; Palit and Hussain 2018; Palit 2018).

Future flow of scientific thoughts in the field of nanotechnology needs to be directed towards scientific and engineering emancipation in the field of nanomaterials and engineered nanomaterials. The other areas of scientific grit and scientific determination in the field of environmental engineering and chemical engineering are the application areas of desalination, novel separation processes, conventional and non-conventional environmental engineering techniques. Human suffering is immense in the field of arsenic groundwater and drinking water contamination in India, Bangladesh and many developing and developed countries around the world. Thus the need of a detailed treatise in the field of arsenic groundwater remediation and the application of green and environmental sustainability. (Barrow 2005; Palit 2017, 2016a, b; Palit et al. 2017; Hussain and Kharisov 2017; Hussain 2017, 2018, 2019, 2020a) Today mankind stands in the midst of deep scientific provenance and deep scientific introspection as global water crisis and climate change destroys the vast scientific landscape. Thus the need of a detailed treatise in the field of green sustainability and arsenic and heavy metal groundwater remediation. Futuristic vision and futuristic flow of scientific thoughts should be targeted in the field of desalination, membrane science, conventional and non-conventional environmental engineering techniques (Hussain 2019, 2020a). A new era in the field of conventional and non-conventional environmental engineering techniques will surely emerge as civilization moves forward.

12 Conclusion, Summary and Scientific Perspectives

Human mankind, civilization, science and technology today are in the path of new scientific rejuvenation and effective regeneration. Drinking water issues and industrial wastewater treatment issues are a burden to science, engineering and civilization today. Arsenic and heavy metal groundwater contamination are destroying the vast scientific landscape today. The utmost needs of the hour are newer environmental engineering techniques and newer innovations in the field of environmental engineering science, nanotechnology and chemical engineering science. Scientific provenance, deep scientific ingenuity and the veritable necessities of human civilization will go a long and visionary way in the true emancipation of environmental engineering science, water remediation and environmental pollution control. The scientific perspectives in the field of groundwater remediation, potable water treatment and industrial wastewater treatment needs to be revisited and re-envisioned with the passage of scientific history and visionary timeframe. Novel separation processes such as membrane science needs to be implemented in every scientific endeavor in the field of environmental engineering science. The authors in this entire treatise delves deep into the field of green sustainability and its application in the field of arsenic groundwater remediation. Today there are fewer applications of sustainable development in the field of arsenic and heavy metal groundwater remediation. The authors deeply pronounces the needs of green and environmental sustainability in the field of environmental and water remediation. Environmental and energy sustainability are the areas of scientific research pursuit which are still not explored. Today these areas of science and engineering needs to be investigated with the passage of history and time. In developing and developed countries around the world, arsenic and heavy metal groundwater and drinking water remediation are today highly neglected. Man's immense scientific vision and mankind's immense scientific and engineering prowess needs to be revisited and re-envisioned as civilization moves forward. The areas of water remediation and industrial pollution control are also highly neglected in developing countries around the world and emerging economies. Thus the need of a detailed treatise in the field of water pollution control, industrial wastewater treatment and environmental remediation. Today, the wonder of science and engineering are the areas of nanotechnology, nanomaterials and engineered nanomaterials. So the application of nanotechnology in water remediation and environmental pollution control will surely usher in a new area in science and technology. Besides the areas of desalination and novel separation processes such as membrane separation processes in water and wastewater treatment will open new avenues in the field of science and engineering. The author validates these scientific and engineering issues with scientific vision and scientific determination. Thus a new era in the field of environmental engineering and nanotechnology will surely emerge with the passage of scientific history, human civilization and time.

References

Abdelbasir SM, Shalan AE (2019) An overview of nanomaterials for industrial wastewater treatment. Korean J Chem Eng 36(8):1209–1225

Annual Report on the Environment in Japan, 2016 (2016) 1–35

Barrow CJ (2005) Environmental management and development. Routledge, Taylor and Francis Group, London, United Kingdom

Chakraborti D, Singh SK, Rashid MH, Rahman MM (2018) Arsenic: occurrence in groundwater. In: Nriagu J (ed) Encyclopedia of environmental health, 2nd edn. Elsevier, Netherlands, pp 1–16

Chatterjee S, Chetia M, Voronina A, Gupta DK (2017) Prospects of combating arsenic: physico-chemical aspects, Chapter-5. In: Gupta DK, Chatterjee S (eds) Book–Arsenic contamination in the environment. Springer International Publishing AG, pp 103–121

Colgan S, Donlon B (2010) Science and sustainability: research-based knowledge for environmental protection. In: STRIVE Programme, EPA STRIVE Programme (2007–2013). Environmental Protection Agency, Ireland

Goni FA, Shukor SA, Mukhtar M, Sahran S (2015) Environmental sustainability: research growth and trends. Adv Sci Lett 21:192–185

Gupta DK, Tiwari S, Razafindrabe BHN (2017) Arsenic contamination from historical aspects to the present, Chapter-1. In: Gupta DK, Chatterjee S (eds) Book- Arsenic contamination in the environment. Springer International Publishing AG, pp 1–12

Hashim MA, Mukhopadhyay S, Sahu JN, Sengupta B (2011) Remediation technologies for heavy metal contaminated groundwater. J Environ Manage 92:2355–2388

Hassan MM (2018) Arsenic in groundwater: poisoning and risk assessment. CRC Press, Taylor and Francis Group, Boca Raton, Florida, USA

Hussain CM (2017) Magnetic nanomaterials for environmental analysis, Chapter-19. In: CM Hussain, B Kharisov (eds) Book–Advanced environmental analysis-application of nanomaterials, vol 1. The Royal Society of Chemistry, Cambridge, United Kingdom, pp 3–13

Hussain CM (2018) Handbook of nanomaterials for industrial applications. Elsevier Amsterdam, Netherlands

Hussain CM (2019) Handbook of environmental materials management. Elsevier, Amsterdam, Netherlands

Hussain CM (2020a) Handbook of functionalized nanomaterials for industrial applications. Elsevier, Amsterdam, Netherlands

Hussain CM, Kharisov B (2017) Advanced environmental analysis-application of nanomaterials, Volume-1, (Book). The Royal Society of Chemistry, Cambridge, United Kingdom

Mazumder DNG, Haque R, Ghosh N, De BK, Santra A, Chakraborty D, Smith AH (1998) Arsenic levels in drinking water and the prevalence of skin lesions in West Bengal India. Int J Epidemiol 27:871–877

Mishra S, Dwivedi S, Kumar A, Chauhan R, Awasthi S, Mattusch J, Tripathi RD (2016) Current status of groundwater arsenic contamination in India and recent advancements in removal techniques from drinking water. Int J Plant Environ 2(1–2):1–15

Palit S (2016a) Filtration: Frontiers of the engineering and science of nanofiltration-a far-reaching review. In: Ortiz-Mendez U, Kharissova OV, Kharisov BI (eds) CRC Concise Encyclopedia of Nanotechnology. Taylor and Francis, pp 205–214

Palit S (2016b) Nanofiltration and ultrafiltration-the next generation environmental engineering tool and a vision for the future. International Journal of Chem Tech Research 9(5):848–856

Palit S (2017) Advanced environmental engineering separation processes. In: Editors- CM, Hussain B (eds) environmental analysis and application of nanotechnology–a far-reaching review, Chapter-14, Book- Advanced environmental analysis–application of nanomaterials, Volume-1. Kharisov, The Royal Society of Chemistry, Cambridge, United, Kingdom, pp 377–416

Palit S (2017) Application of nanotechnology, nanofiltration and drinking and wastewater treatment-a vision for the future, Chapter-17. In: Grumezescu AM (ed.) Book-Water purification. Academic Press, USA, pp 587–620

Palit S (2018) Recent advances in the application of engineered nanomaterials in the environment industry, Chapter-47. In: Hussain CM (ed.) Book-Handbook of nanomaterials for industrial applications. Elsevier, Netherlands, pp 883–893

Palit S, Hussain CM (2018) Engineered nanomaterial for industrial use, Chapter-1. In: Hussain CM (ed.) Book- Handbook of nanomaterials for industrial applications. Elsevier, Netherlands, pp 3–12

Palit S, Misra K, Mishra J (2019) Arsenic contamination in South Asian regions : the difficulties, challenges and vision for the future, Chapter-5. In: Ahuja S (ed.) Book-Evaluating water quality to prevent future disasters, Separation Science and Technology, vol 11. pp 113–123

Saikia J, Gogoi A, Baruah S (2019) Nanotechnology for water remediation, Chapter-7. In: Dasgupta N, et al. (eds) Book-Environmental nanotechnology, environmental chemistry for a sustainable world. Springer Nature Switzerland Ag, pp 195–211

United States Environmental Protection Agency Report (2011) Sustainability and U.S. EPA, Committee on Incorporating Sustainability in the US Environmental Protection Agency. National Research Council, USA

Zamora-Ledezma C, Negreta-Bolagay D, Figueroa F, Zamora-Ledezma E, Ni M, Alexis F, Guerrero VH (2021) Heavy metal water pollution: a fresh look about hazards, novel and conventional remediation methods. Environ Technol Innov 22(101504):1–26

Bio-Based Materials Used in Food Packaging to Increase the Shelf Life of Food Products

Neeta Shivakumar, Mounika Sri Ramesh Babu, Shravya Vasudeva, and H. Akshay

1 Introduction

Food packaging is the enclosing of food products to protect them from spoilage, environmental factors or also tampering. This deterioration of food is primarily caused by oxidation, microbial spoilage, and metabolism, which are all influenced by environmental contamination and other factors like temperature, humidity, light, physical damage, microorganism, odours, shocks, and dust. Food packaging is used to maintain food quality, assure food safety, and increase the shelf life of foods. Different types of foods have different storage and transportation requirements. For example, the preservation of fruits and vegetables necessitates a reduction in respiration and transpiration rates, which is usually achieved by controlling humidity, temperature, light, and gas environment, among other things. The packaging depends on the type of food being packed and varies over a wide range of materials and forms. Most commonly used packaging materials include petrochemical-based polymers like polypropylene, polyethylene, and polyamide among many others (Peelman et al. 2016). These materials are typically used because of certain desirable properties like that of good material performance, good barrier properties, and also their cheaper

N. Shivakumar (✉)
Department of Biotechnology, R.V. College of Engineering, Bengaluru, India
e-mail: neeta@rvce.edu.in

M. S. Ramesh Babu · S. Vasudeva
Department of Electronics and Communication Engineering, R.V. College of Engineering, Bengaluru, India
e-mail: shravyav.ec19@rvce.edu.in

S. Vasudeva
e-mail: akshayh.cs18@rvce.edu.in

H. Akshay
Department of Computer Science and Engineering, R.V. College of Engineering, Bengaluru, India
e-mail: mounikasrirb.ec19@rvce.edu.in

A. K. Mishra and C. M. Hussain (eds.), *Biobased Materials*,
https://doi.org/10.1007/978-981-19-6024-6_9

costs. However, such materials can be harmful to the environment when we're looking at the long-term effects. As a response to this, industries are looking into using packaging made of renewable resources such as biodegradable and bio-based materials. Examples of existing bio-based materials include paper made from wood fibres, bio-PE made from sugarcane, bio-PET, starch-based plastics, etc. These bio-based materials are already used for packing various food items, for example, starch-based packaging is commonly used for chocolates, cellulose-based packaging for organic foods, nuts, herbs, and bread, PLA-based packaging for fresh produce, etc. Research on expanding the use of bio-based materials is being conducted, but generally, the findings of such research aren't reaching the companies and are, therefore, not widely implemented.

There are certain parameters that help us classify whether a material is good for packaging, or not with some of the notable examples being gas barrier requirements and water barrier requirements. Other ideal packaging characteristics include:

- Stability over a wide range of temperatures.
- Sufficient space for rapid cooling of the contents inside.
- Protection from oxygen, light, and moisture.
- Compatibility with the food product.
- Protection from adulteration.
- Closure characteristics like opening, sealing, and resealing.
- Aesthetics and strong marketing appeal.
- Low cost and availability.

Nanotechnology is also an upcoming field expected to improve food packaging. Indeed, new nano-based food packaging materials possess unique characteristics including antimicrobial potential, oxygen scavengers, barriers to gas or moisture, etc. The implementation of such nanomaterials in food packaging increases the shelf life of food without causing any undesirable alteration in its quality. The usage of nanomaterials in food packaging is still in the embryonic stage, and hence, the present review focuses on recent advances and an overview of the current status in the field.

2   Shelf Life and Its Determination

Shelf life is defined as the time period, under certain defined storage conditions, during which food remains safe, retains desired sensory, chemical, physical, and biological characteristics as well as complies with any label declaration. Shelf life, in other words, means the duration for which a product may be stored without becoming unsuitable for consumption and it generally depends on the degradation mechanism of the specific product. It is legally mandatory to produce a product that is chemically and microbially safe and when no products are introduced to the market, their shelf life is tested. The same happens with products with unconventional processing regimes, for example, sauces. This is done so by testing the products against a wide range of spoilage organisms over a predetermined time period until the levels except for allowable levels (Smith and Stratton 2007). However, this can be costly given that the range of testing and the time required for it are quite high.

Products tested for shelf life should be taken from trial production runs instead of test kitchen samples due to the different environments and differences in production methods. Some testing services also offer accelerated shelf life wherein the product is held at above normal temperatures and humidity. The processor is legally obliged to produce a synthetically and microbially safe product. A few processors test the product shelf life when presenting a new product or a product with an unconventional processing routine such as minimally processed sauces. This is done by testing arranged products for a range of spoilage organisms over a predetermined time span till the levels observed exceed allowable levels (Smith and Stratton 2007). This can be expensive, considering the range of testing and the length of time required for testing. Products tested for shelf life should be taken from trial production runs instead of test kitchen samples, due to the differences in production methods and conditions. Few testing services offer accelerated shelf life testing where the product is kept at above normal temperature and/or humidity.

These studies are helpful for testing package integrity over time and can provide data on worst-case scenarios that may occur in product holding environments. However, these should not be completely relied upon for determining the normal shelf life of a product. Conventional thermal processes usually are more reliable and cost-effective for entrepreneurs who wish to market sauces, condiments, and salsas, and when a conventional thermal process is usually used to produce these products, there is no need for expensive shelf life testing.

2.1 Common Methods of Increasing the Shelf Life Extension

Shelf life extension relies on changing the storage conditions or the product packaging to inhibit microbial growth. Some of the common methods are:

(i) *Modified Atmosphere Packaging (MAP)*

Actively or passively managing or adjusting the environment surrounding the product within a box consisting of various types and/or combinations of films is what modified atmosphere packaging entails. A modified atmosphere is one that has been particularly constructed by changing the natural distribution and composition of atmospheric gases. When used in packaging, this entails altering or managing the composition of gases contained within each package to offer optimal conditions for extending the shelf life of perishable food and beverage goods while reducing oxidation and spoiling.

Passive and active modified environment packaging are the two types available. The FDA distinguishes between active modified atmosphere packaging and passive modified atmosphere packaging by defining active modified atmosphere packaging as the displacement of gases in the package, which is then replaced by a desired mixture of gases. Depending on the product, various MAP technologies can be implemented, for example, gas flushing, barrier packaging films, scavenger packs, on package valves, etc. (Sarkar and Aparna 2020).

(ii) *Controlled Atmosphere Packaging (CAP):*

Here, the atmosphere is regulated in controlled atmosphere storage by regulating the levels of three distinct gases—oxygen, carbon dioxide, and nitrogen—each of which has a specialised purpose as described below:

– Oxygen helps to keep meat's red colour.
– Carbon dioxide: Prevents mould and germs from growing.
– Nitrogen is an inert gas that stops the products from breathing and enzymatic activity, as well as gas diffusion through the film.

(iii) *Vacuum packaging*

This is a method of packaging that has been in use for many years. Here, the food is packed in a material with low oxygen permeability, and then vacuum is created by drawing out the air in the package prior to sealing. The package collapses around the product once the air is evacuated. This packaging technique mainly aims to extend the shelf life of food products. Products with high fat content become rancid due to the reaction with oxygen present in the package. Thus, in order to reduce the oxygen available within the package, vacuum packaging proves to be a promising technique.

(iv) *Bioactive packaging*

This is one of the new packaging technologies which has a direct impact on the health of the consumer by producing healthier packaged foods. The biopolymers involved in this packaging technology are capable of withholding desired bioactive principles in optimal conditions until their eventual release into the food product which is employed by enzyme encapsulation, enzyme immobilisation, microencapsulation, and nanoencapsulation. This technology helps in maintaining the bioactive substance until its controlled or fast release within the packed food during the storage period, or prior to its consumption by taking into consideration the specific product or functional substance characteristics or requirements (Sarkar and Aparna 2020).

(v) *Smart or intelligent packaging*

This innovative packaging technology is an integrated and multidisciplinary approach that involves expertise from the fields of physics, chemistry, biochemistry, electronics, and food science and technology. Smart or intelligent packaging uses a variety of chemical sensors or biosensors to monitor the quality and safety of food products. These sensors can suitably monitor the food quality and safety and provide a signal about the freshness, pH, oxygen, carbon dioxide, pathogens, gas leakage, ripeness regulators and indicators, bioprobes, radio frequency indicators and toxin indicators, and time temperature indicator of the product (Sarkar and Aparna 2020).

2.2 Limiting Attributes and the End of the Shelf Life

2.2.1 Identification of Limiting Attributes

Chemical, biochemical, physical, and microbiological changes occur during the shelf life of a product. Distinguishing a single critical attribute that can unequivocally determine the end of a product's shelf life is difficult, if not impossible. Attributes that characterise the end of a product's shelf life are product, consumer, and market-specific. Furthermore, various criteria are used not only to choose a limiting quality attribute, but also to select the threshold level that marks the end of the shelf life (Corradini 2018).

Loss of nutritional value in foods with an intermediate and long shelf life is normally employed as the main criteria for rejection since the degradation of labile vitamins and bioactive compounds may fall beneath the guaranteed content in a product's label before the beginning of defects in sensory quality. Then again, in exceptionally perishable food products, the end of the shelf life is usually signalled by the loss of sensory attributes or the appearance of organoleptic defects that are easily recognisable by the consumer, like bad odours or discoloration. The progressive weakening of specific sensory attributes or overall sensory acceptability is monitored using sensory analysis during storage. In this context, the survival analysis methodology is often used to evaluate the probability that a consumer will reject the product at each stage of testing. When this methodology is applied, the end of the shelf life of a product is determined as the time required to attain a consumer rejection probability level, for example, 25% or 50%. This value represents the level or amount of rejection of a product at the end of its nominal shelf life that a producer considers bearable.

2.2.2 A Broader Assessment of the End of Shelf Life: Single Versus Multiple Attributes

Due to the complex nature of foods, it is difficult to select a single limiting attribute and might not provide an accurate representation of loss of acceptability. The choice of a marker of the end of the shelf life can be worked with by comparing the performance of several attributes at the same time under hypothetical but realistic storage scenarios (Corradini 2018). Researchers have proposed and developed an interactive programme to evaluate the progression of two shelf life limiting criteria simultaneously, which can contribute to more accurate identification of limiting attributes.

The compilation of several restricting factors within a single index has also been suggested as a feasible approach to obtain a more realistic representation of the actual shelf life of a product. The use of a global stability index (GSI) can consolidate sensory, chemical, and microbiological attributes into a single index by selecting representative markers of decay and assigning weighting factors to each of them during the GSI estimation. This approach has been used to determine fresh produce and fish quality. Standard procedures to assign weight factors have not been consistently reported, and in some cases, the weights for each attribute were selected whimsically, which is a current limitation of its application. Despite its flaws, this approach has the potential to provide a more comprehensive estimation of the overall quality and shelf life of highly perishable products.

The determination of shelf life based on fingerprinting kinetics has been advanced to provide more realistic estimations of shelf life by simultaneously using multiple markers associated with deteriorative reactions. The authors propose key quality differences in food during storage, identify markers in terms of their relevance towards deteriorative processes, and use multivariate analysis to connect the selected markers to specific food reactions. This can aid in:

(a) identifying significant quality attributes by reducing the number of markers needed for effective shelf life prediction
(b) determining if a combination of attributes is enough to determine the end of shelf life
(c) potentially acquiring mechanistic insights into the general deterioration of a product.

2.2.3   Open Dating of Foods: Limitations and Relevance

Choosing a limiting quality attribute and a corresponding threshold level beyond which the product is no longer acceptable allows estimating cut-off time values, i.e., expiration dates, for a product. An expiration date depicts the time needed for the chosen quality attribute to reach its unacceptable level under specific storage conditions, which usually includes a randomly selected safety margin. It is stamped on the product's packaging to provide guidance to retailers and consumers in the form of "use by date" or "best before" labels for perishable or shelf-stable food products, respectively.

Recent surveys on temperature distributions of refrigerators and products within the food chain have shown a high prevalence of deviations from the recommended conditions, both at commercial and consumer levels. Improper settings, faulty refrigeration units, and packing of the product so that it blocks airflow and disturbs heat transfer are some of the reasons associated with the prevalence of temperatures above the recommended levels. The inherent static nature of open dating labels does not allow to factor in handling conditions that might decrease the shelf life of a product, e.g., temperature abuse. Temperature changes, or other deviations from the recommended storage conditions, may result in a discrepancy between the actual remaining shelf life and the nominal shelf life on the product's label, causing uncertainty and confusion among consumers and retailers.

In a comprehensive review on consumer perceptions and open labelling, the authors emphasise the limited understanding that consumers and also other stakeholders have of food labelling terminology and the interrelation between expiration dates, food safety, and food quality. This has led to the use of the popular rule "when in doubt, throw it out" to assist in the decision-making process, which can cause unnecessary food waste. This poses doubts about the utility of fixed expiration date labelling to assure product quality and safety. The reliability and relevance of fixed expiration date labels reduce as technological advances that can provide a real-time shelf life estimation for food batches or individual items gain popularity and/or become affordable (Corradini 2018).

3 Bio-Based Materials in Food Packaging and Their Composites

Bio-based materials often used in packaging include bio-based polymers, bio-based nanomaterials, bio-based fibres, and their composites. Bio-based polymers can be divided into four categories:

(1) biomass-derived polymers such as cellulose, hemicellulose, and chitin
(2) polymers made from biomass monomers, such as polylactic acid (PLA) and bio polyethylene (BioPE),
(3) polymers made by microorganisms, such as polyhydroxyalkanoates (PHAs) and bacterial cellulose, and
(4) biodegradable polymers made from petrochemical monomers, such as poly(caprolactone) (PCL), poly(butylene succinate-co-adipate) (PBSA), poly-butylene (PPC).

Bio-based nanomaterials primarily consist of cellulose nanocrystals, cellulose nanofibers, chitin nanocrystals, and other biomass nanomaterials. Natural fibres created by animals, plants, and geological processes are referred to as bio-based fibres. Because food packaging materials must typically meet a number of criteria, including mechanical qualities, permeability, and antibacterial capabilities, the real-world applications are numerous. Some of these bio-based materials and their packaging properties are discussed in detail below.

3.1 Bio-Based Polymers

Cellulose—The most abundant biopolymer in nature is cellulose, which is a linear semi-rigid polymer with D-glucose as the repeating unit. Because of the presence of intermolecular and intramolecular hydrogen interactions, it is difficult to dissolve cellulose directly for utilisation. In the real world, cellophane, which is created from regenerated cellulose, has been widely utilised for food packaging. Cellophane films are transparent, mechanically rigid, and have great dimension stability. They're commonly used as candy wrappers, but they're also used to package food products like cheese, cookies, coffee, and chocolates. Antimicrobial properties of cellophane integrated with nisin have also been found for meat packing. UV-shielding and antibacterial capabilities of cellulose/lignin composite transparent films have been described, with the potential to be employed as food packaging materials.

Chitin/Chitosan—After cellulose, chitin is the most abundant biopolymer on the planet. Chitosan is a cationic biopolymer made from chitin that has been deacetylated. Chitosan's structure contains amino and hydroxyl groups, allowing it to have antibacterial properties against both gram-positive and gram-negative bacteria. Chitosan films have antibacterial and antioxidant properties that make them ideal for food packaging and it has also been mixed with (1) other biopolymers, like polysaccharides and proteins, (2) some synthetic polymers, such as polyvinyl alcohol and poly lactic acid, (3) some functional extracts, like that of beeswax and honeysuckle flower extract, and (4) nanomaterials, such as metal and metal oxide nanomaterials, graphene oxide, montmorillonite, silica, CNF among others to adjust the mechanical, thermal, and barrier properties (Wang et al. 2022). Chitosan was able to be further functionalized thanks to these reactive groups, allowing for a wide range of derivatives that were found to have better food packaging capabilities. Alkyl chitosan, quaternary chitosan, and carboxymethyl chitosan are some examples of water-soluble chitosan derivatives that have been employed as efficient additions for food packaging materials with increased barrier qualities.

Lignin—Lignin is a complex polymer with a cross-linked structure found mostly in plant cell walls. Lignin includes a variety of active groups, and its antibacterial, UV-shielding, and anti- oxidation qualities have made it popular in food packaging. The antibacterial characteristics of lignin were examined using HPMC-lignin and HPMC-lignin-chitosan composite films, and it was discovered that the concentration of lignin is crucial for their antimicrobial activity. Antibacterial characteristics were obtained by combining lignin with PVA/gelatin composites (Wang et al. 2022). The hydroxyl groups of phenolic compounds in lignin interact with the cell membrane, causing cell lysis in the end. Lignin can be included in a variety of polymer matrices to improve the mechanical and wettability of the products. Various approaches have been used to use lignin as an antioxidant agent in active food packaging.

Starch—Because starch is a semi-crystalline polymer, it is not flexible enough for food packing. Chemical or physical changes, plasticization, and enzymatic treatment were utilised to enhance the flexibility of starch films. The use of citric acid and gelatin has been shown to improve the characteristics and structure of starch films. Certain starch derivatives, such as acetylated starch, were used as paper coatings and have demonstrated superior water vapour barrier and gas barrier qualities for food packaging. The sulphur hexafluoride plasma treatment was used to counteract starch's hydrophilicity, and the contact angle reached 140°. Furthermore, various additives, such as CNF, ZnO, MgO, and nanoclay, can be combined with starch to create antibacterial and UV- shielding coating materials and improve barrier properties.

3.2 Bio-Based Nanomaterials

Cellulose Nanofibres—In comparison to paper, cellulose nanofibers (CNF) were manufactured into films and their water vapour transport capabilities were examined. Soybean oil-based polymers were coated on the surface of cellulose nanofibers (CNFs) films to increase their water vapour barrier characteristics. Sol–gel technique, layer by layer assembly, electrospinning, and composite extrusion have all been utilised to increase CNF films' water vapour barrier and water resistance. To create composites for food packaging, the CNFs were mixed with polymers such as polylactic acid (PLA) and polyethylene glycol (PEG). To achieve good compatibility with PLA, cellulose nanofibers were chemically treated to add hydrophobic properties. The CNFs were coated on paper substrates and their properties, such as the release of active compounds through the nanoporous networks of CNFs, water retention value, air resistance, and tensile strength have been investigated (Wang et al. 2022).

Cellulose Nanocrystals—Cellulose nanocrystals (CNC) can be extracted from a variety of sources using a variety of techniques. The CNC was widely used as a reinforcing agent for a variety of materials, often with chemical changes to improve compatibility. To make food packaging materials, it was mixed with polyethylene, polypropylene, polylactic acid (PLA), PVA, PET, and carboxymethyl cellulose (CMC). As an example, the resultant composite films from PVA/CNC/CMC were demonstrated to be enhanced with mechanical, barrier, and thermal properties as well as better transparency for food packaging. PLA/nanocellulose composite materials may have antibacterial capabilities if they are combined with antimicrobial ingredients (Wang et al. 2022).

Bacterial Cellulose—Bacterial cellulose is a kind of cellulose manufactured by bacteria rather than plants. It has high purity, high degree of polymerization, high strength, and high water holding capacity. Bacterial cellulose functionalized with bovine lactoferrin was employed as antimicrobial edible packaging. To make food packaging, antimicrobial peptides were immobilised in bacterial cellulose. Food packaging has also benefited from the development of composite films made of bacterial cellulose and polymers. Food packaging materials made of PVA/gelatin reinforced with bacterial cellulose nanowhiskers were also studied. Their nanowhiskers composite films, which were coated with electrospun polyhydroxyalkanoates/cellulose nanowhiskers and had improved mechanical and barrier properties, were investigated for food packaging applications. Bacterial cellulose films containing ZnO nanoparticles were found to have better mechanical, barrier, and antibacterial properties. Furthermore, the release of ZnO may be controlled, making active packaging a viable option. Because essential oils function as plasticizers, blending some essential oil in bacterial cellulose films reduces tensile strength and increases elongation at break for edible films.

3.3 Bio-Based Fibres

Paper—In comparison to other synthetic polymers, paper is made from cellulose fibres and has advantages such as recyclability, biodegradability, and compostability as food packaging materials. Paper and card/paperboard packaging exist, and well-known and long-established examples of paper-based food packaging include tea bags, egg trays, and juice boxes. Softwood, hardwood, and/or agro-residual fibres derived through chemical and mechanical pulping processes are used in these products. As a result, their composition and fibre morphology are critical for the targeted end use and client demands. Tea bags, for example, have a high fibre coarseness to produce a smoother texture. Because agro-residual fibres are shorter and contain a certain proportion of non-fibrous components, they cannot completely replace wood-based fibres. Paper's hydrophilicity and barrier characteristics are inferior. Due to hygroexpansion, a high relative humidity (RH) can severely affect product qualities including compressive strength and dimensional stability. Coatings such as synthetic polymers and lamination with aluminium foil, which fall under the category of Food Contact Materials, were employed to overcome these obstacles (FCM). The use of coatings or inks as FCM in paper-based food packaging might be problematic due to the presence of dangerous chemicals that can move and contaminate the food inside. Inks, for example, were found to have genotoxic or antagonistic effects on oestrogen and androgen receptors, among other things. Furthermore, there are still questions about sustainability, recyclability, and biodegradability with these approaches. As a result, the paper surfaces were either covered with bio-based polymers, such as PLA or PHBV, or chemically grafted with proteins. Dip-coating or combining guanidine-based antimicrobial polymers with antibacterial paper has also been described. The modified beeswaxes were found to have dual roles of increasing both the antibacterial and water barrier characteristics of paper for packaging by grafting the guanidine-based polymer on the surface of beeswaxes particles (Wang et al. 2022). To increase the water vapour barrier qualities and water resistance of packaging paper, beeswaxes–chitosan emulsions and beeswax mixed emulsions were also applied on the surface of the paper. Paper using a cellulase-assisted refining process has a better barrier and grease resistance. Polysaccharides have also proven to be efficient coatings for paper packaging, owing to their wide availability and excellent film-forming capabilities. The barrier and mechanical characteristics of the coated paper improved as a result of the process. Because of its high crystallinity, high hydrophilicity, difficulty in dissolving, and poor film-forming capabilities, cellulose cannot be utilised directly as a covering for paper packaging (Wang et al. 2022). Cellulose ethers, such as carboxymethyl cellulose (CMC), methyl cellulose (MC), hydroxypropyl cellulose (HPC), and hydroxypropyl methyl cellulose (HPMC), and cellulose esters, such as cellulose acetate, can be used as food packaging materials with good mechanical, barrier, and film-forming properties. The HPC and cellulose acetate are also beneficial due to their nontoxicity, edibility and thus direct contact with food.

3.4   Other Natural Fibres and Their Composites

The utilisation of lignocellulosic wastes such as almond shell, rice husk, and seagrass to reinforce biodegradable polymers for food packaging was examined. The addition of these fibres to the PHB matrix was demonstrated to reduce thermal stability and water barrier performance. As a result of the high water permeability, the material strength is lowered, the life lifetime is shortened, and the susceptibility to microbial development is raised. In contrast to some of the other types of fibres, fibres derived from almond shells produced the most balanced material qualities. The usage of betel nut waste fibres was also examined. Their research revealed improved material qualities with up to 10% fibre content and significantly lower water vapour permeability comparable with food packaging cardboard. In a review paper, sources of lignocellulosic fibres for biocomposites were listed, including fibres from rice straw or husks, sugarcane bagasse, barley straw or husks, and maize. Adding nanoclay to lignocellulosic fibre reinforced bioplastics is one way to avoid increased water permeability. The use of nanoclay results in intercalated and exfoliated formations with improved barrier characteristics. Plant extracts, particularly those containing phenolic and flavonoid components, or the oregano essential oil mentioned above, can be added to increase stability against microbial development and, as a result, extended shelf life, food safety, and quality.

4   Smart Nanopackaging for the Enhancement of Food Shelf Life

Nanotechnology-derived food packaging materials have the biggest application in the food sector. It is fascinating to note that the use of nanomaterials has increased in the last decade to a great extent. Nanotechnology-based food packaging has been partitioned into two types, i.e. , improved packaging and active packaging. Improved packaging contains the addition of nanomaterials into polymer matrix to enhance the gas barrier properties such as polymer and clay nanocomposites, while active packaging nanomaterials can directly interact with food or environment, which enable the proper protection of the food (Rai et al. 2018).

4.1   Diversity in Nanopackaging Materials

Various types of nanopackaging materials include:

1. **Active packaging materials**: The nano-based active packaging includes the incorporation of nanomaterials in the packaging system, which can improve and maintain the quality of food. Some nanomaterials are used as additives and

can be consolidated directly into packaging material or introduced into separate sachets or embedded into food contact material. Mostly, nanomaterials or active substances used in packaging improve the internal environment of food by absorbing ethylene, oxygen, carbon dioxide, moisture, flavour, and other gaseous compounds and thereby promote food preservation.

2. **Smart packaging material**: Smart packaging materials are the substances or materials that monitor the condition of packed food or the environment surrounding the food. Smart packaging can also be defined as inexpensive labels attached to essential packaging such as pouches, trays, and bottles or to the shipping container, which can help to communicate throughout the supply chain. Engineered platinum nanoparticles can be utilised to measure the changes in pH of food packaging material. The most promising use of smart packaging was observed to be the detection of oxygen and moisture content in food.

3. **Edible coating**: Edible coating or films can be described as a thin layer of edible material present on the surface of food, which acts as a barrier to mass transfer. These edible coatings provide a barrier to oxygen, moisture, gas, etc. They are prepared from edible ingredients, that can be easily eaten and prevent pollution. Nanoparticles incorporated edible films help in improving physicochemical characteristics such as strength, mechanical properties, and flexibility. It enhances texture, colour, and flavour of food.

4. **Biodegradable packaging material**: Biodegradable polymers, due to its eco-friendly nature, can be easily decomposed by bacteria and other microorganisms. Biodegradable films are created from biopolymers such as proteins, polysaccharides, lipids, and their combination. Nanocomposites film produced from biopolymers is called bionanocomposites. It is a combination of polymer matrix with the filling of organic and inorganic material, and their dimension is less than 100 nm.

4.2   Nano Antimicrobials for Extending Shelf Life of Food

The food contaminated with microorganisms leads to its deterioration and reduces shelf life. Incorporation of antimicrobial agents into food packaging materials can inhibit the microbes and extend the shelf life of the product. Various metal and metal oxide nanoparticles are known for their antimicrobial activity and can be used in food packaging (Rai et al. 2018).

A few experiments were conducted under this topic, to get a deeper insight:

(1) Emamifar et al. (2010) proposed the formulation of low-density polyethylene (LDPE) polymer matrix incorporated with silver nanoparticles and zinc oxide nanoparticles and assessed antimicrobial activity of nanocomposite film against food spoilage microorganisms. The effect of nanocomposite film combined with nanoparticles was evaluated for the preservation of orange juice. It was discovered that LDPE film extends the shelf life of orange juice, and hence, it can be used in the preservation of fruit juice.

(2) Motlagh et al. and Yang et al. (2010) studied the impact of silver nanoparticles–LDPE packaging material for the preservation of barberries and strawberries. It was demonstrated that LDPE–silver nanoparticles film enhanced the sensory and physiological qualities of barberries and strawberries as compared to the fruits packed in normal polyethylene bags.

(3) Mahdi et al. (2012) studied the effect of silver nanoparticles–polyvinyl chloride nanopackaging on minced beef stored at 4 °C. After a week, it was found that silver nanoparticles-incorporated packaging material shows inhibitory effect. The bacterial load was very less in the meat which was coated with nano packaging material in comparison with packed with common food packaging.

(4) Metak formulated polyethylene film containing silver nanoparticles and titanium dioxide nanoparticles and looked into the effect of film on fresh apples and slices of bread, carrots, fresh orange juice, etc. It was observed that the film inhibited food spoilage microorganisms such as *Staphylococcus aureus*, *E. coli*, *L. monocytogenes*, and coliforms.

(5) Martinez-Abad et al. (2012) prepared nanocomposite film of ethylene alcohol that was incorporated with silver nanoparticles and evaluated antimicrobial activity of food in contact with fresh-cut fruits, fresh juice, and vegetables. Authors found that film inhibited the activity of food spoilage microorganisms, consequently enhancing the shelf life of the food products.

5 Future Scope

Traditional food preservation methods like physical and chemical have existed for centuries to prolong the product shelf life. However, consumers with increasing awareness regarding synthetic preservatives' health concerns are keen on adopting preservatives-free products. Industry players are keen to develop innovative food preservation solutions that extend product shelf life while maintaining their freshness.

Enlisted below are two attractive, innovative shelf life extending solutions in the market that would play an important role in the coming future of the food packaging industry:

Sani Concentrate by Parx Materials

- Broad-spectrum antimicrobial (Gram-positive and Gram-negative bacteria), antifungal, and antiviral performance (includes SARS-CoV-2/Covid-19, Influenza, Human Corona 229E, etc.) helps prevent biofilm formation.
- Safe food contact materials prolong shelf life and prevent contamination.
- Applicable for meat, poultry, and chicken.

SAF-D by Bluwrap

- Shelf life preservation and delivery service technology.
- It helps extend the shelf life of perishable proteins.
- Applicable for perishable food products like seafood, pork, and beef.

These days, the market for active, intelligent, and smart food packaging is rapidly increasing. However, such systems are still barely applied due to their high cost compared to traditional packaging materials and the limited possibility of integration within the existing packages. It is critical to make sure that intelligent packaging is sustainable in terms of application, design, and production. This is important, not only to improve the management of the food supply chain, but also to reduce global food waste and environmental pollution. Specifically, the increasing environmental burden related to petroleum-derived plastics imposes the enrichment of packaging products with eco-friendly features. However, till date, intelligent devices such as colorimetric and optical tags and especially RFID and sensors are often not yet designed with respect to recyclability and/or sustainability mechanisms, since their main purpose is to better food shelf- life by being as cheap and functional as possible. Thus, such naturally occurring polymers based on renewable resources, such as cellulose and chitosan, may represent considerable areas of innovation and exciting opportunities for the inexpensive, safe, and sustainable fabrication of eco-friendly intelligent packaging materials (Dodero et al. 2021).

Nanotechnology's usage in food science and study has progressed tremendously. Nanotechnology assists in the identification of pollutants, bacteria, pesticides and, and also the preservation of food safety by monitoring, tracing, and recording. Carbon nanotubes are gaining a lot of popularity as a means to track harmful food spoilage, microorganisms, and proteins by integrating them into packaging materials. More-over, carbon nanotubes (CNTs) have the ability to turn future FP materials into intelligent devices. Several possible hazards and toxicity issues associated with the use of nanomaterials were identified as a consequence of the never-ending studies on the topic, and these concerns should be resolved. Due to the variations in their properties over time, the effect of these nanoscale particles on the atmosphere, live-stock, and humans is uncertain. Such nanoparticles may also pass across biological walls, e.g., the blood– brain barrier, to reach different cells and organs (Hutapea et al. 2021).

6 Conclusion

Bio-based materials, such as bio-based polymers, bio-based nanomaterials, bio-based fibres, and bio-based composites, have been shown to be effective food preservation technologies. Coatings, films, hydrogels, aerogels, and emulsions made from bio-based materials were used in food packaging. These bio-based materials' mechanical qualities, gas barrier performance, and antibacterial capabilities were all studied in relation to their food packaging applications. The performance of active packaging and smart packaging, as well as the characteristics of these bio-based materials, were also concluded. However, as globalisation drives up demand for food preservation and transportation, further research into bio-based materials for food packaging will need to improve their performance even more. Despite the fact that bio-based mate-rials for packaged foods have made significant development, there are still several barriers to their commercialization. One of the most important considerations is how to reduce the cost and energy required for the production and processing of these bio-based materials, which includes biopolymer extraction, isolation of bio-based nanomaterials and natural fibres, and the preparation of bio-based materials for food packaging. Commercial processes and equipment that are more suited to the large-scale manufacture of bio-based products are in high demand. However, from the technical point of view, these bio-based materials show a promising future for the food packaging industry and also for prolonging the shelf life of the products (Wang et al. 2022).

Previous studies' results also suggest that nanotechnology is becoming increasingly essential and is now being used in the food industry. Nanotechnology has significant prospects for novel food packaging advances that will benefit both consumers and business. Nanoparticles have been shown to exhibit broad-spectrum antibacterial action, and as a result, they can be utilised with food packaging material to limit the growth of common food contaminating microbes, extending the shelf life of food.

References

Corradini M (2018) Shelf life of food products: from open labeling to real-timemeasurements. Ann Rev Food Sci Technol, 9.https://doi.org/10.1146/annurev-food-030117-012433

Dodero A, Escher A, Bertucci S, Castellano M, Lova P (2021) Intelligent packaging for real-time monitoring of food-quality: current and future developments. Appl Sci 11:3532. https://doi.org/10.3390/app11083532

Emamifar A, Kadivar M, Shahedi M, Solaimanianzad S (2010) Evaluation of nanocomposite packaging containing Ag and ZnO on the shelf life of fresh orange juice. Innovative Food Sci Emerg Technol 11(4):742–748. ISSN 1466-8564

Hutapea S, Ghazi Al-Shawi S, Chen T-C, You X, Dmitry Bokov WalidKamal Abdelbasset, Suksatan W (2021) Study on food preservation materials based on nano-particle reagents. 20 Sept 2021.https://doi.org/10.1590/fst.39721

Martínez-Abad A, Berglund J, Toriz G, Gatenholm P, Henriksson G, Lindström M, Wohlert J, Vilaplana F (2017) Regular motifs in xylan modulate molecular flexibility and interactions with cellulose surfaces. Plant Physiol 175(4):1579–1592. https://doi.org/10.1104/pp.17.01184

Peelman N, Ragaert P, Verguldt E, Devlieghere F, Meulenaer B (2016) Applicability of biobased packaging materials for long shelf-life food products. Packaging Res, 1. https://doi.org/10.1515/pacres-2016-0002

Rai M, Ingle A, Gupta I, Pandit R, Paralikar P, GadeA, Chaud M, Santos C (2018) Smart nanopackaging for the enhancement of food shelf life. Environ Chem Lett, 17.https://doi.org/10.1007/s10311-018-0794-8

Sarkar S, Aparna K (2020) Food packaging and storage. https://doi.org/10.22271/ed.book.959

Smith DA, Stratton JE (2007) Food preservation, safety and shelf life extension. https://extension publications.unl.edu/assets/html/g1816/build/g1816.htm

Wang J, Euring M, Ostendorf K, Zhang K (2022) Biobased materials for food packaging. J. Bioresources Bioproducts 7(1):1–13, ISSN 2369–9698. https://doi.org/10.1016/j.jobab.2021.11.004

Yang S et al. (2010) Paradigm for industrial strain improvement identifies sodium acetate tolerance loci in Zymomonas mobilis and Saccharomyces cerevisiae. Proc Natl Acad Sci USA 107(23):10395–10400

Bio Polymers and Sensors Used in Food Packaging—Present and Future Prospects

Neeta Shivakumar, Sinchana Raj, Shahbaaz Ahmed, and M. Rajeswari

1 Introduction

Globalization, Urbanization and dynamism indicate an increase in demand for healthy, organic, fresh ingredients and food. There is a growth in awareness of health, safety and sustainability of the environment (Dobrucka 2013). Food is our most basic necessity, its processing, packaging, production and Transport involves a lot of intricacies with respect to a lot of aspects and we need to check a lot of criteria in terms of factors of health, (Fuertes et al. 2016) economy and environment. The safety of foods is the main objective of food law. There should be quality control in food manufacturing which can be achieved from technology, chemical composition, physical and sensory features of the produce and the microbiological safety and nutritional value (Lim et al. 2014). Protection, marketing, containment (means ensuring accurate proportions of ingredients to avoid spills), safety, ergonomics (is related to the minimisation of physical effort for transportation, storage, usage and disposal of containers) (Azzi et al. 2012) and communication with the user (this function is regulated by the law. The information in the communication must include details like weight, nutrition value, origin, mode of transport, ingredients, precautions, recycling and disposal) play an important role in packaging. One way to handle the problems of food safety and food waste is by selecting and developing

N. Shivakumar (✉) · M. Rajeswari
Department of Biotechnology, R.V. College of Engineering, Bengaluru, India
e-mail: neeta@rvce.edu.in

M. Rajeswari
e-mail: rajeswarim@rvce.edu.in

S. Raj · S. Ahmed
Department of Computer Science and Engineering, R.V. College of Engineering, Bengaluru, India
e-mail: sinchanaraj.cs19@rvce.edu.in

S. Ahmed
e-mail: shahbaazahmed.cs18@rvce.edu.in

© The Author(s), under exclusive license to Springer Nature Singapore Pte Ltd. 2023
A. K. Mishra and C. M. Hussain (eds.), *Biobased Materials*,
https://doi.org/10.1007/978-981-19-6024-6_10

proper food packaging. The main functions of food packages have always been to maintain hygiene, protect food during transportation and storage and increase the shelf life of the food. With the advent of technology, there are smart features introduced in packaging like environment-friendly packaging materials with enhanced properties that are also biodegradable, and smart sensors that have the ability to monitor the quality of food. Several such examples include Time temperature indicators, nitrogen sensors detecting food decay, fruit ripening indicators, CO_2 and O_2 monitoring sensors (Fuertes et al. 2016), pathogen sensors and a lot of such smart features.

The most used materials in food packaging are plastics which consist of PP, LDPE, LLDPE, PET, etc. Low cost, strength, light weight, stability, flexibility, ability to getconverted from thin films to rigid containers, impermeability to gasses and numerous solvents and sterilization without affecting food quality are some of the reasons for the success of plastics as the materials for food packaging.

Plastics have innumerable benefits but are equally problematic for both the environment and the human body. The annual production of plastics is around 350 million tons of which only 1% is bio-based and the rest is fossil-derived having a large carbon footprint. In addition to it, plastic has adverse effects on the environment causing plastic pollution where polymers don't degrade and break down into smaller bits which end up in the air particles, soil and water. Plastic also contributes to global warming and also poses a threat to flora and fauna (Halonen et al. n.d).

The solution to deal with harmful plastics is to introduce more sustainable bio-based materials that can enhance the environment, human health and economy. The requirement here is not just bio-based materials but smart bio-based food packaging materials which monitor the physical, chemical and biological conditions of food (Kerry et al. n.d).

The smart can be brought in through sensors. Sensors are components composed of chemical and biological receptors that are designed for the recognition of a target analyte and a transducer which converts the process of recognition into a measurable electrical signal which generates a qualitative and quantitative output. Biosensors rely more on the properties of the biomolecule analyte like enzymes, aptamers and antibodies. These biosensors are integrated with a transducer to increase the selectivity of the targeted analyte. Hence, these sensors enhance the shelf life and measurement capabilities of food products (Mannelli et al. 2003).

2 Biopolymers in Food Packaging

Biopolymers are made up of a family of materials with very different features, properties and areas of and applications. Bio-based polymers are polymers that are produced from renewable resources. These polymers include both natural and synthetic polymers synthesized from monomers that are obtained from biological sources. There are many ways of the production of biopolymers. Some of them are (Torres-Giner n.d):

(1) Direct extraction of biopolymers like starch and cellulose with further molding to make thermoplastic starch polymers or using additional functionalization as in acetylation, carboxymethylation and phosphorylation to produce CA, carboxymethyl cellulose and cellulose diphenyl-phosphate which are then polymerized further or used as additives in those polymers.
(2) Hydrolysis of sugars followed by bacterial synthesis of polyesters, examples being PHAs also including PHB.
(3) Conversion to sugars which undergo the process of fermentation to lactic acid which is followed by the direct polycondensation or ring-opening condensation of lactide to PLA.
(4) Chemical conversion to monomers which is followed by polymerization examples being amino acids which are obtained by hydrolysis are polymerized with esters of lactonized unsaturated fatty acids in PEA synthesis.

Bio-based polymers can be biodegradable but sometimes not all biodegradable polymers are bio-based. Synthetic biodegradable polymers could be developed from bio-based monomers in the near future. Biopolymers are advantageous in terms of the conservation of fossil resources by the use of biomass which is regenerating (Ashter 2016).

2.1 Food Packaging Materials Based on Biopolymers

Since there is increasing awareness about going environmentally friendly and sustainable, there have been developments in the production of different kinds of biopolymeric packaging materials. Like conventional plastics, biopolymeric packaging also provides physical protection during transportation and storage and adequate physical and chemical conditions for quality, safety and shelf life of food (Rhim et al. 2013).

2.1.1 Polylactide

PLA in recent times is one of the most researched biopolymers and also a commercialized polymer seen as a fitting substitute for the conventional polymers used as packaging materials as it's compostable and bio-based (Rhim et al. 2013). It shares the same properties as conventional polymers like Polycarbonate, PET and PS. The outstanding properties of PLA are good transparency, rigidity, printability, heat sealability and melting processability. Injection molding, blow molding, thermoforming and extrusion are some of the large-scale production methods. There are some properties limiting it like low glass temperature, which hinders the utilization over 55 °C, and also has low ductility and tensile strength. The limitations can be improved by varying the ratio of the isomers, modifying its stereochemistry and enhancing its mechanical and thermochemical properties by mixing with other polymers and fillers (Babu et al. 2013).

PLA is efficient for fresh products which don't require protection against oxygen. It can be used in bottles, candy wraps, cups and food trays. Also, its extreme permeation to water makes it ideal for packaging applications like increasing the shelf life of fresh fruits and bread. It is also used as coating which acts as a barrier to reduce the biopolymer permeability (Relinque et al. 2019).

2.1.2 Polyhydroxyalkanoates

PHA's currently are one of the most important alternatives to fossil-derived polymers having the most potential to substitute polyolefins in the packaging industry. Factors like its biocompatibility and physical properties aid in making it a potential substitute. PHA's come from a family of biopolyesters which are produced by a large range of microorganisms as carbon storage material. It has similar properties of biodegradability, mechanical strength (Reddy et al. 2003), thermoplasticity and water resistance like PP and PS. But the production is highly expensive due to the high costs of downstream process and fermentation. The common methods of producing PHA's are injection molding, extrusion, thermoforming, film blowing, etc. (Keshavarz and Roy 2010). PHA's are ideal for different areas of food packaging like blow molded bottles, containers, milk cartons, etc. (Khosravi-Darani and Bucci 2015). The use of nanofillers like antimicrobial and antioxidant substances imbibed in a PHA-based material can enhance the shelf life of the packaged food (Requena et al. 2017) and increase the protection and sensory features.

2.1.3 Bio-Based Polyethylene

Also called 'Microbial' or green polyethylene is produced by the catalytic dehydration of bioethanol which is obtained by microbial fermentation which is followed by normal polymerization to produce polyethylene. But Bio-PE is nonbiodegradable and shares the same properties as that of conventional PE. Currently, bio-PE is produced on an industrial scale from bioethanol which is derived from renewable feedstocks like beet and sugarcane, maize, wood, wheat, corn and such plant wastes through biological fermentation process and microbial stain (Chen et al. 2007; Torres-Giner et al. 2017; Braskem. 2014). Bio-PE has the potential to replace all the packaging applications involving conventional fossil-derived PE due to its good lifetime performance, recyclability and low price (LyondellBasell. 2019). Current applications are found in yogurt cups by Danone, fruit juice bottles by Odwalla and plastic caps and closures for paperboard cartons by Tetra Pak (Robertson 2015).

2.1.4 Cellulose and Derivatives

Cellulose is one of the most abundant polymers on earth. Wood is a very major source of cellulose. It's a naturally occurring linear polymer made of 1,4-linked-β-d-anhydro glucopyranose units that are covalently linked via acetal functions between the equatorial –OH group of C4 and C1 carbon atom (Dong et al. 2019). Since neat cellulose is unsuitable for the production of films due to its high crystalline nature and insoluble in water due to strong intramolecular and intermolecular hydrogen bonding between the individual chains, it's dissolved in a mixture of NaOH and CS2 and recast into H_2SO_4. This treatment results in the manufacture of cellophane film with good mechanical properties. It is usually coated with nitrocellulose wax or PVDC to enhance moisture sensitiveness. This is later used for fresh products, baked foods, processed meat, candy and cheese (Petersen et al. 1999). Though the gas and moisture barrier properties of cellulose acetate are not ideal for the packaging of foods, the film is really good for products that require high moisture as it allows respiration and reduces fogging (Stein et al. 2010).

2.1.5 Thermoplastic Starch

Starch is a naturally occurring polysaccharide that is obtained by a number of crops like cassava and corn. It is considered as one of the most potential biopolymers which can be used for food packaging due to its properties of availability, biodegradability and low cost (Chan et al. 2018). Starch is made of a mixture of two polymers: amylose and amylopectin. Amylose is a linear polysaccharide composed entirely of d-glucose units joined by α-1,4-glycosidic linkages. Amylopectin is a branched-chain polysaccharide made of glucose units linked primarily by α-1,4-glycosidic bonds but with occasional α-1,6-glycosidic bonds (Zhu 2015). Amylopectin is responsible for native dry starch which makes it limited because of its high brittleness, high viscosity, poor melt processability, retrogradation and insolubility in cold water. Films developed from native starch show moderate oxygen barrier properties but poor moisture barrier. TPS can be blended with other polymers and fillers to enhance the mechanical properties and attained water resistance (Kaseem et al. 2012). TPS blends show chemical resistance, low cost and enhanced flexibility and toughness (Wang et al. 2009). The use of TPS in biopolymer blends is particularly relevant to obtain materials with high elongation at break properties in food packaging (Liu et al. 2009).

2.1.6 Proteins

Protein-based films have become a new research topic because of their cohesiveness, low cost, biodegradability features and film-forming capacity. Proteins often act as a good barrier against oxygen and other gasses and smell. They also show high vapor permeability due to their hydrophilic nature (Andrade et al. 2015; Azevedo et al. 2015). Protein films have been developed from various other substances like gelatin (Gelatin is a water-soluble protein that is prepared by thermal denaturation of collagen in the presence of dilute acid (gelatin type A) or alkali (gelatin type B). Collagen is found in animal skins and bones such as connective tissues, skin and bones), soy protein (Soy protein is an inexpensive renewable resource, sustainable, abundant, and functional, constituted by different globulins with mainly polar amino acids also including acidic and basic amino acids and nonpolar amino acids fractions such as 2S, 7S, 11S, and 15S.), milk protein (Biodegradable films can also be formed from milk proteins), casein(comprising 80% of total milk protein, consists of three main components, α, β, and γ with Mws in the 19–25 kDa range) (Ahmad et al. 2012; Martucci and Ruseckaite 2010; Tongnuanchan et al. 2014), etc.

3   Present Status of Active and Functional Food Packaging

Normal packaging methods have been long used not only to support product handling, but also to conserve the nutrition value, decrease food spoilage and elongate the shelf life of the food. In recent times, efforts are made to design and develop smart packaging (Mustafa and Andreescu 2018) which has the ability to provide features like detection of spoilage and communication to the consumer about the said spoilage when it occurs along with preservation of the product. Some examples of active packaging are sachets and pads that absorb and emit. The absorbing sachets contain O_2 scavengers (Busolo and Lagaron 2012) to decrease fat oxidation, ethylene scavengers to reduce the ripening of fruits and vegetables, odor and humidity absorbers and antimicrobial growth obstructors. Some emitting sachets are antimicrobial growth inhibitors in meat and antioxidant releasers to decrease the oxidation of oil and fat.

Smartness in packaging can involve sensing platforms to give information about food quality throughout the entire food chain like composition, bacterial growth and storage facilities and conditions. Some present systems include temperature indicators, Integrity and freshness indicators and Radio frequency Identification (RFID) (Zabala et al. 2015).

Time and Temperature Indicators

TTI's have been the most widely used smart application in recent times due to their simplicity, efficiency, low cost and affordability (Dobrucka 2013). They are used to monitor the quality of food and inform the status of food to the consumer. There are different types of TTI's developed on the enzymatic base, polymeric and biological reactions (Kim et al. 2016). It is significant to monitor the changes in the temperature and time parameters in order to ensure the quality and safety of food products that have certain temperature constraints such as chilled or frozen food, where the cold control point is critical during the time of transport and distribution (Salgado et al. n.d).

Antimicrobials and Antioxidants

An increase in microbial growth increases the risk of foodborne diseases, accelerates the change in color, smell and texture of food, thus resulting in a shorter shelf life of food. That's why there is a requirement for the (Lastname et al. n.d) development of antimicrobial packaging. They can be classified as either those that contain antimicrobial agents which migrate to the food surface or those that work against surface microbes. Some substances have been added to biopolymers for food packaging for activation of antimicrobial properties. Those are organic salts and acids like sorbic, propionic and ascorbic acids (Lastname et al. n.d), which are strong antimicrobial agents as they have the power to alter membrane transport and permeability and reduce intracellular pH. Ascorbic, citric, lactic, tartaric, gallic and vanillic acids are also used as preservative agents. Some polymers like chitosan are inherently antimicrobial, and had been used as films and coatings. Ethanol is also used as an antimicrobial agent, particularly effective against fungi although it can also inhibit yeast and bacterial growth (Lastname et al. n.d).

Modified Atmosphere Packaging

In MAP, there is an alteration of the gas atmosphere in the food package to preserve the quality of food and elongate the shelf life (Lastname et al. n.d). Thereby increasing the conditions of fresh foods and can be applied to perishable food products as well. MAP reduces microbial growth and odor development, and delays ripening. O_2, N_2 and CO_2 are some of the most important gasses used in MAP. According to numerous studies, there has been successful incorporation of MAP in fresh fruits and vegetables, meat and fish. It was reported that packaged carrots showed enhanced properties for the combination of chitosan coating and MAP. There are also various applications of various polysaccharides in MAP systems (Lastname et al. n.d), such as alginate, cellulose, chitosan, gellan gum, pectin, carrageenan and xanthan. MAP, when coupled with natural or chemical biopolymers, additives and technologies like nanotechnology, increases its potential uses.

Oxygen Scavengers

Oxygen increases the oxidative spoilage of food, nasty smells, loss of aroma and flavor, growth of aerobic microbes and nutritional value of food products by decreasing the general stability and shelf life of the food product (Lastname et al. n.d), thus it's necessary to control the levels of oxygen into packages to reduce different spoilage reaction rates in food. Oxygen scavengers are developed to oxidize quickly to reduce the level of free oxygen in hermetically packaged food systems and to decrease oxygen permeation rates through the walls of the container. This can be integrated as a part of multilayer films sandwiched between inert layers or can be coated as a surface of the packaging material. This is how oxygen scavengers based on biopolymers were designed. Some of them include protein edible films (Realini and Marcos n.d) added with ascorbic acid, ascorbyl palmitate-β-cyclodextrin inclusion complexes, cast excluded PLA films with α-tocopherol microparticles. These systems have a better consumer acceptance than the traditional sachets or packets used in packages.

Packages with Self Heating and Self Cooling

These systems are oriented toward ready-to-eat foods and don't aim to extend the shelf life of packaged foods (Lastname et al. n.d). With the help of self heating packages, food can be heated without external sources of energy using calcium or magnesium dioxide and water which produces an exothermic reaction. They are in use in coffee packets, military portions and so on. Likewise, self cooling packages involve endothermic reactions (Wu et al. 2011).

4 Contribution of Nanotechnology to the Monitoring of Food Security

Nanotechnology is the study, design, synthesis, creation, manipulation and application of materials, devices whose dimensions are in nanoscale (Neethirajan and Jayas 2011). The future of packaging is dependent on the development of nanomaterials and nanoparticles like fullerenes, nanofibres and nanocylinders. The electronic and optical features of nanomaterials enable the development of a new generation of smart devices. Nanotechnology can be exploited as an interdisciplinary powerful tool for developing intelligent packaging technology. For the designing and development of intelligent packaging, it is important for the integration and the technological advancement of the sensors, nanosensors and indicators. A sensor/nanosensor computes only some of the aspects, while an indicator integrates both the measurement and display of the particular matter in the study. The sensors and nanosensor must be connected to a device for signal transduction (Brody et al. 2011) of the receptor, whereas an indicator directly provides the qualitative or semiquantitative information of the quality for a visible change. Nanotechnology facilitates the application of nanosensors in food packaging to control the quality during various stages of food processing and packaging and to ensure the product quality to the consumer. Nanotechnology coupled with intelligent packaging can (Liu et al. 2008) provide authentication, tracking and locating product features to avoid adulteration, falsification and prevention in the diversity of the food products intended for a particular market. Still, there are many concerns like the uncertainty of the nanoparticle behavior in the body and the toxicity toward humans. So it's necessary to impose a set of rules and regulations on the security of food of IP implications.

4.1 Nanosensors

Sensors and nanosensors are used generally in food packaging to control the internal and external conditions of food. From a microbiological perspective, the main purpose of the nanosensor is to reduce the pathogen's detection time from days to hours to minutes (Zhang et al. 2010). The nanosensors are used in the detection of molecule gasses, microorganisms and detection by surface enhanced Raman Spectroscopy. Some examples include nanosensors in raw bacon packaging for detecting oxygen, electronic tongue which holds an array of nanosensors sensitive to gasses released by spoiled foods giving a result if the food is fresh or not, use of fluorescent nanoparticles for detecting the toxins and pathogens in food and crops which is based on the functionalized quantum dots coupled with immunomagnetic separation in apple juice and milk, nanosensors used to detect the temperature changes, humidity, escherichia coli (Horner et al. 2006) in food sample by measuring the scattering of light by cellular mitochondria, organophosphate pesticide residues in food, biosensors for detecting Salmonella, CO_2, pathogen in food, Bacillus cereus. Research and development in nanotechnology and nanosensors have led to the scientific advances that facilitate a new generation of nanosensors and nanomachines (Prescott et al. 2013).

4.2 Biosensors in Food Analysis

Smart food packaging is a growing field in the research community. The use of bionanocomposites in smart food packaging materials can provide the promising benefit of mechanical, thermal and gas barrier properties. Bionanocomposites with antimicrobial nanoparticles can inhibit the growth of microorganisms and improve the mechanical, thermal and electrical properties of packaging materials (Mustafa and Andreescu 2018). Although the advantage of bionanocomposites has been described for food packaging, a range of factors is needed to be considered when developing commercial smart food packaging, such as product price, health responsiveness and diffusion properties of bionanocomposites from packaging materials to foods.

The biosensor is a very advanced technology in recent years and is considered a very effective tool for developing smart food packaging. It is a quick, accurate and dependable method, but the application of biosensors with packaging materials has some limitations. It is required to consider the sensitivity, microstructure, specificity and the limit of detection of biosensors for developing smart food packaging. Biomolecular integration such as enzymes, immunosystems, tissues, organelles, or whole cells, with a variety of transduction methods, such as thermal, electrical or optical signals, has enabled the development of a wide variety of biosensing devices. The development of biosensors is described with examples for detecting pathogens, allergens and other toxicants such as pesticides and mycotoxins. In this section, we briefly highlight the application of biosensors for allergens, toxicants and pathogen detection.

4.2.1 Biosensors for Food-Allergen Detection

Allergens present in food products like milk, eggs, seafood (crustaceans), nuts (peanuts), soybeans, gluten and so on have become an increased concern of safety (Mustafa and Andreescu 2018). At least about 10% of preschool children are undergoing clinical food allergies (Yman et al. 2006). There have been developments in a variety of immune-based and DNA sensors fabricated for allergen detection. A sensor has been developed by NIMA company for the detection of peanut allergens in ppm by incorporating antibody-based detection and magnetite beads. (Arah1) peanut allergens were detected by surface plasmon resonance immune-based biosensors in chocolate candy bars (Moura et al. 2012). A colorimetric, silicon-based, optical thin-film biosensor chip with PCR amplification was developed which had the ability to identify 8 food allergens found in wheat, peanuts, cashews, soybeans, chicken, beef, fish and shrimp all at a time. A label-free immunological electrochemical sensor for β-lactoglobulin, an allergen usually found in milk, was developed with a LOD of 0.85 pg/mL.

4.2.2 Biosensors for Bacterial Pathogens Detection

Rapid detection of pathogenic bacteria plays a very significant role in food analysis. The main methods of detection of pathogens are based on Polymerase Chain Reaction (PCR) and plate counting which needs a long analysis time and sample enrichment. Biosensors can be represented as an alternative for pathogen detection as they are portable and have a potential for onsite detection. Most of the biosensors used for bacterial pathogens detection are those based on immune and DNA recognition, but these need extensive involve labeling, preparation procedures, multiple washing steps and specialized facilities (Mustafa and Andreescu 2018). As an alternative, synthetic antimicrobial peptides have been proposed as recognition agents, which enable the detection and quantification of four bacterial strains: *Escherichia coli*, (Ping et al. 2012) *Pseudomonas aeruginosa, Staphylococcus aureus* and *Staphylococcus epidermidis*. Colorimetric biosensor strips fabricated with peptides immobilized on a gold chip were also used for detecting *Listeria monocytogenes* in milk and meat samples.

Very recently, a fluorescent DNAzyme probe that binds *E. coli* was developed and printed on a cyclo-olefin polymer transparent package.

4.2.3 Biosensors for Food Adulteration, Authenticity and Toxicity Assessment

Food adulteration is based on the change of the composition of food for the purpose of financial profit regardless of the concern for safety. The adulteration of food is becoming an increasing concern creating a necessity for reliable analysis methods. A serious issue of safety is the adulteration of milk products with melamine due to its property to improve the level of protein content as determined by protein quantification assay. An antibody-based biosensor that is optical was developed for melamine detection (Mustafa and Andreescu 2018). A polyclonal antibody was immobilized on the surface of an SPR biosensor, resulting in a sensitive platform measuring IC50 of 67.9 ng/mL. The sensor was designed selective for melamine but it also showed a cross reactivity with cyromazine, an insecticide that decays to form melamine (Niranjana Prabhu and Prashantha 2016). A label-free AgNP colorimetric sensor was developed to detect melamine in milk. The AGNPs produced a tinge of Yellow–red color in the presence of melamine which was a result of aggregation.

5 Future Trends

The manufacture of food packaging using biopolymers and sensors also contributes to a circular economy and a bioeconomy concept, along with the Sustainable Development Goals of the UN (Taherimehr n.d) and with the European Green Deal which is set by the European Commission. Encapsulating the pigments can immobilize the compound inside the solid particles or liquid cavities to maintain structure, stability and also to control the release of the pigment. The usage of nanoparticles, such as ZnO, as ultraviolet light blocking agents, is also a potent mechanism to protect and immobilize pigments. Nanoclays, such as montmorillonite (MMT), commonly used for reinforcement purposes, do show promising results as pigments stabilizers. The combination of two different pigments to increase the sensor's stability and the variety of color changes can be a novel strategy. By mixing alcohol-soluble pigments and water-soluble pigments, the possibilities to use as indicators are enhanced, becoming suitable for both water-soluble and fat-soluble foods. The use of bilayers is also a strategy that can be applied to help stabilize the pigments. In the biopolymers, and to test them in situ, to certain types of food, to understand its potential as smart food packaging. There is a necessity to develop bio-based sensors that can be incorporated into fruits and vegetables, since these edibles are easily perishable and present an economic and environmental loss. Biopolymer-based packaging systems that can detect leaks in packages or that can detect pathogens, toxins, allergens (Sharma et al. 2017) and other chemical markers, or that can determine how long a temperature breach has lasted across a cold chain, would be acknowledged by the food industries, once the quantity of food waste produced could be reduced and the sustainability of the chain could be improved.

The application of natural pigments and biopolymers for smart and intelligent food packaging systems still shows some constraints and a more extensive study is required to understand better how the bio-based sensors interact with the food products. Even though most of the indicators developed are not in direct contact with the food, some migration and toxicological studies are necessary. Another key point is that in spite of the innovation, design and development of new bio-based smart packaging materials, there are still some constraints and restrictions. According to the research on literature and market, no suitable solutions are currently available commercially. One main reason is the increase in the packaging cost and the fact that existing solutions are not compatible with digital platforms.

As most of the current bio-based sensors provide a visual response, this would be also interesting to develop systems that can provide different response information that can be emitted through a signal, which is readable and transmitted through app notifications or LED (light-emitting diode) messages. Extremely sensitive, low cost paper-based eco-friendly, electrical sensors could be developed to spread the use of intelligent packaging. This would contribute to minimizing food waste and providing value to byproducts from lignocellulosic industrial crops. More focused research is required in bringing more value such as manufacturing the packaging material simpler yet making it smarter, where consumers are able to assess and analyze the quality, safety, shelf life and nutritional values of the contents of packet with effective cost (Ching et al. 1993). The benefits however should not come at the cost of curing environmental issues and should be eco-friendly.

Intelligent or smart packaging doesn't tell about the safety and quality of food, instead, it tells about the environment around the package and the condition of the package. On the other hand, active packaging helps in elongating the shelf life of the food by releasing active compounds into the package. The combination of both of these technologies can be applied in the future to get a product with superior quality. The concerns of safety related to these novel packaging include migration, human ingestion and accidental leakage of active or intelligent components from packaging material. The food packaging industry should focus majorly on reducing waste and its adverse effect on the environment by 2050.

Case study in Europe

The carbon footprint of food not eaten is approximately 495 million tons of CO_2. In Europe, around 23 million tons of plastic packaging are formed annually and if the current trend continues, by 2050, this will become 1,124 million tons. By following the new packaging innovations, by 2050, 50% of food waste can be reduced. If the method of sustainable packaging is followed effectively, at the European level, a net reduction of 370 million tons of CO_2 can be expected. The future of food packaging will support more of a circular economy than a linear economy (Tharanathan 2003). Recycling is not an exclusive solution to the plastic economy. Alternative methods should also be considered. Perhaps by 2050, 50% of the European food packaging material can be made from renewable resources with the up-cycling of organic wastes and the 50% oil-based materials left will be recycled in a closed loop. By moving toward the circular economy, a profit of about 0.6 trillion can be generated annually in Europe.

A case study of India

An intelligent packaging system should be a developer in the future that mainly focuses on food security and reduces the spoilage and wastage of food products. In India, there is a very big percentage of food that is spoiled and wasted during transportation and storage of food.

Advancements in nanosensors and their application in food containers will be introduced in the future as much (Rhim et al. 2013) research is going on in this field. In a new intelligent packaging system, there should be an assistant at every level of food manufacturing and monitoring, recording, management and communication at every step of the food supply chain from harvesting from the field to the consumer's hands.

6 Conclusion

The solution to the global challenges related to food safety, carbon neutrality and renewable sources would be bio-based smart food packaging. Majorly, there are three main components of bio–based smart food packaging that have been covered in this review. They are a) the packaging materials that are responsible for providing a safe coverage for the products, b) advanced additives and coating to preserve food products and c) sensors which are renewable enabling technologies that can be able to detect food quality. The current advancement of nanotechnology has a very abundant benefit to the food industry. Implementing chemical sensors, biological sensors and indicator labels in smart and active packaging, as well as the development of methods enabling freshness investigation of food products and crops, is a growing scientific field of study. Intelligent packaging is an emerging field that is bound to grow exponentially. The help of nanosensors and biosensors can create a breakthrough in the field of intelligent packaging solutions. This new packaging system can help in the detection, tracking, recording, monitoring and communication throughout the supply chain. It is expected that smart and intelligent packaging will reach a low cost relative to the food product. Although smart and intelligent packaging is becoming more relevant and growing in the scientific field, it is important to assess the life cycle, economic and user perception analysis and socioeconomic and environmental impact of each of those solutions. Most sensors that have been reportedly developed are not yet fully commercialized. These are still in their initial design or proof of concept stage and they still need more effort to further be developed into a marketable product to meet the concerns in terms of health and regulation. Additionally, integration of materials in smart packaging and sensors should be done with total consideration of safety regulations due to the potential migration and contact with food.

References

Ahmad M, Benjakul S, Sumpavapol P, Nirmal NP (2012) Quality changes of sea bass slices wrapped with gelatin film incorporated with lemongrass essential oil. Int J Food Microbiol 155(3):171–178

Andrade R, Skurtys O, Osorio F (2015) Drop impact of gelatin coating formulated with cellulose nanofibers on banana and eggplant epicarps. LWT Food Sci Technol 61(2):422–429

Ashter SA (ed) (2016) In plastics design library, introduction to bioplastics engineering. William Andrew Publishing, pp 81–151

Azevedo VM, Silva EK, Gonçalves Pereira CF et al (2015) Whey protein isolate biodegradable films: influence of the citric acid and montmorillonite clay nanoparticles on the physical properties. Food Hydrocolloids 43:252–258

Azzi A, Battini D, Persona A, Sgarbossa F (2012) Packaging design: general framework and research agenda. Packag Technol Sci 25(8):435–456

Babu RP, O'Connor K, Seeram R (2013) Current progress on bio-based polymers and their future trends. Prog Biomater 2(1):8

Braskem (2014) I'm green polyethylene. Innovation and differentiation for your product

Brody AL, Zhuang H, Han JH (2011) Modified atmosphere packaging for fresh-cut fruits and vegetables. In: Modified atmosphere packaging for fresh-cut fruits and vegetables. Wiley, pp 1–7

Busolo MA, Lagaron JM (2012) Oxygen scavenging polyolefin nanocomposite films containing an iron modified kaolinite of interest in active food packaging applications. Innov Food Sci Emerg Technol 16:211–217. https://doi.org/10.1016/j.ifset.2012.06.008

Chan CM, Vandi L-J, Pratt S et al (2018) Composites of wood and biodegradable thermoplastics: a review. Polym Rev 58(3):444–494

Chen G, Li S, Jiao F, Yuan Q (2007) Catalytic dehydration of bioethanol to ethylene over TiO_2/γ-Al_2O_3 catalysts in microchannel reactors. Catal Today 125(1):111–119

Ching C, Kaplan D, Thomas E (eds) (1993) Biodegradable polymers and packaging. Technomic Publishing Company Inc., Lancaster

Çoban et al (2016); Noshirvani et al (2017); Ribeiro-Santos et al (2017); Sirocchi et al (2017); Mir et al (2018); Bouarab Chibane et al (2019)

De Moura MR, Mattoso LH, Zucolotto V (2012) Development of cellulose-based bactericidal nanocomposites containing silver nanoparticles and their use as active food packaging. J Food Eng 109:520–524. https://doi.org/10.1016/j.jfoodeng.2011.10.030

Dobrucka R (2013) The future of active and intelligent packaging industry. Logforum 9(2):103–110

Dong Y, Novo DC, Mosquera-Giraldo LI et al (2019) Conjugation of bile esters to cellulose by olefin cross-metathesis: a strategy for accessing complex polysaccharide structures. Carbohyd Polym 221:37–47

Fuertes et al (2016); Ghaani et al (2016); Ahmed et al (2018); Badia-Melis et al (2018); Galstyan et al (2018); Mustafa, Andreescu (2018); Yousefi et al (2019)

Fuertes G, Soto I, Carrasco R, Vargas M, Sabattin J, Lagos C (2016) Intelligent packaging systems: sensors and Nanosensors to monitor food quality and safety. J Sensors, Article ID 4046061, 8 p. https://doi.org/10.1155/2016/4046061

Gaikwad et al (2018); Dey and Neogi (2019)

Halonen N, Pálvölgyi PS, Bassani A, Fiorentini C, Nair R, Spigno G, Kordas K. https://doi.org/10.3389/fmats.2020.00082

Horner SR, Mace CR, Rothberg LJ, Miller BL (2006) A proteomic biosensor for enteropathogenic E. coli. Biosens Bioelectron 21(8):1659–1663

Johnson et al (2018); Santos et al (2018)

Kaseem M, Hamad K, Deri F (2012) Thermoplastic starch blends: a review of recent works. Polym Sci, Ser A 54(2):165–176

Kerry J, O'grady M, Hogan S (2006) Past, current and potential utilisation of active and intelligent packaging systems for meat and muscle-based products: a review. Meat Sci 74:113–130. https://doi.org/10.1016/j.meatsci.2006.04.024

Keshavarz T, Roy I (2010) Polyhydroxyalkanoates: bioplastics with a green agenda. Curr Opin Microbiol 13(3):321–326

Khosravi-Darani K, Bucci DZ (2015) Application of poly(hydroxyalkanoate) in food packaging: improvements by nanotechnology. Chem Biochem Eng Q 29(2):275–285

Kim JU, Ghafoor K, Ahn J et al (2016) Kinetic modeling and characterization of a diffusion-based time-temperature indicator (TTI) for monitoring microbial quality of non-pasteurized angelica juice. LWT—Food Sci Technol 67:143–150

Lim SAH, Antony J, Albliwi S (2014) Statistical Process Control (SPC) in the food industry—a systematic review and future research agenda. Trends Food Sci Technol 37(2):137–151

Liu S, Yuan L, Yue X, Zheng Z, Tang Z (2008) Recent advances in nanosensors for organophosphate pesticide detection. Adv Powder Technol 19(5):419–441

Liu H, Xie F, Yu L et al (2009) Thermal processing of starch-based polymers. Prog Polym Sci 34(12):1348–1368

LyondellBasell (2019) Circulen and Circulen Plus

Mannelli I, Minunni M, Tombelli S, Mascini M (2003) Quartz crystal microbalance (QCM) affinity biosensor for genetically modified organisms (GMOs) detection. Biosens Bioelectron 18:129–140. https://doi.org/10.1016/S0956-5663(02)00166-5

Martucci JF, Ruseckaite RA (2010) Biodegradable three-layer film derived from bovine gelatin. J Food Eng 99(3):377–383

Mustafa F, Andreescu S (2018) Chemical and biological sensors for food-quality monitoring and smart packaging. Foods 7(10):168. Published 2018 Oct 16. https://doi.org/10.3390/foods7100 168Rodriguez-Aguilera R, Oliveira JC (2009) Review of design engineering methods and applications of active and modified atmosphere packaging systems. Food Eng Rev 1:66–83. https://doi.org/10.1007/s12393-009-9001-9

Neethirajan S, Jayas DS (2011) Nanotechnology for the food and bioprocessing industries. Food Bioprocess Technol 4(1):39–47

Niranjana Prabhu T, Prashantha K (2016) A review on present status and future challenges of starch based polymer films and their composites in food packaging applications. Polym Compos. https://doi.org/10.1002/pc.24236

Oliveira et al (2015); Belay et al (2016); Saini et al (2017)

Petersen K, Væggemose Nielsen P, Bertelsen G et al (1999) Potential of biobased materials for food packaging. Trends Food Sci Technol 10(2):52–68

Ping H, Zhang M, Li H, Li S, Chen Q, Sun C, Zhang T (2012) Visual detection of melamine in raw milk by label-free silver nanoparticles. Food Control 23:191–197. https://doi.org/10.1016/j.foodcont.2011.07.009

Prescott SL, Pawankar R, Allen KJ, Campbell DE, Sinn JK, Fiocchi A, Ebisawa M, Sampson HA, Beyer K, Lee B-W (2013) A global survey of changing patterns of food allergy burden in children. World Allergy Organ J 6:1. https://doi.org/10.1186/1939-4551-6-21

Realini and Marcos (2014); Biji et al (2015)

Reddy CSK, Ghai R, Rashmi, Kalia VC (2003) Polyhydroxyalkanoates: an overview Bioresource Technol 87(2):137–146

Relinque JJ, de León AS, Hernández-Saz J et al (2019) Development of surface-coated polylactic acid/polyhydroxyalkanoate (PLA/PHA) nanocomposites. Polymers 11(3)

Requena R, Vargas M, Chiralt A (2017) Release kinetics of carvacrol and eugenol from poly(hydroxybutyrate-co-hydroxyvalerate) (PHBV) films for food packaging applications. Eur Polymer J 92:185–193

Rhim J-W, Park H-M, Ha C-S (2013) Bio-nanocomposites for food packaging applications. Prog Polym Sci 38(10):1629–1652

Robertson GL.(2015) Trends in food packaging. J Inst Food Sci Technol

Salgado Pablo R, Di Luciana G, Musso Yanina S, Mauri Adriana N. https://doi.org/10.3389/fsufs.2021.630393

Sharma C, Manepalli PH, Thatte A, Thomas S, Kalarikkal N, Alavi S (2017) Biodegradable starch/PVOH/Laponite RD-based Bionanocomposite films coated with graphene oxide: preparation and performance characterization for food packaging applications. Colloid Polym Sci 295:1695–1708. https://doi.org/10.1007/s00396-017-4114-9

Tharanathan RN (2003) Biodegradable films and composite coatings: past, present, and future. Trends Food Sci Technol 14:71–78. https://doi.org/10.1016/S0924-2244(02)00280-7

Torres-Giner S, Figueroa-Lopez KJ, Melendez-Rodriguez B, Prieto C, Pardo-Figuerez M, Lagaron JM, Emerging Trends in Biopolymers for Food Packaging

Tongnuanchan P, Benjakul S, Prodpran T (2014) Structural, morphological and thermal behaviour characterisations of fish gelatin film incorporated with basil and citronella essential oils as affected by surfactants. Food Hydrocolloids 41:33–43

Torres-Giner S, Torres A, Ferrándiz M et al (2017) Antimicrobial activity of metal cation-exchanged zeolites and their evaluation on injection-molded pieces of bio-based high-density polyethylene. J Food Saf 37(4):e12348

Taherimehr M, YousefniaPasha H, Tabatabaeekoloor R, Pesaranhajiabbas E, Trends and challenges of biopolymer-based nanocomposites in food packaging. https://doi.org/10.1111/1541-4337.12832

Vom Stein T, Grande P, Sibilla F et al (2010) Salt-assisted organic-acid-catalyzed depolymerization of cellulose. Green Chem 12(10):1844–1849

Wang Z-F, Peng Z, Li S-D et al (2009) The impact of esterification on the properties of starch/natural rubber composite. Compos Sci Technol 69(11):1797–1803

Wu Y, Lin Y-M, Bol AA et al (2011) High-frequency, scaled graphene transistors on diamond-like carbon. Nature 472(7341):74–78

Yman IM, Eriksson A, Johansson MA, Hellens K-E (2006) Food allergen detection with biosensor immunoassays. J AOAC Int 89:856–861

Zabala S, Castán J, Martínez C (2015) Development of a time-temperature indicator (TTI) label by rotary printing technologies. Food Control 50:57–64

Zhang H, Li Z, Wang W, Wang C, Liu L (2010) Na+-doped zinc oxide nanofiber membrane for high speed humidity sensor. J Am Ceram Soc 93(1):142–146

Zhu F (2015) Composition, structure, physicochemical properties, and modifications of cassava starch. Carbohyd Polym 122:456–480

Biofunctional Textiles: Functional Polymer-Carriers with Antiviral, Antibacterial, Antifungal, and Repellent Activity

Ma Alejandra Martinez, Liesel B. Gende, Vera A. Alvarez, and Jimena S. Gonzalez

1 Introduction

Consumer demand for biofunctional textiles has been growing in the last two decades. Biofunctional textiles are fibrous substrates that have been modified by the presence of a substance or component in order to obtain new properties and to add value; those make them more interesting and attractive to the consumer and for marketing (Bezerra et al. 2016). Biofunctional textiles have a specific function when stimulated; they are especially useful when the textile begins to have functions that allow it to interact with the skin, being suitable for use in the administration of active substances (Specos et al. 2010). In recent years, several advanced technologies such as nanotechnology have been developed, among them, microencapsulation techniques applied to textiles offer the possibility of functionalizing fibers, threads and fabrics and producing innovative products with many advantages over traditional textile products. Such advantages are associated with microencapsulation, through this technology different active compounds can be isolated, such as antimicrobials, vitamins, and essential oils, among others, in this way they can be protected from environmental factors (oxygen, temperature, humidity, and light), the release of active compounds is controlled, it can mask unwanted properties of active ingredients as well as it can convert liquid substances into solids (Valdes et al. 2018).

M. A. Martinez (✉) · V. A. Alvarez · J. S. Gonzalez
Materiales Compuestos Termoplásticos (CoMP), Instituto de Investigaciones en Ciencia y Tecnología de Materiales (INTEMA), Consejo Nacional de Investigaciones Científicas y Técnicas (CONICET), Universidad Nacional de Mar del Plata (UNMdP), Colón 10890, 7600 Mar del Plata, Argentina
e-mail: alejandra.martinez@intema.gob.ar

L. B. Gende
Instituto de Investigaciones en Producción, Sanidad y Ambiente (IIPROSAM), Facultad de Ciencias Exactas y Naturales (FCEyN), Universidad Nacional de Mar del Plata, CONICET, CP7600 Mar del Plata, Argentina

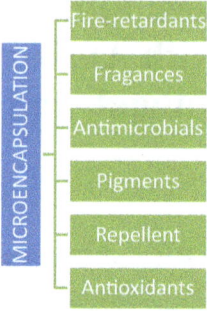

Fig. 1 Functionalities that microencapsulation of EOs can introduce on textiles

Microencapsulation technique allows to introduce new and important attributes into garments and fabrics, such as the breathability of the fabric, softness, durability, and also to incorporate new properties such as water repellency, anti-odor, UV protection, fragrances, repellent against mosquitoes, textile thermoregulation, fire retardant, antimicrobial properties, among others (Marinković et al. 2006) see Fig. 1. Microencapsulation allows a controlled release of functional agents and also protects against reactions with humidity, light and oxygen (Specos et al. 2010).

Some of the textile finishes currently on the market are potentially toxic to humans and the environment (Zheng et al. 2019), widely used in personal care, first-skin clothing and household cleaning products (Benavides Portilla 2017). This has led to the need of developing finishes that are effective, safe and biodegradable for humans and the environment finishes (Reshma et al. 2018).

The demand for aromatic and medicinal plants grows continuously as consumers are increasingly informed about the potential of their benefits (Hanif et al. 2019). Essential oils (EOs) are aromatic and volatile liquid extracts obtained from various parts of plants such as barks, seeds, flowers, husks, fruits, roots, leaves, and wood, fruits, among others (Aziz et al. 2018). As natural products, they have interesting physicochemical characteristics with high added values respecting the environment. EOs have diverse and relevant biological activities as a result of their complex chemical composition, their structures are often made up of more than 100 different terpenic and/or phenolic compounds. They are usually used in the medical field as a result of their biocidal activities (bactericidal, virucidal and fungicidal) and for their medicinal properties, encompassing additional fields such as agriculture and the environment (Reshma et al. 2018), see Fig. 2.

In this context, the microencapsulation of EOs in textiles could offer a biodegradable and ecological alternative to traditional and toxic finishes. This technique preserves the functional and physicochemical properties of the EOs, such as physical stability, protection of encapsulated material from degradation, and controlled release (Beşen 2019; Asbahani et al. 2015).

The release rate of encapsulated volatile essential oils depends on the materials used (core and shell) for the development of the microcapsules and the environmental

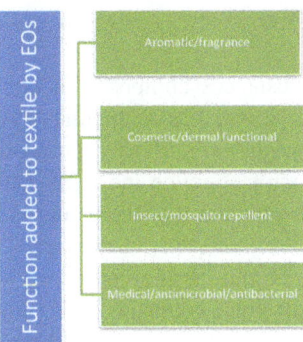

Fig. 2 Uses of EOs to add specific function to textile

conditions (temperature, pH and humidity) (Asbahani et al. 2015). Different types of biomaterials can be used to encapsulate EOs; biopolymers are the most widely used ones, due to their interesting characteristics such as high solubility in water, biodegradability and easy availability. They can be obtained from plants, animals, marine, and microbial sources (Shaaya and Rafaeli 2007), in some cases to formulate encapsulated EOs it is necessary to alter some characteristics such as porosity, polarity, stability, and permeability, among others for their processing (Asbahani et al. 2015). Several techniques have been used to encapsulate essential oils such as ionic gelation, coacervation, nano-precipitation, nano-emulsion, and drying processes (Lee et al. 2001; Shaaya and Rafaeli 2007; Tworkoski 2002). The last step is how to apply the microcapsules to the textile substrate. The microcapsules can be applied to textile fabrics by using the conventional finishing treatments such as: padding, in which the fabric is dipped in microcapsules suspension and squeezed through rollers; bath exhaustion process, where the fabric is soaked in the microcapsule suspension for a given time under controlled temperature and pH; spraying of the capsules onto the fabrics; through screen printing with an appropriate binder and thickener onto the fabrics (Marinković et al. 2006). The processes mentioned above usually require a subsequent curing step for setting. In some cases, when it is necessary to increase the affinity between the tissue and the microcapsule, an activation treatment can be performed to improve the anchoring of the microcapsules on the tissue surface (Massella et al. 2019).

The aim of this review is to describe biofunctional systems reported, formed by vehicles of natural origin that transport active compounds (natural substances) to inhibit microorganisms. In particular, essential oils, which have antibacterial, anti-fungal and antiviral activities in order to use them as micro and nano-finishes in woven and non-woven textiles oriented to the field of bios health are described here.

2 Essential Oils (EOs)

Essential oils are natural substances composed of complex mixtures of volatile compounds (Jianu 2014). These volatile compounds have diverse ecological functions, as defensive substances against microorganisms and herbivores, also important to attract insects for the dispersion of pollen and seeds (Bakkali et al. 2008). In nature, play an important role in the protection of the plants as antibacterials, antifungals, antivirals, and insecticides.

They are liquid, volatile, limpid, lipid soluble, and soluble in organic solvents that can be synthesized by all plant organs as buds, flowers, leaves, stems, twigs, seeds, fruits, roots, wood, or bark, and are stored in secretory cells, cavities, canals, epidermal cells or glandular trichomes (Bakkali et al. 2008).

They are commonly used as flavouring agents in food products, drinks, perfumeries, pharmaceuticals, and cosmetics (Burt 2004).

2.1 Extraction Methods

Essential oils are complex mixtures of low molecular compounds weight extracted by steam distillation, hydrodistillation, solvent extraction (Nakatsu et al. 2000), solvent-free microwave extraction (Okoh et al. 2010), and supercritical fluid extraction (Carvalho et al. 2020), among others of lesser importance (Fig. 3).

However, each technique has an effect on the extraction yield as well as on the oil properties. The elevated temperatures on steam and hydrodistillation can cause chemical modifications to the oil components, as loss of the most volatile molecules. When using the solvent extraction, it becomes difficult to obtain totally free of that product and there is also a loss of the highly volatile components. In contrast, extraction by supercritical fluids generally leads to high-quality and solvent-free products (Rezzoug et al. 2005).

Fig. 3 Different methods of extracting essential oils

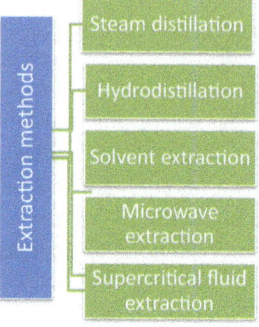

Besides, the extraction product can vary in quality and in composition according to climate, soil composition, plant organ, age, and vegetative cycle stage. Many factors including genetic variation, variety or ecotype, nutrition, application of fertilizers, geographic location, climate, seasonal variations, hydric stress, and also the type of drying and storage, affect the chemistry of Eos (Angioni et al. 2006).

2.2 Taxonomy of Essential Oil

Plants producing EOs belong to various genera distributed in 60 botanical families, such as Alliaceae, Apiaceae, Asteraceae, Lamiaceae, Myrtaceae, Poaceae, and Rutaceae (Carson and Hammer 2010). Almost all essential oils are rich in terpenic compounds, regardless of the botanical family to which the plant belongs. While plant families like Apiaceae (Umbelliferae), Lamiaceae, Myrtaceae, Piperaceae, and Rutaceae contain phenylpropanoids more frequently (Raut and Karuppayil 2014). Some examples are shown in Table 1.

Table 1 Publications related with EOs botanical origin and majority chemical composition

Botanical family	Species	Common name	Principal compound	References
Lauraceae	*Laurus nobilis* *Cinnamomum zeylanicum*	Laurel Cinnamon	1,8 cineole Cinnamaldehyde	Álvarez et al. (2019) Brenda Gende et al. (2008)
Myrtaceae	*Syzygium aromaticum* *Eucalyptus globulus*	Clove Eucalyptus	Eugenol 1,8 cineole	Maggi et al. (2011) Gende et al. (2010)
Lamiaceae	*Mentha piperita/Mentha pulegium/Mentha arvensis Lavandula angustifolia Origanum vulgare Thymus vulgaris*	Mint Lavender Oregano Thyme	Menthol/piperitone oxide Linalool Carvacrol Thymol	Brenda Gende et al. (2014) Martucci et al. (2015) Borugă et al. (2014)
Umbelliferae	*Pimpinella anisum Foeniculum vulgare*	Anise Fennel	Anethol	Gende et al. (2011)

2.3   Chemical Composition

Essential oils of plants can contain many different isomers, and it is difficult to identify these compounds accurately. Gas chromatography-mass spectrometry (GC–MS) turns out to be a useful technique for the qualitative and quantitative analysis of essential oils (Meng et al. 2021). There are mixtures which can contain various components at quite different concentrations. They can be made up of 20–100 different plant secondary metabolites belonging to a variety of chemical classes. Terpenoids and phenylpropanoids (aromatic compounds) form the major constituents of the essential oils. In addition, a few aromatic and aliphatic constituents are also present (Alonso 1998).

The components include two groups of distinct biosynthetical origin (Pichersky et al. 2006). The main group is composed of terpenes and terpenoids and the other of aromatic constituents (Fig. 4).

Thin layer chromatography (TLC), and other physicochemical properties such as density, refraction and acid indexes, and UV and IR spectra can be useful to characterize essential oils, these can be analyzed with the aim of establishing the quality, purity and chemical stability of the essential oil (Brenda Gende et al. 2008).

As mentioned above, regarding their composition, essential oils are complex mixtures of numerous molecules, and their biological effects could be the result of synergism of all of them or reflect only those of the main present at the highest levels according to gas chromatographical analysis (Bakkali et al. 2008).

According to literature, generally, the major components are found to reflect quite well the biophysical and biological features of the essential oils from which they were isolated (Ipek et al. 2005), the amplitude of their effects being just dependent on their concentration when they were tested alone or comprised in essential oils.

Various biological effects depend on the reactivity, this is related to the electronic structure and can be modeled with quantum descriptors (Netzeva et al. 2005). The quantitative structure–activity relationship (QSAR) studies a correlation between the different physicochemical parameters of the compounds studied and their biological activity (Chang et al. 2007). The data structuring proved that there exists a relationship between the molecular structures of the essential oil compounds and their biological activity (Voda et al. 2004).

2.4   Biological Activities of Eos

2.4.1   Antibacterial

Plant molecules are well known for their antimicrobial properties. Especially plant EOs have been shown to exhibit broad spectrum inhibitory activities against various Gram positive and Gram negative bacterial pathogens (Lang and Buchbauer 2012; Teixeira et al. 2013).

Terpenic compounds

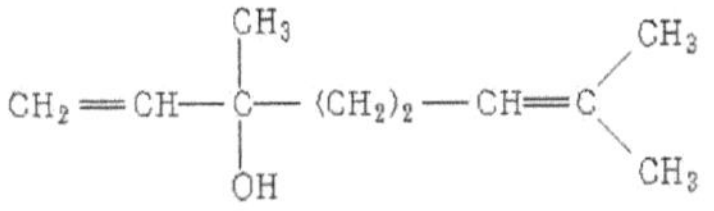

Linalool

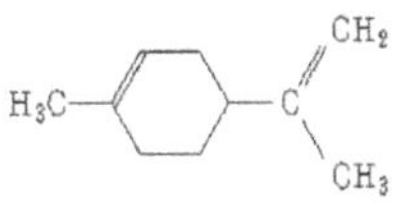

Limonene

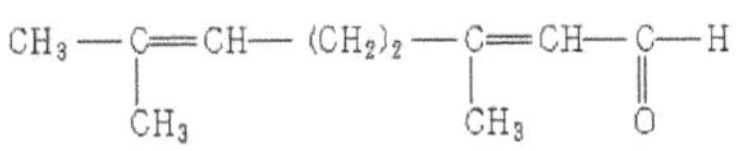

Citral

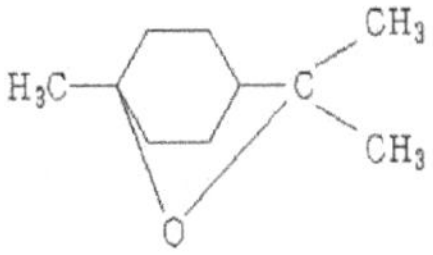

1,8 Cineol

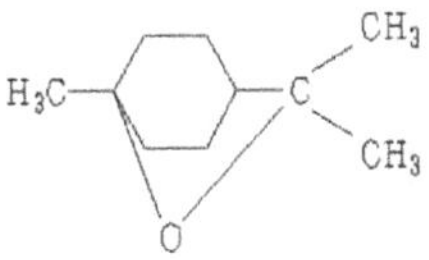

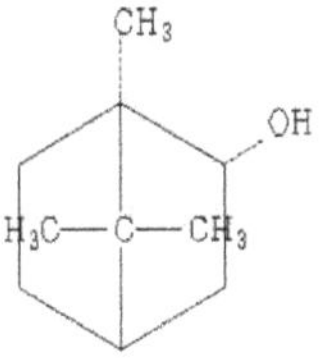

Aromatic compounds

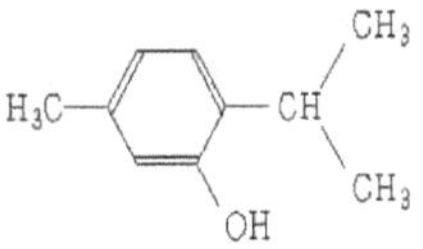

Thymol

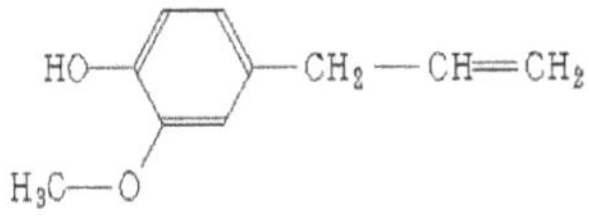

Eugenol

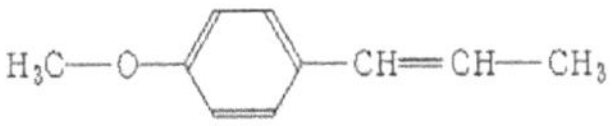

Anetol

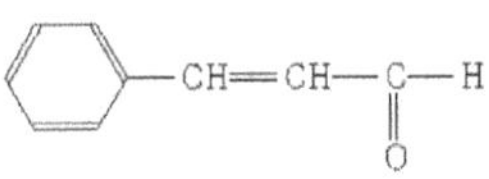

Cinnamaldehyde

Fig. 4 Chemical structures extracted from: SDBSWeb: http://riodb01.ibase.aist.go.jp/sdbs/ (National Institute of Advanced Industrial Science and Technology, date of access)

The techniques commonly used to determine antibacterial activity are serial dilution in liquid or solid media and bioautography. In the first case, by determination of the minimal inhibitory concentration (MIC) defined as the lowest concentration of the antimicrobial agent in which there is no macroscopic bacterial growth. In the case of bioautography, the technique is used to define the active constituents (Işcan et al. 2002).

It has been seen that the presence of components with phenolic structures, such as thymol, carvacrol, eugenol, and cinnamaldehyde, were greatly active against microorganisms, also, the importance of the hydroxyl group in the phenolic structure was confirmed (Dorman and Deans 2000; Sharifi-Rad et al. 2017). Activity amongst the phenols and alcohols is at least partly due to the hydroxyl group, which has intrinsic antimicrobial activity and contributes to the relatively greater solubility of these components in biological membranes. Furthermore, the ability of components to release or accept protons has been postulated as an important factor in antimicrobial activity (ben Arfa et al. 2006).

In the work of Nazzaro et al. (2017) were able to determine an important relationship between the chemical structure of the compounds and the antimicrobial activity. They found that the bioactivity of the compounds is modified by the presence and position of oxygen contained in the functional group, as well as an additional hydroxyl group. Similarly, the position of the double bond affects the antimicrobial activity, being higher for alpha pinene (endocyclic double bond) than for beta pinene (exocyclic double bond). Also, the derivatives with the amino group present greater activity, being the terpenic amines more active than the oxygenated terpenoids.

2.4.2 Antifungal

Natural products with antifungal properties are also an interesting new therapeutic alternative to synthetic drugs. Many studies have demonstrated the effectiveness of essential oils against fungal species (de Cruz et al. 2013).

The cell wall of fungi (eukaryotic organisms) may be considered as the prime target for selectively toxic antifungal agents because of its chitin structure (Nazzaro et al. 2017). EOs can be one of the most promising natural products for fungal inhibition (Hu et al. 2017; Kalemba and Kunicka 2005). In fact, many of them obtained from different plants or herbs exhibited intense antifungal properties (Bakkali et al. 2008; Lang and Buchbauer 2012).

Phytochemicals and plant essential oils possess antimicrobial activities that are able to prevent food spoilage due to fungi (e.g., *Aspergillus, Penicillium*) and intoxications (due to mycotoxins). The antifungal activity of substances such as terpenoids, polyphenols and thiols, as well as some of the mechanisms of action were observed to affect especially mycelial growth and germination (Redondo-Blanco et al. 2020).

The antifungal activity of essential oil might be caused by the properties of the compounds, that due to their highly lipophilic nature and low molecular weight are capable of disrupting the cell membrane, causing cell death or inhibiting the sporulation and germination of fungi. Although several in vitro tests indicate that terpenes/terpenoids show ineffective antimicrobial activity when used as individual compounds, compared to the complete EO (Tian et al. 2013).

2.4.3 Antioxidant

The antioxidant potential of an essential oil depends as well as on its composition. The antioxidant capacity of EOs can be revealed through their reducing power by Ferric reducing antioxidant power (FRAP) assay and DPPH radical scavenging efficiency (Martucci et al. 2015).

The greatest antioxidant power is in accordance with the presence of electron donor chemicals such as carvacrol and thymol, which can react with free radicals and turn them into more stable products, and so terminate radical chain reactions (Burt 2004). The phenolic compounds are free radical acceptors that delay or inhibit the autoxidation initiation step or interrupt the autoxidation propagation step (jmbfs Kacaniova 2022). Numerous studies have demonstrated the antioxidant properties of essential oils. Terpenoids and phenolic constituents of EOs exhibit significant antioxidant effects (Tomaino et al. 2005; Sánchez-Vioque et al. 2013).

Also, it has been shown that the free radical scavenging effect exhibit a dose-dependent in line with the results reported by jmbfs Kacaniova (2022) and, in many cases, higher concentrations of EOs not induce significant changes in the reducing power, indicating a saturation level (Martucci et al. 2015).

2.4.4 Antiviral

Essential oils are shown to possess notable antiviral properties (Raut and Karuppayil 2014). Inhibition of viral replication is believed to be due to the presence of monoterpene, sesquiterpene and phenylpropanoid constituents of EOs. Particularly, phenylpropanoids and sesquiterpenes compounds, such as trans-anethole, eugenol, β-eudesmol, farnesol, β-caryophyllene, and β-caryophyllene oxide, which are present in many essential oils, were evaluated for their antiviral activity against herpes simplex virus type 1 (HSV-1) on in vitro assays. It was observed that phenylpropanoids inhibited their infectivity by about 60–80% and sesquiterpenes suppressed virus infection by 40–98% and the essential oil of star anise inhibited the virus in a percentage >99% (Schnitzler et al. 2011). Both, star anise essential oil and all isolated compounds exhibited anti-HSV-1 activity by direct inactivation of free virus particles in viral suspension assays. It was also observed that all drugs interacted in a dose-dependent manner.

It has also been seen in Effect of Essential Oils on the Enveloped Viruses (2022) that antiviral activity was confined to the ability to interfere with viral envelope structures, so that adsorption or entry of the virus into the host cells is prevented.

Closing this section, as previously indicated, some components of essential oils are effective against a variety of microorganisms such as bacteria, fungi, viruses, as well as parasites (Elissondo et al. 2008; Albanese et al. 2022), microsporidian parasites (Porrini et al. 2017), mites (Damiani et al. 2009; Brasesco et al. 2017), among others.

Thin layer chromatography (TLC), and other physicochemical properties such as density, refraction and acid indexes, and UV and IR spectra can be useful to characterize essential oils, these can be analyzed with the aim of establishing the quality, purity and chemical stability of the essential oil (Brenda Gende et al. 2008).

As mentioned above, regarding their composition, essential oils are complex mixtures of numerous molecules, and their biological effects could be the result of synergism of all of them or reflect only those of the main present at the highest levels according to gas chromatographical analysis (Bakkali et al. 2008).

According to literature, generally, the major components are found to reflect quite well the biophysical and biological features of the essential oils from which they were isolated (Ipek et al. 2005), the amplitude of their effects being just dependent on their concentration when they were tested alone or comprised in essential oils.

Various biological effects depend on the reactivity, this is related to the electronic structure and can be modeled with quantum descriptors (Netzeva et al. 2005). The quantitative structure–activity relationship (QSAR) studies a correlation between the different physicochemical parameters of the compounds studied and their biological activity (Chang et al. 2007). The data structuring proved that there exists a relationship between the molecular structures of the essential oil compounds and their biological activity (Voda et al. 2004).

3   Components of Capsules (Encapsulated EOs)

The **shell material** of capsules can be formulated by using a wide variety of materials including natural and synthetic polymers, depending on the chemical characteristics and intended use of the core, the conditions under which the product is stored, and the processing conditions, which the microcapsules are exposed, as well as their cost and availability (Jamil et al. 2016a). The shell is the most effective approach to improve stability by forming a barrier between the active principle and the external environment. Then, the wall, shell, or encapsulating agent, of the microcapsule, in addition to the structuring function, serves to protect and isolate the compound from the external environment. It is desirable that the shell material has no reactivity with the active ingredient, be inexpensive, and show consistent properties during storage. Depending on the kind of **wall material** and its inherent characteristics, the compound is released from the wall material via various mechanisms such as swelling, dissolution, or degradation. Different materials can be selected from synthetic or natural polymers, such as waxes and lipids, proteins, carbohydrates, gums, and other polymer classes. The choice of wall materials depends upon several factors including expected product objectives and requirements; the nature of the core material; the process of encapsulation; economics and whether the coating material is appropriate for this use (Soliman et al. 2013b).

Crosslinkers are compounds applied to improve the physical properties and stability of the microcapsules and the efficiency of encapsulation (Chen et al. 2019). However, crosslinkers are also used for the connection of the microcapsules with the textile substrate. Some crosslinking agents used for microcapsules are: Tripolyphosphate glutaraldehyde, genipin, transglutaminase, tannic acid, urea, and so on, (Yoplac et al. 2021) and for the connection between the capsules and the textile, citric acid, and 1–4-butane tetracarboxylic are employed (Hanif et al. 2019; Ahn et al. 2008; Kolanowski et al. 2006).

Finally, there are **surfactants** widely used in the preparation of microcapsules that present amphiphilic behavior, which are Tween (Hanif et al. 2019; Burt 2004; Ipek et al. 2005; Nguyen et al. 2021), Span (Burt 2004; Nguyen et al. 2021; Jafari et al. 2008), polyvinyl alcohol, polyglycerol polyricinoleate, lutensol, among others. In the system, surfactants, when used, produce micelles, which are supramolecular arrangements possessing a hydrophobic central core and a hydrophilic crown. The entropy reduction with thermodynamically unfavorable interactions between the lipophilic surfactant tails and water molecules causes self-assembly in larger molecular organizations when the surfactant concentration exceeds the "critical micelle concentration (Nguyen et al. 2021). The surfactant can control the particle size and agglomeration (Bakry et al. 2016a), also it helps to reduce molecular interactions of chemical groups in the particle surface (Van der Waals, hydrogen bonding, or hydrophobic interactions) (Hosseini et al. 2013). All these are summarized in Fig. 5.

Biopolymers Used to Encapsulate Essential Oils

Different kinds of biomaterials can be used to encapsulate EOs; biopolymers are the most widely used. Polysaccharide's lipids and proteins are the most frequently used as encapsulating materials for essential oils. They have interesting characteristics that make them useful for such applications, mainly, high water solubility, biodegradability, and easy availability. They can be obtained from plants (starch, cellulose, and gluten), animals (dextrin, chitosan, and casein), and marine (alginates and xanthan) and microbial sources (Massella et al. 2019) and also, they are classified in this way (Fig. 6).

Some of these polymers can be directly used to formulate encapsulated EOs whereas others need chemical modification or processing to alter some characteristics such as porosity, polarity, stability, and permeability, among others.

The Table 2 summarized the chemical structures and some relevant characteristics of biopolymers commonly used to encapsulate EOs.

3.1 Oil Encapsulation Benefits

The selection of a polymer for a specific EO encapsulation depends on several parameters as their safety, applicability, biocompatibility, availability, and cost.

The reasons to encapsulate EOs are summarized in Fig. 7 (Okoh et al. 2010; Carvalho et al. 2020; Rezzoug et al. 2005; Angioni et al. 2006).

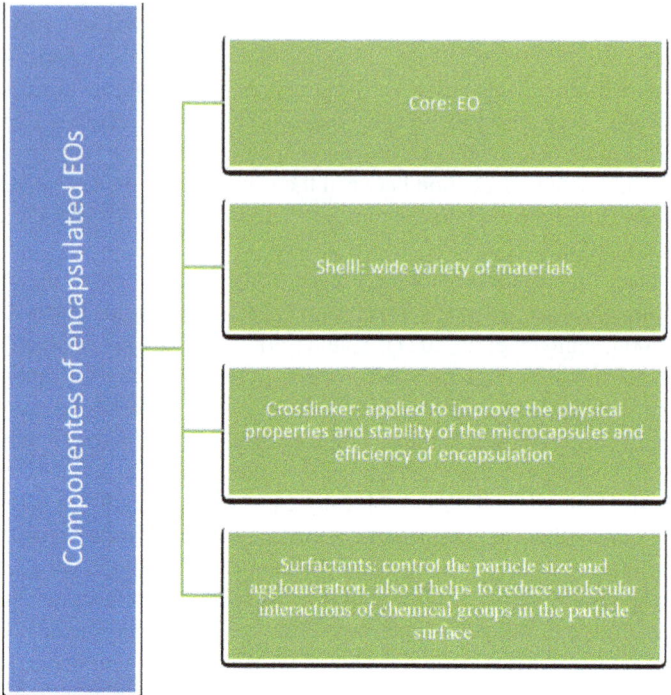

Fig. 5 Capsules components

3.2 Techniques and Processes Used to Encapsulate Essential Oils

Overall, encapsulation can be defined as a method by which a material (or a mixture of materials) called active principle (or core material) is entrapped within or coated with another material (Massella et al. 2019) known as shell (or wall) material; the combination of both is identified as encapsulate or carrier.

Several encapsulation techniques (Fig. 8) have been used to encapsulate essential oil (Lee et al. 2001; Shaaya and Rafaeli 2007; Tworkoski 2002).

3.3 Physicochemical Characterization of Encapsulated EOs

Morphological, chemical and physical characteristics of encapsulated EOs are crucial to assess the variables that affect the optimization of each process and product. Several characterization techniques are used for EOs based formulations (Fig. 9).

Microscopic techniques, like OM (Optical Microscopy), SEM (Scanning Electron Microscopy) and TEM (Transmission Electron Microscopy) are commonly used to

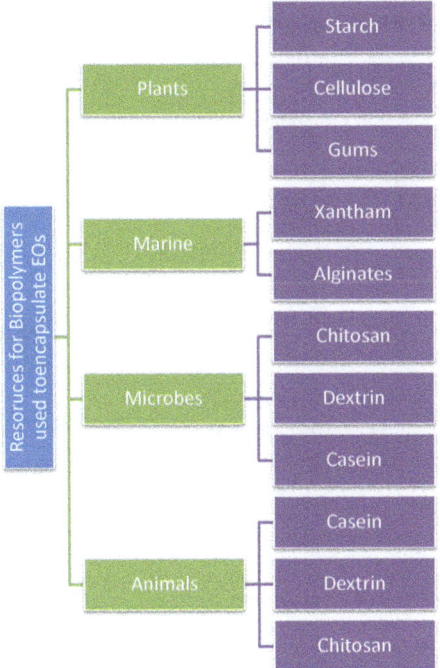

Fig. 6 Examples of biopolymers used to encapsulate EOs and its origins

analyze the morphology of encapsulated Eos (Okoh et al. 2010; Raut and Karuppayil 2014) whereas the analysis of particle size and also the surface charge are commonly carried out by dynamic laser scattering (DLS) technique with Zeta Potential analyzer (de Oliveira et al. 2014).

Thermal behavior can be used to define changes in the physical characteristics, such as melting, evaporation, sublimation, decomposition of EOs, and of the biopolymer used for encapsulation and loaded encapsulates before and after the production process. It is also useful to analyze the retention/release properties and the thermal stability of encapsulated Eos (Shaaya and Rafaeli 2007; Brenda Gende et al. 2008).

The determination of loading capacity and encapsulation efficiency is required to determine the quantity of loaded EO inside the encapsulate. The encapsulation efficiency can be defined as the ratio between the mass of EO inside the encapsulate and the initial mass of EO whereas the loading capacity can be calculated as the ratio of the weight of loaded EO and mass of prepared formulation (Liao et al. 2021).

Table 2 Chemical structures and some relevant characteristics of biopolymers commonly used to encapsulate EOs

Biopolymer	Characteristics	Chemical structure
Starch	Available polymer of plant origin. Currently, modified forms of starch are extensively used in the encapsulation. The aim of starch modification is to alter the structure and affect the hydrogen bonding in order to enhance its applicability in the encapsulation (Campos et al. 2015)	 Amylose Amylopectin
Casein	Is the chief protein in milk and the essential ingredient of cheese. In pure form, it is an amorphous white solid, tasteless and odorless. Casein has been used as an encapsulation material for different types of substances, mainly hydrophobic ingredients (Chen et al. 2019)	
Cellulose	Most abundant carbohydrate present in nature. It is insoluble in water. It is an important structural component of the primary cell wall of green plants, many forms of algae and the oomycetes. Cellulose is a semicrystalline material, has no taste, is odorless, is hydrophilic is chiral and is biodegradable. Cellulose and derivatives have been used to encapsulate several kinds of active principles (Nguyen et al. 2021)	
Chitosan	Chitin is the second most abundant polymer in nature after cellulose. It is the principal part of the exoskeleton of crustaceans such as crabs, shrimps, prawns, lobsters, and cell walls of some fungi such as *Aspergillus spp*, Zygomicetes and *Mucor spp*. Chitosan is a deacetylated derivative of chitin. Chitosan has been employed to encapsulate different Eos (Jamil et al. 2016b)	
Alginates	Anionic biopolymer is produced by marine algae and is used in the preparation of nanocapsules, beads, and nanoparticles due to its sensible biocompatibility, biodegradability, non-toxicity, mucoadhesion, gelation, and film formation properties. Alginates have been used to encapsulate EOs with antifungal and antibacterial activities (Soliman et al. 2013c)	

(continued)

Table 2 (continued)

Biopolymer	Characteristics	Chemical structure
Dextrin	Maltodextrins, cyclodextrins, and dextrin whites (DE value 20–100) are the major hydrolysates derived using acid treatment or enzyme modification of the starch and have been widely used for the encapsulation of EOs and other flavouring agents (Yoplac et al. 2021)	

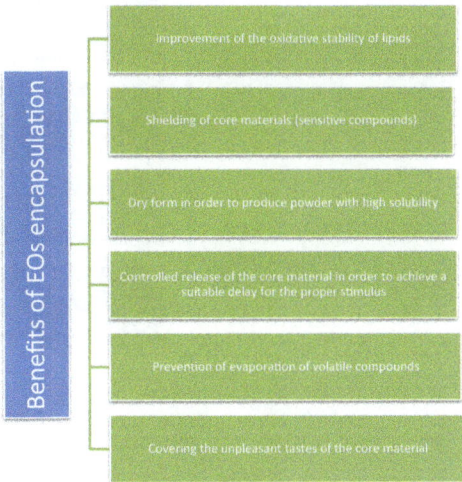

Fig. 7 Advantages of encapsulation of EOs

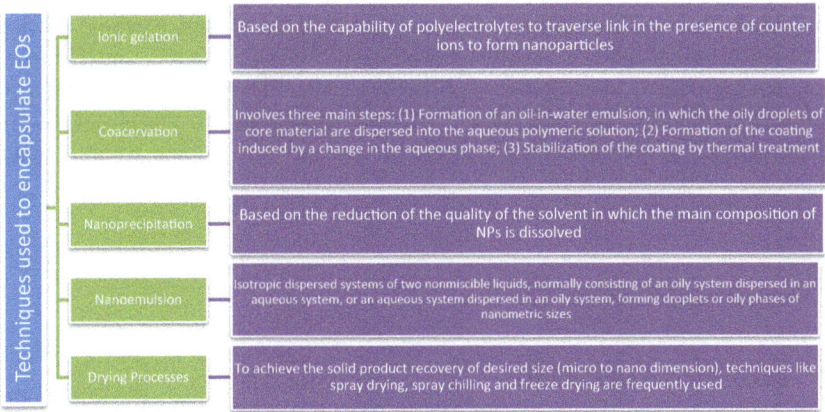

Fig. 8 Different techniques use for encapsulated EOs in biopolymers

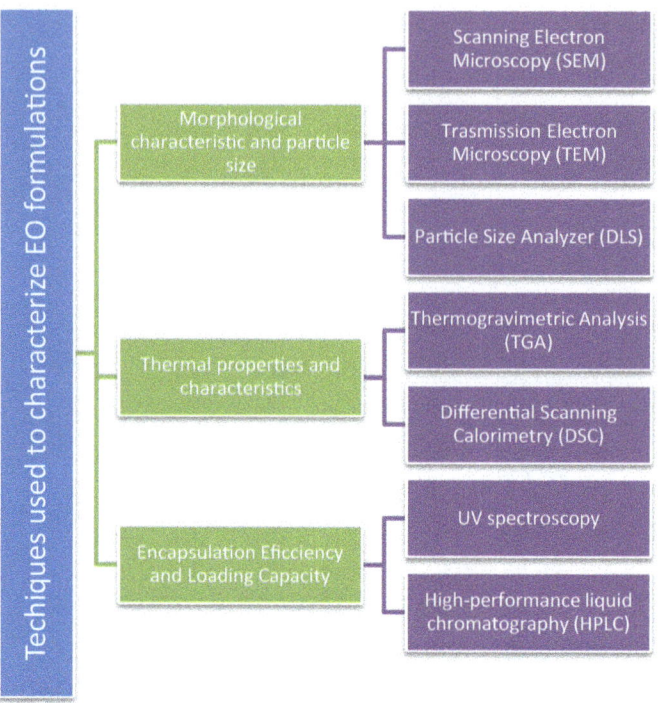

Fig. 9 Characterization techniques of EOs based formulations

3.4 Release Characteristics of Encapsulated EOs

The release of EOs from biopolymers is controlled by several alternate mechanisms: diffusion, polymer relaxation (swelling/shrinking), erosion, and fragmentation, among others (Matalanis et al. 2011) (Fig. 10). The mentioned mechanisms rely on the properties of encapsulated EO, the used polymer, the physicochemical characteristics of obtained encapsulates and the surrounding environmental conditions.

4 Functional Textile

4.1 Finishing Treatments

Different techniques have been developed to functionalize textile materials with microcapsules. The fixation of the microcapsules can be done in several ways: within the fiber structure permanently, to embed them in a binder, mix them in foam or graft

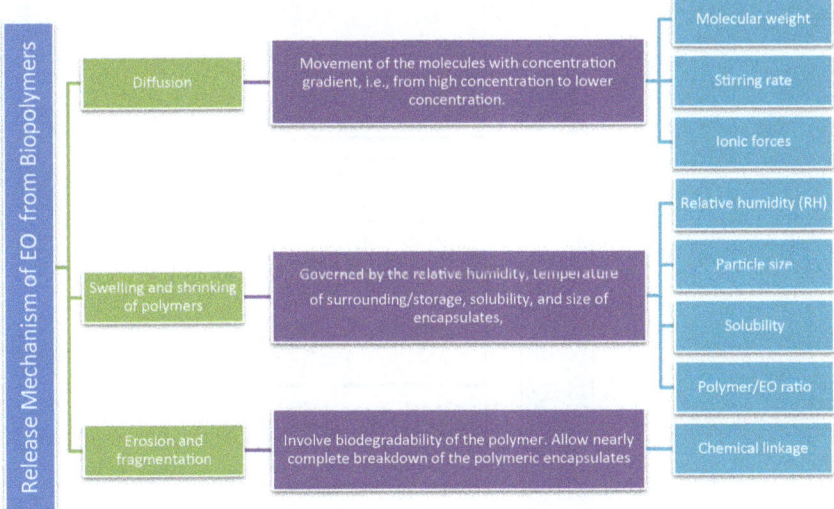

Fig. 10 Mechanisms involved in EOs release from formulations

them according to the expected end use and coating properties of the microcapsules (Massella et al. 2019). The selection of the textile finishing process to functionalize the fabrics depends on several factors (Yip and Luk 2016) (Fig. 11).

The chemical affinity between the textile substrate and the microcapsule plays a very important role in all cases in controlling the release behavior (Arias et al. 2018). The finishing processes chemical and physical–chemical are superseding the high energy methods. Chemical finishes can use microcapsules that can be applied to the textile by several methods similar to dyeing processes, followed by curing for fixation (Valle et al. 2021). In some cases, before finishing, an activation treatment by plasma method can be carried out on the surface to increase its affinity with the capsules. It should also be noted that higher temperatures and longer cure times generally promote better adhesion of these capsules on textiles, but in turn, there will be a greater load loss of active compounds at the same time (Yip and Luk 2016).

Microcapsules can be grafted onto textile fabric by using any one of the following techniques (Fig. 12).

1. Padding, the fabric that has been previously immersed into the microcapsule solution followed by curing for fixation (Gende et al. 2011)
2. Exhaustion method (applied after dyeing) consists of soaking the fabric for a controlled duration of time and temperature (Sannapapamma et al. 2018).
3. Spraying, the microcapsules are applied to the surface of the textile, then a fixing process is applied (Pichersky et al. 2006)
4. Screen printing where the microcapsules are used with an appropriate binder and thickener onto the fabrics. Other printing techniques are photography, electrostatics, pressure transfer, thermal transfer, and ink jet printing (Hozić and

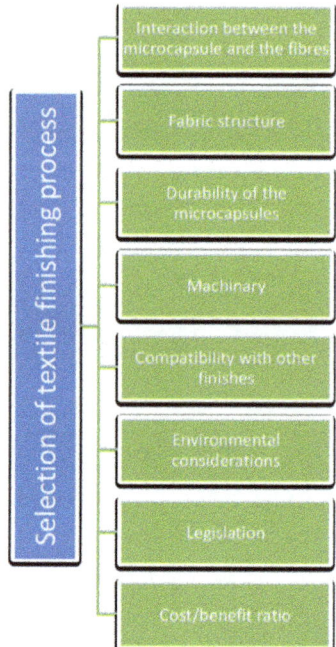

Fig. 11 Factors influencing the selection of textile finishing process

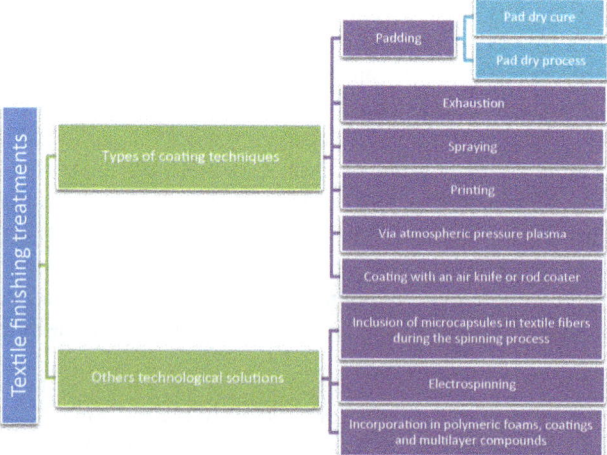

Fig. 12 Textile finishing treatments

Kert 2019; Aqueous pigmented inks for ink jet printers1990; Coatings and Laminates—Google Libros 2022).

5. Via atmospheric pressure plasma, the tissue surface is modified to then embed the microcapsules using one of the techniques listed above, followed by thermal fixation with a fixing agent with a monomeric or oligomeric crosslinker (Ramos Montilla and Vestuario 2018; Almeida et al. 2006).

6. Coating with an air knife or rod coater (Netzeva et al. 2005; Lang and Buchbauer 2012).

In addition to coating techniques, other technological solutions have been developed and used for the incorporation of microcapsules in textile products, such as:

7. Inclusion of microcapsules in textile fibers during the spinning process (Iqbal and Sun 2014).

8. Electrospinning, a technique for producing fibers with the microcapsules inside them, from polymer melts and solutions using an electric field (Guan et al. 2018).

9. Incorporation in polymeric foams, coatings and multilayer compounds, to be used in selected parts of textile garments or footwear (Dorman and Deans 2000; Sharifi-Rad et al. 2017).

Microcapsules can be applied by conventional finishing techniques. The most common technique applied to textile chemical finishes is padding (Fig. 13). It is a method, the fabric is continuously dipped in liquor and squeezed to a certain degree by passing between a pair of rollers, followed by a dried (removed water) and cured (heated to cause a chemical reaction) for fixation. This is one of the finishing methods used to embed microcapsules on the textile surface when there is no affinity between the microcapsules and the textile substrate. In these cases, the microcapsule solutions usually require a dispersant (to promote their diffusion through the textile material), followed by the addition of a crosslinking agent to adhere them to the fabric. Another common method to apply chemical finishes used is exhausted or immersion, in this case, the method has strong affinities for fiber surfaces; consists of a batch process where is soaked the fabric in microcapsule solution composition (salts, pH, surfactants, etc.) for a controlled time and temperature. Later the liquor is drained and the textile material is dried without washing, followed by curing for fixation.

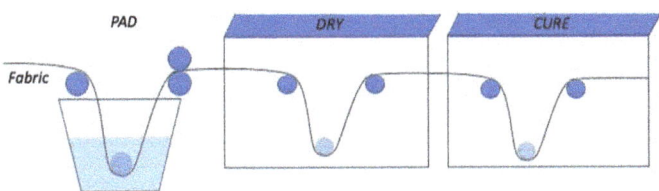

Fig. 13 Padding process

4.2 Physicochemical Characterization of the Microcapsule-Treated Fabrics

Several techniques are used to characterize the functionalized textiles, some of them already described previously in the physical–chemical characterization of the encapsulated EO section.

The structure of the tissues treated with microcapsules can be characterized with various physicochemical techniques. Some of them are listed below (Sánchez-Silva et al. 2012).

Mechanical Tests

- *Maximum force and elongation* using the strip method (Sánchez et al. 2010). Another standard with which the resistance and elongation properties at the break of the treated samples can be determined is the ASTM D 5034-95 standard, in this method a tensile test equipment is used for the measurement (Cheng et al. 2010).
- *Tear Strength* (ASTM D 1242), this test method determines the force required to propagate a single tear from a cut in a fabric. To evaluate the test samples, it is used an Elmendorf Tear tester (UNE-EN 1998).
- *Water vapor permeability* (UNE-EN ISO 15496 2018), determines the vapor permeability of textiles, in order to provide the manufacturer with a simple method for quality control (Method and for Breaking Strength and Elongation of Textile Fabrics (Grab Test) 2022).
- *Permeability to air* is an important factor in textile materials applied with special finishing. Air permeability is the amount of air passing through a specific area in a given time; the higher the value of Pka.s/m, the higher the air resistance of a fabric will be. The air permeability method (applicable to most types of fabrics), in terms of fabric air resistance can be performed with the instrument KES-F8-AP1, Kato Tech (ISO 9237 1995) (Method and for Tearing Strength of Fabrics by Falling-Pendulum (Elmendorf-Type) Apparatus 2022).

Durability of Textiles Treatment

- *Durability to Washing*, determines the color fastness in textiles of any nature against domestic or commercial washing procedures (UNE-EN ISO 105-C06 2010 and AATCC 61) (UNE-EN ISO 2018; ISO-ISO 1995).
- *Durability to Abrasion*, using this test method (ASTM D4966), fabrics of all kinds can be tested. The treated fabrics were subjected to the Martindale Abrasion Tester for 10, 50 and 100 abrasion cycles respectively. Then the degree of rupture of the microcapsules present in the textiles was observed under SEM (UNE-EN ISO 2010).

Scanning Electron Microscopy (SEM) and Fourier transform infrared (FTIR)

- SEM and FTIR analyzes show distribution and gradual decrease of microcapsules on the surface of textile samples after washing and abrasion cycles, respectively (AATCC 2022).

Comfort Properties

- Using Kawabata Fabric Evaluation System (KES-F) (Kato Tech Co., Ltd Japan) the target hand of the textiles can be evaluated after treatments with microcapsules. Total value of the hand is generated with respect to the tensile, bending, shear, surface, color, and compression properties of the fabric. In addition, the geometrical smoothness of the surface and the frictional smoothness are measured, because the rough surface of the fingerprint shape is sensitive to the roughness of the cloth surface (Method and for Abrasion Resistance of Textile Fabrics (Martindale Abrasion Tester Method) 2022).

Biological Safety Tests on Skin Test Systems

Biological test systems allow the scientific determination of interactions between textiles and skin accurately and to recognize and assess the potential benefits and risks. For this, the biological evaluation standard for medical devices UNE-EN ISO 10993 (Grgac et al. 2020) is applied, which determines which test methods should be used according to the type and duration of bodily contact with the textile.

Cytotoxicity Test, tissue compatibility, this standard is particularly important in looking at whether the finished textile is potentially toxic to cells, i.e. bioactive leaching substances could be released from the material during normal use. In the cytotoxicity test, an extract of the textile is prepared using artificial perspiration and the effects of this on L 929 fibroblasts and HaCaT keratinocytes are observed in the human epidermis, in this way it is determined if it contains potentially toxic components for the cells (Sánchez-Silva et al. 2012).

Potential for sensitization and irritation, by means of the Draize test, evaluates the potential of the test textile material to produce dermal irritation by applying it to the skin of rabbits, as a suitable animal model (Kawabata and Niwa 1995).

Evaluating Antimicrobial Textiles, using the agar diffusion and suspension tests in vitro, it is possible to evaluate the antimicrobial action of textiles.

The agar diffusion test is recommended in textile research as it is a rapid and preliminary qualitative method to distinguish between active and passive antimicrobials. It is a semi-quantitative analysis. Also, this type of method is effective for testing textiles with antifungal activity (UNE-EN ISO 2021).

With suspension tests it is possible to evaluate the degree of antimicrobial activity of textiles in a quantitative way. It can record the degree of inhibition possessed by the product textile against the growth of test bacteria (Pruebas de toxicología—Prueba de irritación cutánea aguda (Prueba de Draize para piel) (Acute Dermal Irritation test) 2021).

Listed below are some of the most popular standards that test the antimicrobial efficacy of high-performance textiles: SN 195920-1992; SN 195921-1992; EN

14119:2003-12; ASTM E 2149-01; JIS Z 2801; JIS L 1902-2002 (Pruebas de toxi-cología—Prueba de irritación cutánea aguda (Prueba de Draize para piel) (Acute Dermal Irritation test) 2021).

Antiviral test, this standard provides a quantitative test method to assess the antiviral performance of textile products, determining whether they are capable of reducing the number of infectious viral particles that come into contact with the textile Surface (ISO 18184 2019) (UNE-EN ISO 2021; Gulati et al. 2021).

Repellent Activity Tests

Laboratory repellent bioassays based on the World Health Organization Method (WHO) protocol, determine the protection time after treatment with a single dose of repellent or the percentage of protection in relation to the dose of repellent. The WHO method consists of about five variations that meet different testing needs (Höfer 2006).

Indoor tests, this procedure complies with a protocol which was developed for the Stiftung Warentest, where the application tests are given under constant conditions: air-conditioned test rooms, the number, age, and species of mosquitoes used in the study can be controlled. This type of test is more realistic than cage tests, and is also a good alternative to field tests (ISO-ISO 2019). Cono test is another test of an insecticide-treated textile that is based on introducing mosquitoes into a standard-ized cone for a defined period of time, then they are removed and determines the knockdown rates. Another test is cage tests of repellent-treated textiles, for such a test volunteers are required to cover for a certain time an area of their forearms with the treated textiles to later expose themselves in front of the mosquitoes in the cage. This test proves to be a quick and cost-effective way to determine the repellency in textiles treated against long-term insects.

Field test is the most significant evidence to prove the efficacy of textiles treated with mosquito repellent. Treated and controlled (untreated) samples are carried out in areas where mosquitoes are abundant. The effectiveness of the treated textile is measured as a function of the protection time (Barnard et al. 2006).

4.3 Examples of Antimicrobial and Repellent Textiles

There are several publications related with antimicrobial and repellent textiles prepared by the incorporation of microencapsulated EOs to textiles, some of them are summarized in Table 3.

Table 3 Publications related with antimicrobial, antibacterial and repellent textiles prepared by the incorporation of microencapsulated EOs and/or its main components

Purpose	EO or its main components	Findings	References
Antibacterial	Tea tree	β-cyclodextrin were applied to viscose fabric, showing antibacterial activity against *E. coli* and *S. aureus*	Beşen (2019)
	Lime	The growth of *E. coli*, *B. cereus*, *S. typhimurium*, and *S. aureus* on the surface of cotton fabrics containing lime EO microcapsules was inhibited before and after washing	Wijesirigunawardana et al. (2018) (2) (PDF) Textiles repelentes de mosquitos (2022)
	Vanillin and limonene compounds	Application of limonene and vanillin encapsulated in chitosan/gum Arabic inhibited *E. coli* and *S. aureus* growth	Sharkawy et al. (2017) Murugesan (2017)
	Peppermint	Encapsulated peppermint EO in alginate was applied on cotton fabric. The EO was slowly released and exhibited inhibitory activity against *S. aureus* and *E. coli*	Wijesirigunawardana and Perera (2018)
Mosquito repellent	Limonene and permethrin compounds	Cotton fabrics containing limonene and permethrin capsules showed repellency after 20 washing cycles	Sharkawy et al. (2017)
	Lemongrass	Cotton fabrics containing lemongrass encapsulates had resistance against mosquito bites and repellent durability after 30 wash cycles	Vinayagamoorthy et al. (2017) Ghayempour and Montazer (2016)
	Chrysanthemum, eucalyptus, lavender, citronella	Textile finishing on cotton substrates, with mosquito repellency, based on gelatin microcapsules (wall material) and essential oils (active substances)	Türkoğlu et al. (2020)

(continued)

Table 3 (continued)

Purpose	EO or its main components	Findings	References
Antimicrobial	Thyme	Microcapsules of thyme EO and gelatin/gum Arabic were applied on non-woven fabric and showed inhibitory activity against *S. aureus*, *E. coli* and *C. albicans*	Karagonlu et al. (2018) Vinayagamoorthy et al. (2017)
	Litsea, lemon	Treatment of polyester and cotton fabrics from Lemon/Litsea essential oil microemulsion, with strong antimicrobial activity against skin-associated microorganisms *E. coli*, *S. aureus*, *S.* epidermidis and *T. rubrum* and as well as possible repellent properties of mosquitoes	Teli and Chavan (2016)
	Red pepper seed	Microcapsules incorporating ozonized red pepper seed oil (ORPSO) with antimicrobial properties, anchored in non-woven fabrics to prepare functional textiles	Karagonlu et al. (2018)
	Thymus leptobotrys and other components	Cotton textile fibers with antifungal activity, functionalized with microcapsules of gum arabic and essential oil, intended for the area of medical textiles	Soroh et al. (2021)

5 Conclusions and Final Remarks

The uses of EOs are exponentially growing in many industries (pharmaceutical, medical and food). The nature of EOs brings products with antimicrobial, antioxidant, antitumor, and anti-inflammatory properties.

EOs Microencapsulation for functional textile have a bright future prospect not only because this technology effectively provides controlled release of EOs under predetermined conditions besides, the EOs conserve their effects and are protected from environmental erosion or degradation.

Nowadays, the textile industry is one of the most contaminant industries (the second in the Word, following the oil industry). Therefore, the industries and the researchers are working to stop using hazardous substances and switch to natural materials. Hence, biopolymers and EOs are presented as a market attractive as functionalizing agents for textiles.

In this scenario, it is the urge to apply new strategies for prepared natural and ecofriendy textiles. One of the possible solutions is the combination of biopolymers with EOs.

Due to the pandemic, the use of antimicrobial textiles (face shields, personal protective equipment, among others) has increased substantially, it is possible that it will be considered an essential product in the future.

References

(2) (PDF) Textiles repelentes de mosquitos: una visión general textiles repelentes de mosquitos: una visión general. https://www.researchgate.net/publication/326508235_MOSQUITO_REPELL ENT_TEXTILES_AN_OVERVIEW_Mosquito_repellent_textiles_An_overview. Accessed 20 Feb 2022

AATCC 61 (2022) Test method for colorfastness to laundering: accelerated. https://global.ihs.com/ doc_detail.cfm?&input_doc_number=&input_doc_title=&document_name=AATCC%2061& item_s_key=00255811&item_key_date=791231&origin=DSSC. Accessed 6 Feb 2022

Ahn JH, Kim YP, Lee YM, Seo EM, Lee KW, Kim HS (2008) Optimization of microencapsulation of seed oil by response surface methodology. Food Chem 107(1):98–105

Albanese AA, Elissondo MC, Gende LB, Eguaras (2022) Google Académico. https://scholar. google.es/scholar?hl=es&as_sdt=0%2C5&q=+Albanese+A.A.%2C+Elissondo+M.C.%2C+ Gende+L.B.%2C+Eguaras+M.J.%2C+Denegri+M.+2009.+Echinococcus+granulosus%3A+ in+vitro+efficacy+of+Rosmarinus+officinalis+essential+oil+on+protoscoleces.+Int+J+of+ess ential+oil+therapeutics.+3%3A69-75.&btnG=. Accessed 7 Mar 2022

Almeida L, Carneiro N, Souto AP. Aplicación del plasma en la industria textil. http://repositorium. sdum.uminho.pt/handle/1822/18647. Accessed 6 Jan 2022

Alonso JR (1998) Tratado de fitomedicina: bases clínicas y farmacológicas, p 1038

Álvarez B, Damiani N, Czerner M, Martucci J, Gende L. Application of active and edible films of sodium caseinate to extend the shelf-life of Argentine Pategrás Cheese

Angioni A, Barra A, Coroneo V, Dessi S, Cabras P (2006) Chemical composition, seasonal variability, and antifungal activity of Lavandula stoechas L. ssp. stoechas essential oils from stem/leaves and flowers. J Agricult Food Chem 54(12):4364–4370. https://pubs.acs.org/doi/abs/ 10.1021/jf0603329. Accessed 7 Mar 2022

Aqueous pigmented inks for ink jet printers (1990 Apr 11)

Arias MJL, Coderch L, Martí M, Alonso C, Carmona OG, Carmona CG et al (2018) Vehiculation of active principles as away to create smart and biofunctional textiles. Materials 11(11)

Asbahani A, Addi E, Miladi K, ... AB-J of C (2014) Undefined. Preparation of medical cotton textile activated by Thymus leptobotrys essential oil colloidal particles: evaluation of antifungal properties. ingentaconnect.com. https://www.ingentaconnect.com/contentone/asp/jcsb/2014/000 00003/00000003/art00006. Accessed 28 Feb 2022

Asbahani A el, Miladi K, Badri W, Sala M, Addi EHA, Casabianca H et al (2015) Essential oils: From extraction to encapsulation. Int J Pharmaceutics. Elsevier B.V., 483:220–243

Aziz ZAA, Ahmad A, Setapar SHM, Karakucuk A, Azim MM, Lokhat D et al (2018) Essential oils: extraction techniques, pharmaceutical and therapeutic potential—a review. Curr Drug Metab 19(13):1100–1110

Bakkali F, Averbeck S, Averbeck D, Idaomar M (2008) Biological effects of essential oils—a review. Food Chem Toxicol 462008:446–475

Bakry AM, Abbas S, Ali B, Majeed H, Abouelwafa MY, Mousa A et al (2016a) Microencapsulation of oils: a comprehensive review of benefits, techniques, and applications. Comprehens Rev Food Sci Food Saf 15(1):143–182. https://onlinelibrary.wiley.com/doi/full/10.1111/1541-4337.12179. Accessed 8 Mar 2022

Bakry AM, Abbas S, Ali B, Majeed H, Abouelwafa MY, Mousa A et al (2016b) Microencapsulation of oils: a comprehensive review of benefits, techniques, and applications. Comprehens Rev Food Sci Food Saf 15(1):143–182

Barnard D, Bernier U, Xue R, Debboun M (2006) Standard methods for testing mosquito repellents. Insect Repellents 25:103–110

ben Arfa A, Combes S, Preziosi-Belloy L, Gontard N, Chalier P (2006) Antimicrobial activity of carvacrol related to its chemical structure. Lett Appl Microbiol 43(2):149–154. https://onlinelib rary.wiley.com/doi/full/10.1111/j.1472-765X.2006.01938.x. Accessed 7 Mar 2022

Benavides Portilla KE (2017) Acabado antibacterial en calcetines de acrílico con triclosán. http://repositorio.utn.edu.ec/handle/123456789/6615. Accessed 2 Jan 2022

Beşen BS (2019) Production of disposable antibacterial textiles via application of tea tree oil encapsulated into different wall materials. Fib Polym 20(12):2587–2593. https://link.springer.com/art icle/10.1007/s12221-019-9350-9. Accesesd 4 Jan 2022

Bezerra FM, Carmona OG, Carmona CG, Lis MJ, de Moraes FF (2016) Controlled release of microencapsulated citronella essential oil on cotton and polyester matrices. Cellulose 23(2):1459–1470

Borugă O, Jianu C, Mişcă C, Goleţ I, Gruia AT, Horhat FG (2014) Thymus vulgaris essential oil: chemical composition and antimicrobial activity. J Med Life 7(Special Issue 3):56. /pmc/articles/PMC4391421/. Accessed 8 Mar 2022

Brasesco C, Gende L, Negri P, Szawarski N, Iglesias A, Eguaras M et al (2017) Assessing in vitro acaricidal effect and joint action of a binary mixture between essential oil compounds (Thymol, phellandrene, eucalyptol, cinnamaldehyde, myrcene, carvacrol) over ectoparasitic mite varroa destructor (Acari: Varroidae). J Apicult Sci 61(2):203–215

Brenda Gende L, Floris I, Fritz R, Javier Eguaras M (2008) Antimicrobial activity of cinnamon (Cinnamomum zeylanicum) essential oil and its main components against Paenibacillus larvae from Argentine. Bull Insectol 61(1):1–4. www.bulletinofinsectology.org. Accessed 7 Mar 2022

Brenda Gende L, Mendiara S, Jorgelina Fernández N, van Baren C, Leo Lira A di, Bandoni A et al (2014) Essentials oils of some Mentha spp. and their relation with antimicrobial activity against Paenibacillus larvae, the causative agent of American foulbrood in honey bees, by using the bioautography technique. Bull Insectol 67(1):13–20

Burt S (2004) Essential oils: their antibacterial properties and potential applications in foods—a review. Int J Food Microbiol 94(3):223–253

Campos EVR, de Oliveira JL, Fraceto LF, Singh B (2015) Polysaccharides as safer release systems for agrochemicals. Agron Sustain Dev 35(1):47–66

Carson CF, Hammer KA (2010) Chemistry and bioactivity of essential oils

Carvalho DN, Lopez-Cebral R, Sousa RO, Alves AL, Reys LL, Silva SS, et al. Marine collagen-chitosan-fucoidan cryogels as cell-laden biocomposites envisaging tissue engineering. Biomed Mater (Bristol) 15(5)

Chang HJ, Kim HJ, Chun HS (2007) Quantitative structure-activity relationship (QSAR) for neuro-protective activity of terpenoids. Life Sci. 80(9):835–41. https://pubmed.ncbi.nlm.nih.gov/171 66521/. Accessed 7 Mar 2022

Chen H, Wooten H, Thompson L, Pan K (2019) Nanoparticles of casein micelles for encapsulation of food ingredients. Biopolym Nanostruct Food Encapsulat Purposes 1:39–68

Chen H, Wooten H, Thompson L, Pan K (2019) Nanoparticles of casein micelles for encapsulation of food ingredients. Biopolym Nanostruct Food Encapsulat Purposes. Elsevier Inc. 39–68 p

Cheng SY, Yuen CWM, Kan CW, Cheuk KKL, Tang JCO (2010) Systematic characterization of cosmetic textiles. Text Res J 80(6):524–536

da Cruz CL, Fernández Pinto V, Patriarca A (2013) Application of plant derived compounds to control fungal spoilage and mycotoxin production in foods. Int J Food Microbiol 166(1):1–14

Damiani N, Gende LB, Bailac P, Marcangeli JA, Eguaras MJ (2009) Acaricidal and insecticidal activity of essential oils on Varroa destructor (Acari: Varroidae) and Apis mellifera (Hymenoptera: Apidae). Parasitol Res 106(1):145–152. https://pubmed.ncbi.nlm.nih.gov/19795133/. Accessed 7 Mar 2022

de Oliveira EF, Paula HCB, de Paula RCM (2014) Alginate/cashew gum nanoparticles for essential oil encapsulation. Colloids Surf, B 113:146–151

Dhar P, Ayala U, Andarge E, Morisseau S, Snyder-Leiby T (2004) Study of the structural changes on the antimicrobial activity of [3.1.1.]-bicyclics. J Essential Oil Res 16(6):612–616

Dorman HJD, Deans SG (2000) Antimicrobial agents from plants: antibacterial activity of plant volatile oils. J Appl Microbiol 88(2):308–316. https://onlinelibrary.wiley.com/doi/full/10.1046/j.1365-2672.2000.00969.x. Accessed 7 Mar 2022

Effect of Essential Oils on the Enveloped Viruses: Antiviral Activity of Oregano and Clove Oils on Herpes Simplex Virus Type 1 and Newcastle Disease Virus|Semantic Scholar. https://www.semanticscholar.org/paper/EFFECT-OF-ESSENTIAL-OILS-ON-THE-ENVELOPED-VIRUSES-%3A-Siddiqui-Ettayebi/4fc47b3d923ef53458b9633bb503dcd017712907. Accessed 7 Mar 2022

Elissondo MC, Albani CM, Gende L, Eguaras M, Denegri G (2008) Efficacy of thymol against Echinococcus granulosus protoscoleces. Parasitol Int 57(2):185–190

Gende L, Maggi M, van Baren C, di Leo L, Bandoni A, Fritz R et al (2010) Antimicrobial and miticide activities of Eucalyptus globulus essential oils obtained from different Argentine regions. Span J Agricult Res 8(3):642–650. https://sia.revistas.inia.es/index.php/sjar/article/view/1260. Accessed 8 Mar 2022

Gende LB, Maggi MD, Fritz R, Eguaras MJ, Bailac PN, Ponzi MI (2011) Antimicrobial Activity of Pimpinella anisum and Foeniculum vulgare essential oils against Paenibacillus larvae 21(1):91–93. https://www.tandfonline.com/doi/abs/10.1080/10412905.2009.9700120. Accessed 8 Mar 2022

Ghayempour S, Montazer M (2016) Micro/nanoencapsulation of essential oils and fragrances: Focus on perfumed, antimicrobial, mosquito-repellent and medical textiles. J Microencapsul 33(6):497–510

Grgac SF, Tarbuk A, Dekanic T, Sujka W, Draczynski Z (2020) The Chitosan implementation into cotton and polyester/cotton blend fabrics. Materials 13(7):1616. https://www.mdpi.com/1996-1944/13/7/1616/htm. Accessed 6 Feb 2022

Guan Y, Zhang L, Wang D, West JL, Fu S (2018) Preparation of thermochromic liquid crystal microcapsules for intelligent functional fiber. Mater Des 5(147):28–34

Gulati R, Sharma S, Sharma RK (2021) Antimicrobial textile: recent developments and functional perspective. Polymer Bulletin 1–25. https://link.springer.com/article/10.1007/s00289-021-03826-3. Accessed 28 Feb 2022

Hanif MA, Nisar S, Khan GS, Mushtaq Z, Zubair M (2019) Essential oils. Essential Oil Res 3–17. https://link.springer.com/chapter/10.1007/978-3-030-16546-8_1. Accessed 4 Jan 2022

Höfer D (2006) Antimicrobial textiles—evaluation of their effectiveness and safety. Curr Probl Dermatol 33:42–50

Hosseini SM, Hosseini H, Mohammadifar MA, Mortazavian AM, Mohammadi A, Khosravi-Darani K et al (2013) Incorporation of essential oil in alginate microparticles by multiple emulsion/ionic gelation process. Int J Biol Macromolecules 62:582–588. https://doi.org/10.1016/j.ijbiomac.2013.09.054

Hozić N, Kert M (2019) Influence of different colourants on properties of cotton fabric. Printed with Microcapsules of Photochromic Dye Vpliv Različnih Kolorantov Na Lastnosti Bombažne Tkanine, Potiskane z Mikrokapsulami Fotokromnega Barvila 62(3):208–218

Hu Y, Zhang J, Kong W, Zhao G, Yang M (2017) Mechanisms of antifungal and anti-aflatoxigenic properties of essential oil derived from turmeric (Curcuma longa L.) on Aspergillus flavus. Food Chem 220:1–8

Ipek E, Zeytinoglu H, Okay S, Tuylu BA, Kurkcuoglu M, Baser KHC (2005) Genotoxicity and antigenotoxicity of Origanum oil and carvacrol evaluated by Ames Salmonella/microsomal test. Food Chem 93(3):551–556

Iqbal K, Sun D (2014) Development of thermo-regulating polypropylene fibre containing microencapsulated phase change materials. Renew Energy 1(71):473–479

Işcan G, Kirimer N, Kürkcüoğlu M, Başer KHC, Demirci F (2002) Antimicrobial screening of Mentha piperita essential oils. J Agricult Food Chem 50(14):3943–3946. https://pubs.acs.org/doi/abs/10.1021/jf011476k. Accessed 7 Mar 2022

ISO-ISO 9237 (1995) Textiles—determination of the permeability of fabrics to air. https://www.iso.org/standard/16869.html. Accessed 4 Feb 2022

ISO-ISO 18184 (2019) Textiles—determination of antiviral activity of textile products. https://www.iso.org/standard/71292.html. Accessed 19 Feb 2022

Jafari SM, Assadpoor E, He Y, Bhandari B (2008) Encapsulation efficiency of food flavours and oils during spray drying. Drying Technol 26(7):816–835

Jamil B, Abbasi R, Abbasi S, Imran M, Khan SU, Ihsan A et al (2016a) Encapsulation of cardamom essential oil in chitosan nano-composites: in-vitro efficacy on antibiotic-resistant bacterial pathogens and cytotoxicity studies. Front Microbiol 7:1580

Jamil B, Abbasi R, Abbasi S, Imran M, Khan SU, Ihsan A et al (2016b) Encapsulation of cardamom essential oil in chitosan nano-composites: in-vitro efficacy on antibiotic-resistant bacterial pathogens and cytotoxicity studies. Front Microbiol 7:1–10

Jianu C (2014) Thymus vulgaris essential oil: chemical composition and antimicrobial activity. J Med Life 7

jmbfs Kacaniova A. J Microbiol Biotechnol Food Sci. http://www.jmbfs.org/jmbfs-kacaniova-a/?issue_id=1307&article_id=19. Accessed 7 Mar 2022

Kalemba D, Kunicka A (2005) Antibacterial and antifungal properties of essential oils. Curr Med Chem 10(10):813–829

Karagonlu S, Başal G, Ozyıldız F, Uzel A (2018) Undefined. Preparation of thyme oil loaded microcapsules for textile applications. neliti.com. https://www.neliti.com/publications/263122/preparation-of-thyme-oil-loaded-microcapsules-for-textile-applications. Accessed 28 Feb 2022

Kawabata S, Niwa M (1995) Objective measurement of fabric hand. In: Modern textile characterization methods, 329–354. https://www.taylorfrancis.com/chapters/edit/10.1201/9780203746684-10/objective-measurement-fabric-hand-sueo-kawabata-masako-niwa. Accessed 5 Feb 2022

Kolanowski W, Ziolkowski M, Weißbrodt J, Kunz B, Laufenberg G (2006) Microencapsulation of fish oil by spray drying—impact on oxidative stability. Part 1. Euro Food Res Technol. 222(3–4):336–342. https://link.springer.com/article/10.1007/s00217-005-0111-1. Accessed 8 Mar 2022

Lam PL, Li L, Yuen CWM, Gambari R, Wong RSM, Chui CH et al (2013) Effects of multiple washing on cotton fabrics containing berberine microcapsules with anti-Staphylococcus aureus activity. J Microencapsul 30(2):143–150

Lang G, Buchbauer G (2012) A review on recent research results (2008–2010) on essential oils as antimicrobials and antifungals. A review. Flavour Fragrance J 27(1):13–39. https://onlinelibrary.wiley.com/doi/full/10.1002/ffj.2082. Accessed 7 Mar 2022

Lee BH, Choi WS, Lee SE, Park BS (2001) Fumigant toxicity of essential oils and their constituent compounds towards the rice weevil, Sitophilus oryzae (L.). Crop Protect 20(4):317–320

Liao W, Badri W, Dumas E, Ghnimi S, Elaissari A, Saurel R et al (2021) Nanoencapsulation of essential oils as natural food antimicrobial agents: an overview. Appl Sci (Switzerland) 11(13)

Maggi MD, Ruffnengo SR, Gende LB, Sarlo EG, Eguaras MJ, Bailac PN et al (2011) Laboratory Evaluations of Syzygium aromaticum (L.) Merr. et Perry essential oil against Varroa destructor 22(2):119–122. https://www.tandfonline.com/doi/abs/10.1080/10412905.2010.970 0278. Accessed 8 Mar 2022

Marinković SŠ, Bezbradica D, Škundrić P (2006) Microencapsulation in the textile industry. Chem Indus Chem Eng Quar 12(1):58–62. http://www.doiserbia.nb.rs/Article.aspx?ID=1451-937206 01058S. Accessed 3 Jan 2022

Martucci JF, Gende LB, Neira LM, Ruseckaite RA (2015) Oregano and lavender essential oils as antioxidant and antimicrobial additives of biogenic gelatin films. Ind Crops Prod 1(71):205–213

Massella D, Giraud S, Guan J, Ferri A, Salaün F (2019) Textiles for health: a review of textile fabrics treated with chitosan microcapsules. Environ Chem Lett 17(4):1787–1800. https://link. springer.com/article/10.1007/s10311-019-00913-w. Accessed 5 Jan 2022

Matalanis A, Jones OG, McClements DJ (2011) Structured biopolymer-based delivery systems for encapsulation, protection, and release of lipophilic compounds. Food Hydrocolloids 25(8):1865–1880

Meng X, Poonia M, Yoo CG, Ragauskas AJ (2021) Recent advances in synthesis and application of Lignin nanoparticles, pp 273–293. https://pubs.acs.org/doi/abs/10.1021/bk-2021-1377.ch011. Accessed 1 Mar 2021

Mohammadi A, Hashemi M, Hosseini SM (2015) Nanoencapsulation of Zataria multiflora essential oil preparation and characterization with enhanced antifungal activity for controlling Botrytis cinerea, the causal agent of gray mould disease. Innov Food Sci Emerg Technol 28:73–80

Murugesan B (2017) Analysis and characterization of Mosquito-Repellent textiles

Nakatsu T, Lupo AT, Chinn JW, Kang RKL (2000) Biological activity of essential oils and their constituents. Stud Nat Prod Chem 21:571–631

Nazzaro F, Fratianni F, Coppola R, de Feo V (2017) Essential oils and antifungal activity, vol 10. Pharmaceuticals. MDPI AG

Netzeva TI, Aptula AO, Benfenati E, Cronin MTD, Gini G, Lessigiarska I et al (2005) Description of the electronic structure of organic chemicals using semiempirical and ab initio methods for development of toxicological QSARs. J Chem Inform Model 45(1):106–114. https://pubs.acs. org/doi/abs/10.1021/ci049747p. Accessed 7 Mar 2022

Nguyen TTT, Le TVA, Dang NN, Nguyen DC, Nguyen PTN, Tran TT et al (2021) Microencapsulation of essential oils by spray-drying and influencing factors. J Food Qual 2021

Okoh OO, Sadimenko AP, Afolayan AJ (2010) Comparative evaluation of the antibacterial activities of the essential oils of Rosmarinus officinalis L. obtained by hydrodistillation and solvent free microwave extraction methods. Food Chem 120(1):308–312

Özyildiz F, Karagönlü S, Basal G, … Uzel A (2013) Undefined. Micro-encapsulation of ozonated red pepper seed oil with antimicrobial activity and application to nonwoven fabric. Wiley Online Libr 56(3):168–179. https://sfamjournals.onlinelibrary.wiley.com/doi/abs/10.1111/lam. 12028. Accessed 28 Feb 2022

Pichersky E, Noel JP, Dudareva N (2006) Biosynthesis of plant volatiles: nature's diversity and ingenuity. Science 311:808–811

Porrini MP, Garrido PM, Gende LB, Rossini C, Hermida L, Marcángeli JA et al (2017) Oral administration of essential oils and main components: study on honey bee survival and Nosema ceranae development 56(5):616–624. https://www.tandfonline.com/doi/abs/10.1080/00218839. 2017.1348714. Accessed 7 Mar 2022

Pruebas de toxicología - Prueba de irritación cutánea aguda (Prueba de Draize para piel) (Acute Dermal Irritation test). ISO 10993-23 (2021) (Evaluación biológica de productos sanitarios. Parte 23: Pruebas de irritación). IVAMI. https://www.ivami.com/es/evaluacion-biologica-de-dispositivos-medicos/6536-pruebas-de-toxicologia-121-prueba-de-irritacion-cutanea-aguda-pru eba-de-draize-para-piel-acute-dermal-irritation-test-iso-10993-23-2021-evaluacion-biologica-de-productos-sanitarios-parte-23-pruebas-de-irritacion. Accessed 5 Feb 2022

Ramos Montilla LC, de Vestuario D (2018) Herramienta para el desarrollo y aplicación de nanopartículas de cobre, aplicadas como recubrimiento hidrófobo y antimicrobiano sobre fibras de poliéster para la construcción de un material textil. https://repository.upb.edu.co/handle/20. 500.11912/4844. Accessed 6 Jan 2022

Raut JS, Karuppayil SM (2014) A status review on the medicinal properties of essential oils. Indus Crops Prod. Elsevier 62:250–264

Redondo-Blanco S, Fernández J, López-Ibáñez S, Miguélez EM, Villar CJ, Lombó F (2020) Plant phytochemicals in food preservation: antifungal bioactivity: a review. J Food Protect 83(1):163–171. https://meridian.allenpress.com/jfp/article/83/1/163/425622/Plant-Phy tochemicals-in-Food-Preservation. Accessed 7 Mar 2022

Reshma A, Brindha Priyadarisini V, Amutha K (2018) Sustainable antimicrobial finishing of fabrics using natural bioactive agents—a review. Int J Life Sci Pharma Res 8(4):10–20. https://doi.org/ 10.22376/ijpbs/lpr.2018.8.4.L10-20. Accessed 4 Jan 2022

Rezzoug SA, Boutekedjiret C, Allaf K (2005) Optimization of operating conditions of rosemary essential oil extraction by a fast controlled pressure drop process using response surface methodology. J Food Eng 71(1):9–17

Sánchez P, Sánchez-Fernandez MV, Romero A, Rodríguez JF, Sánchez-Silva L (2010) Development of thermo-regulating textiles using paraffin wax microcapsules. Thermochim Acta 498(1–2):16–21

Sánchez-Silva L, Rodríguez JF, Romero A, Sánchez P (2012) Preparation of coated thermo-regulating textiles using Rubitherm-RT31 microcapsules. J Appl Polym Sci 124(6):4809–4818. https://onlinelibrary.wiley.com/doi/full/10.1002/app.35546. Accessed 6 Jan 2022

Sánchez-Vioque R, Polissiou M, Astraka K, Mozos-Pascual M de los, Tarantilis P, Herraiz-Peñalver D et al (2013) Polyphenol composition and antioxidant and metal chelating activities of the solid residues from the essential oil industry. Indus Crops Prod 49:150–159

Sannapapamma K, Malligawad Lokanath H, Naikwadi S (2018) Undefined. Antimicrobial and aroma finishing of organic cotton knits using vetiver oil microcapsules for health care textiles. publications.waset.org. https://publications.waset.org/10008658/antimicrobial-and-aroma-finish ing-of-organic-cotton-knits-using-vetiver-oil-microcapsules-for-health-care-textiles. Accessed 6 Jan 2022

Schmidt DG, Buttery HJ, Norbury RJ (1991) Coated perfume particles in fabric softener or antistatic agents

Schnitzler P, Astani A, Reichling J (2011) Screening for antiviral activities of isolated compounds from essential oils. Evid-Based Compl Alter Med 2011

Shaaya E, Rafaeli A (2007) Essential oils as biorational insecticides-potency and mode of action. Insecticides Des Adv Technol 9783540469:249–261

Sharifi-Rad J, Sureda A, Tenore GC, Daglia M, Sharifi-Rad M, Valussi M et al (2017) Biological activities of essential oils: from plant chemoecology to traditional healing systems, vol 22, Molecules. MDPI AG

Sharkawy A, Fernandes IP, Barreiro MF, Rodrigues AE, Shoeib T (2017) Aroma-loaded microcapsules with antibacterial activity for eco-friendly textile application: synthesis, characterization, release, and green grafting. Indus Eng Chem Res 56(19):5516–5526. https://pubs.acs.org/doi/ abs/10.1021/acs.iecr.7b00741. Accessed 28 Feb 2022

Smart textile coatings and laminates—Google Libros. https://books.google.com.ar/books?hl=es& lr=&id=L4hwAgAAQBAJ&oi=fnd&pg=PP1&dq=microcapsules++%2B+Coating+with+an+ air+knife+or+rod+coater+&ots=-cLPnh4keL&sig=EvqhjwhUBloiRF8dtyqZHi9Fphg&redir_ esc=y#v=onepage&q=microcapsules%20%20%2B%20Coating%20with%20an%20air%20k nife%20or%20rod%20coater&f=false. Accessed 6 Jan 2022

Soliman EA, El-Moghazy AY, Mohy El-Din MS, Massoud MA (2013a) Microencapsulation of essential oils within alginate: formulation and in vitro evaluation of antifungal activity. J Encapsulat Adsorpt Sci (1):48–55. http://www.scirp.org/journal/PaperInformation.aspx?Pap erID=29469. Accessed 4 Jan 2022

Soliman EA, El-Moghazy AY, Mohy El-Din MS, Massoud MA (2013b) Microencapsulation of essential oils within alginate: formulation and in vitro evaluation of antifungal activity. J Encapsulat Adsorpt Sci (1):48–55. http://www.scirp.org/journal/PaperInformation.aspx?Pap erID=29469. Accessed 8 Mar 2022

Soliman EA, El-Moghazy AY, El-Din MSM, Massoud MA (2013c) Microencapsulation of essential oils within alginate: formulation and *in Vitro* evaluation of antifungal activity. J Encapsulat Adsorpt Sci 3(1):48–55

Soroh A, Owen L, Rahim N, Masania J, Abioye A, Qutachi O et al (2021) Microemulsification of essential oils for the development of antimicrobial and mosquito repellent functional coatings for textiles. J Appl Microbiol 131(6):2808–2820. https://pubmed.ncbi.nlm.nih.gov/34022108/. Accessed 28 Feb 2022

Specos MMM, Escobar G, Marino P, Puggia C, Tesoriero MVD, Hermida L (2010) Aroma finishing of cotton fabrics by means of microencapsulation techniques. J Ind Text 40(1):13–32

Standard Test Method for Abrasion Resistance of Textile Fabrics (Martindale Abrasion Tester Method). https://www.astm.org/d4966-12r16.html. Accessed 5 Feb 2022

Standard Test Method for Breaking Strength and Elongation of Textile Fabrics (Grab Test). https://www.astm.org/d5034-21.html. Accessed 5 Feb 2022

Standard Test Method for Tearing Strength of Fabrics by Falling-Pendulum (Elmendorf-Type) Apparatus. https://www.astm.org/d1424-21.html. Accesse 5 Feb 2022

Teixeira B, Marques A, Ramos C, Neng NR, Nogueira JMF, Saraiva JA et al (2013) Chemical composition and antibacterial and antioxidant properties of commercial essential oils. Ind Crops Prod 43(1):587–595

Teli MD, Chavan PP (2016) Application of gelatine based microcapsules containing mosquito repellent oils on cellulosic biopolymer. J Bionanosci 10(5):390–395

Tian G, Ji Q, Xu D, Tan L, Quan F, Xia Y (2013) The effect of zinc ion content on flame retardance and thermal degradation of alginate fibers. Fibers and Polymers. 14(5):767–771

Tomaino A, Cimino F, Zimbalatti V, Venuti V, Sulfaro V, de Pasquale A et al (2005) Influence of heating on antioxidant activity and the chemical composition of some spice essential oils. Food Chem 89(4):549–554

Türkoğlu GC, Sarıışık AM, Erkan G, Yıkılmaz MS, Kontart O (2020) Micro- and nano-encapsulation of limonene and permethrin for mosquito repellent finishing of cotton textiles. Iran Polym J 29(4):321–329. https://link.springer.com/article/10.1007/s13726-020-00799-4. Accessed 28 Feb 2022

Tworkoski T (2002) Herbicide effects of essential oils. Weed Sci 50(4):425–431

UNE-EN 12127 (1998) Textiles. Tejidos. Determinación de la masa…. https://en.tienda.aenor.com/norma-une-en-12127-1998-n0010747. Accessed 4 Feb 2022

UNE-EN ISO 105-C06 (2010) Textiles. Ensayos de solidez del color….. https://en.tienda.aenor.com/norma-une-en-iso-105-c06-2010-n0045649. Accessed 5 Feb 2022

UNE-EN ISO 15496 (2018) Textiles. Medición de la permeabilidad a….. https://en.tienda.aenor.com/norma-une-en-iso-15496-2018-n0060848. Accessed 4 Feb 2022

UNE-EN ISO 10993-1 (2021) Evaluación biológica de productos sani…. https://en.tienda.aenor.com/norma-une-en-iso-10993-1-2021-n0067302. Accessed 5 Feb 2022

Valdes A, Ramos M, Beltran A, Garrigos MC (2018) Recent trends in microencapsulation for smart and active innovative textile products. Curr Org Chem 22(12):1237–1248

Valle JAB, Valle R de CSC, Bierhalz ACK, Bezerra FM, Hernandez AL, Lis Arias MJ (2021) Chitosan microcapsules: Methods of the production and use in the textile finishing. J Appl Polym Sci 138(21):50482. https://onlinelibrary.wiley.com/doi/full/10.1002/app.50482. Accessed 5 Jan 2022

Vinayagamoorthy P, Senthilkumar B, Patchiyappan K, Kavitha R (2017) Microencapsulated lemongrass oil for mosquito repellent finishing of knitted cotton wear. Asian J Pharm Clin Res 10(6):303–307

Vishwakarma GS, Gautam N, Babu JN, Mittal S, Jaitak V (2016) Polymeric encapsulates of essential oils and their constituents: a review of preparation techniques, characterization, and sustainable release mechanisms. Polym Rev 56(4):668–701

Voda K, Boh B, Vrtačnik M (2004) A quantitative structure-antifungal activity relationship study of oxygenated aromatic essential oil compounds using data structuring and PLS regression analysis. J Mol Model];10(1):76–84. https://link.springer.com/article/10.1007/s00894-003-0174-5. Accessed 7 Mar 2022

Wijesirigunawardana PB, Perera BGK (2018) Development of a cotton smart textile with medicinal properties using lime oil microcapsules. Acta Chimica Slovenica 65(1):150–159. https://journals.matheo.si/index.php/ACSi/article/view/3727/1587. Accessed 28 Feb 2022

Yip J, Luk MYA (2016) Microencapsulation technologies for antimicrobial textiles. Antimicrobial Text 8:19–46

Yip J, Luk MYA (2016) Microencapsulation technologies for antimicrobial textiles. Antimicrobial Text 1:19–46

Yoplac I, Vargas L, Robert P, Hidalgo A (2021) Characterization and antimicrobial activity of microencapsulated citral with dextrin by spray drying. Heliyon 7(4):e06737

Zheng X, Yan Z, Liu P, Fan J, Wang S, Wang P et al (2019) Research progress on toxic effects and water quality criteria of triclosan. Bull Environ Contam Toxicol 102(6):731–740

Low-Cost and Sustainable Treatment Options for Removal of Cd (II) from Drinking Water Using Indigenous Materials for Rural Communities

S. M. Deepak, M. Rajeswari, and Neeta Shivakumar

1 Introduction

Human health, ecosystem and wealthy economy demand clean and safe water. Maintaining the availability of water and its quality is of great concern (Jagaba et al. 2020). Globally, Cd (II) pollution in drinking water is of huge concern as it is highly carcinogenic to human beings (Villarreal et al. 2020; Li et al. 2019; Tang et al. 2018; Shaikh et al. 2018). Cadmium can create iron deficiency by binding to glutamate, cysteine, histidine and aspartate ligands (Castagnetto et al. 2002). Owing to the similarity in oxidation states, Cadmium can replace zinc from metallothionein and inhibit its free radical scavenging activity. Excess intake of cadmium can cause the formation of renal stones, hypertension, itai-itai diseases and kidney-, liver- and lung-related disorders (Rajeswari et al. 2021, 2013; Hebbani et al. 2021; Johri et al. 2010). Different conventional approaches such as solvent-based extraction, adsorption, chemical settling, electrocoagulation, membrane separation, oxidation and ion exchange are being used for the Cd (II) expulsion (Akinyeye et al. 2020; Isawi 2020; Zhang and Tian 2020). However, such purification techniques have their own share of drawbacks. There could be an ineffective Cd (II) elimination, expensive and production of toxic sludge (Wernke 2020; Monte Blanco et al. 2019; Pirkwieser et al. 2018). Consequently, there is a necessity to look for simple, low-cost and versatile processes. Biosorption has grown as the best option due to its Cd (II) removal efficiency, adaptability, eco-friendliness, simplicity and cost-effectiveness (Giese 2020; Gupta et al. 2019; Ayangbenro and Babalola 2017). Several agricultural by-products including

S. M. Deepak
Department of Mechanical Engineering, P E S University, Bengaluru 560085, India

M. Rajeswari (✉) · N. Shivakumar
Department of Biotechnology, R.V. College of Engineering, Bengaluru 560059, India
e-mail: rajeshwarim@rvce.edu.in

© The Author(s), under exclusive license to Springer Nature Singapore Pte Ltd. 2023
A. K. Mishra and C. M. Hussain (eds.), *Biobased Materials*,
https://doi.org/10.1007/978-981-19-6024-6_12

sugarcane bagasse, coconut fibre, peat, rice husks (Asif and Chen 2017), soya bean, walnut, cotton seeds hull, sawdust, tea waste (Abdolali et al. 2017), potato peel (Bibi et al. 2017), corn cobs and banana peels (Mondal et al. 2018) are currently utilized as biosorbent. In the current study, shelled *Moringa oleifera* seed is used as a natural (without modification or any sort of synthetic treatment), inexpensive and effective biosorbent for Cd (II) removal from drinking water. The Cd (II) ion biosorption efficiency is ascribed to the binding sites of the carboxyl group and hydroxyl groups of Moringa oleifera seed for Cd (II) (Madhuranthakam et al. 2021).

2 Materials and Methods

2.1 Preparation of Moringa Oleifera Biosorbent

Shelled *Moringa oleifera* seeds were cleaned using distilled water to eliminate dust and adhering impurities. Cleaned Moringa seeds were allowed to dry at 55 °C for 24 h in a hot air oven. The dried seed of shelled *Moringa oleifera* was crushed into fine powder using mortar and pestle and sieved through a 600 μm standard sieve.

2.2 Batch Biosorption Studies

Cadmium biosorption experiments using *Moringa oleifera* as the biosorbent were conducted for 100 ml of cadmium solution in 250 ml Borosil Erlenmeyer flasks under various experimental conditions like sorbent dosage, time of contact, temperature, pH and sorbate concentration.

The amount of cadmium ion biosorbed per unit amount of the *Moring oleifera* (Q_e, $\mu g/g$) was evaluated using the following equation:

$$Q_e = \frac{(C_0 - C_e)V}{M} \tag{1}$$

The percentage cadmium ion uptake (% R) was evaluated using Eq. 2

$$\%R = \frac{(c_0 - c_e)}{c_0} \times 100 \tag{2}$$

where Ce is the equilibrium concentration of adsorbate (mg/L), Co is the initial adsorbate concentration (mg/L), V is the volume of solution (L) and M is the adsorbent mass (g).

2.3   Thermodynamic Studies

Batch experiments were performed with cadmium solution separately at various temperatures (15–40 °C) by keeping optimum biosorbent dose (1 g), initial concentration of adsorbate (100 µg/L) and pH 7 constant. The thermodynamic parameters variables such as standard enthalpy change, $\Delta H°$; Gibbs free energy change, $\Delta G°$ and entropy change, $\Delta S°$ were determined using the following Van't Hoff expression (Zhu et al. 2011; Kara and Demirbel 2012; Foroughi-Dahr et al. 2015).

$$\Delta G° = \Delta H° -- T\Delta S° \tag{3}$$

$$ln\,K_d = \frac{\Delta S^0}{R} - \frac{\Delta H^0}{RT} \tag{4}$$

$$Kd = \frac{(Co - Ce)V}{Ce.M} \tag{5}$$

where T is the temperature (K), R is the gas constant (8.314 J mol^{-1} K^{-1}) and K_d is the distribution coefficient (mg g^{-1}).

2.4   Equilibrium Isotherms

Three sorption isotherms models such as Freundlich, Langmuir and Dubinin–Radushkevich were tested to estimate biosorption capacity. The Langmuir isotherm model has one more dimensionless constant known as the separation factor, R_L, which can be estimated by Eq. (6) (Webber and Chakkravorti 1974).

$$RL = 1/(1 + bCo) \tag{6}$$

where b is he Langmuir isotherm constant

R_L value indicates the characteristics of biosorption process to be linear ($R_L = 1$), irreversible biosorption ($R_L = 0$), unfavourable biosorption ($RL > 1$) and favourable biosorption ($0 < RL < 1$) (Chen et al. 2008; Farooq et al. 2010).

2.4.1   The Langmuir Condition Can Be Portrayed in the Form

$$1/\,q_e = (1/\,bq_{max})\,1/C_e + 1/\,q_{max} \tag{7}$$

2.4.2 Linearized Logarithmic Form of the Freundlich Isotherm Condition Has the Following Equation

$$\log q_e = \log k + 1/n \log C_e \tag{8}$$

where qmax is maximum sorbate uptake, q_e is equilibrium sorbate uptake, C_e is equilibrium sorbate uptake ($\mu g/L$), k is adsorption capacity and n is adsorption intensity.

2.4.3 Dubinin–Radushkevich Equation

The logical expression for DR condition is shown in Eq. (9)

$$ln q_e = ln q_{max-k_{DR}} \cdot \varepsilon^2 \tag{9}$$

where k_{Dr} is constant related to adsorption energy and ε is Polanyi potential.

2.5 Biosorption Kinetic Models

The mechanism involved in cadmium removal was assessed by utilizing kinetic models like Lagergren's equation (pseudo-first order) and pseudo-second order.

2.5.1 Pseudo-First-Order Model (Lagergren's Equation)

The pseudo-first-arrange show relates the rate of the technique to the measure of solute evacuated per unit mass of media,
The linearized logarithmic form of Pseudo-first-order model is shown in Eq. (10)

$$Log_{10}(q_{eq} - q_t) = Log(q_{eq}) - \frac{k_1}{2.303}t \tag{10}$$

2.5.2 Pseudo-Second-Order Model

The Pseudo-second-order model equation can be written as Eqs. (11) and (12)

$$\frac{dq_t}{dt} = k_2(q_{eq} - q_t)^2 \tag{11}$$

$$\frac{t}{q_t} = \frac{1}{k_2 q_{eq}^2} + \frac{1}{q_{eq}}t \tag{12}$$

where k_1 is Pseudo-first-order rate constant, k_2 is pseudo-second-order rate constant and t is time in minutes.

3 Results and Discussion

3.1 *Biosorbent characterization*

3.2 *Batch studies*

3.2.1 Effect of pH on Biosorption of Cd (II)

Figure 1 depicts the influence of pH on cadmium removal percentage at a fixed biosorbent dosage of 1 g, 100 ml volume of Cd (II) with 100 µg/L initial concentration, time of 40 min and 30 °C temperature. The percentage of Cd (II) removal was increased sharply at pH 6 to about 64.3% and then slowly increased to a maximum of 91.4% at pH 8. Most of the cadmium expulsion occurred between pH 6 and pH 8. The reduction efficiency of Cd (II) with reducing pH may be due to the presence of high H⁺ ions concentration in the mixture, and also at lower pH, *Moringa oleifera* surface sites especially amino acids with low molecular weight are protonated which results in electrostatic repulsion between the positively charged cadmium ions and protonated low molecular weight amino groups (Eman 2010). As the pH value increases, the carboxylic group of the low molecular weight amino acids is deprotonated; therefore, there exists an electrostatic force of attraction between the positively charged cadmium (II) and negatively charged carboxylate ligands and binding occurs (Agrawal 2007). Since there was no change in percentage biosorption of Cd (II) ions between pH 7 and 8, the value of pH 7 was chosen for the present studies.

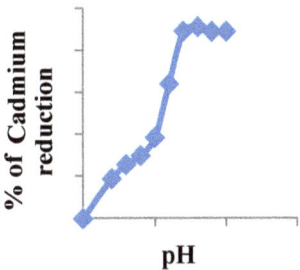

Fig. 1 Influence of pH on the biosorption of cadmium (II)

Fig. 2 Influence of biosorbent dose on the biosorption of cadmium (II)

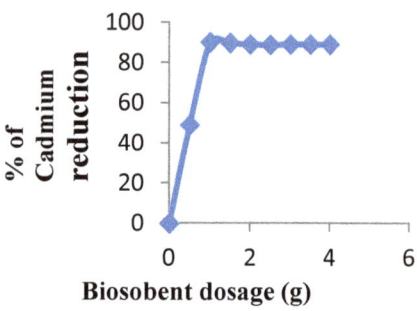

3.2.2 Influence of Moringa Oleifera Seed Dosage on the Evacuation of Cadmium (II)

The influence of Moringa oleifera seed dosage on the evacuation of cadmium (II) was studied by varying amounts of biosorbent from 0.5 to 4 g at pH 7 with initial cadmium (II) concentration of 100 μg/L. Figure 2 depicts that the percentage removal capacity of cadmium (II) increased steadily with increased biomass dosage from 0.5 to 4 g. The maximum biosorption efficiency of 90.2% was achieved at 1 g of biosorbent. These phenomena might be due to the increased surface area with more number of active sites of biosorbent. However, there was no change in the percentage biosorption of cadmium (II) ions. Therefore, the optimum biomass dosage was found to be 1 g. The optimum biosorbent dose in the present study was much lower compared to other studies by Sharma et al. (2006) who stated optimum *Moringa oleifera* dose of 4 g is for 85.01% of cadmium (II) removal using cadmium (II) concentration of 25 μg/mL, test solution of 200 ml and contact time of 40 min at pH of 6.5 (Sharma et al. 2006). Furthermore, Abishekkardam et al. reported optimum *Moringa oleifera* dose of 4 g is for 85.10% removal of Cd (II) ion from 100 μg/mL, at an optimum pH of 6.5 (Karda et al. 2010). The purpose of the current study is to reduce the overall process cost and also to create less volume of sludge after biosorption by using less dose of biosorbent (1 g) for further studies.

3.2.3 Influence of Initial Cadmium (II) Concentration on Biosorption

Figure 3 shows the relation between biosorption efficiency and the initial cadmium (II) concentration at fixed *Moringa oleifera* seed powder of 1 g and initial pH of 7. The initial cadmium concentration was varied from the range of 50–200 μg/L on the percentage removal of cadmium. The removal of cadmium (II) on *Moringa* seed powder increased with raising the initial cadmium (II) concentration reaching to a maximum level of 89.6% at 100 μg/L of cadmium (II) concentration. Later the percentage removal decreased sharply to about 67.2% with a rise in cadmium (II) concentration. This observation could be explained due to the saturation of active sites at higher initial cadmium (II) concentration (Amin et al. 2018).

Fig. 3 Influence of Cd (II) concentration on the biosorption of cadmium (II)

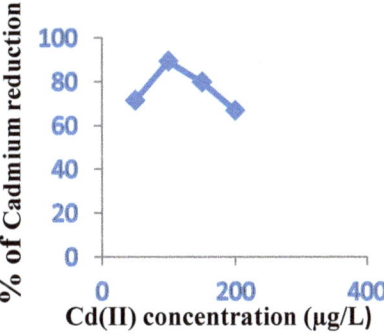

Fig. 4 Influence of contact time on the expulsion of cadmium (II)

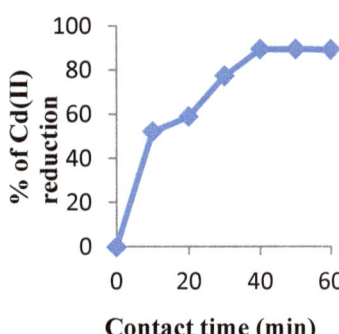

3.2.4   Influence of Time of Contact on Biosorption of Cadmium (II)

The biosorption of cadmium (II) on *Moring oleifera* seed powder with different contact time intervals (10–60 min) at pH value of 7, initial cadmium concentration of 100 µg/L and 1 g of biosorbent is presented in Fig. 4. The removal rate was rapid and steadily increased up to 30 min (77.6% reduction) and finally stabilized as attained equilibrium at 40 min (89.7% reduction). Rapid biosorption during the initial phase could be due to the availability of more number of vacant active sites of the *Moringa oleifera for* the percentage expulsion of cadmium (II) (Das et al. 2014).

3.2.5   Temperature Influence on Biosorption of Cadmium (II)

The temperature influence on biosorption of cadmium (II) by *Moringa oleifera* was investigated at various temperature ranges (20–45 °C) keeping biosorbent dosage (1 g), cadmium (II) initial concentration (100 µg/L), contact time (40 min) and pH 7 constant. As shown in Fig. 5, the rise in temperature led to an increase in biosorption of cadmium (II) percentage. The highest removal was 91.3% for cadmium (II) at 40 °C. However, to reduce the process cost, 30 °C was selected for further experiments which still yield reasonable cadmium (II) reduction of 89.2%. A similar conclusion has been

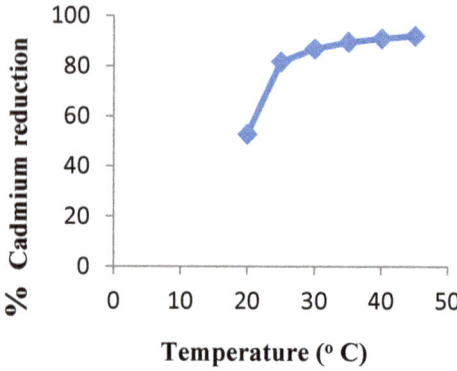

Fig. 5 Influence of temperature on the biosorption of cadmium (II)

given by Temesgen Girma Kebedea et al. (2018) for the reduction of cadmium (II) by *Moringa stenopetala* powder (Kebedea et al. 2018).

3.3 Equilibrium Isotherms for Biosorption of Cd (II) Using Moringa Oleifera

Three sorption isotherms models such as Langmuir, Freundlich and Dubinin–Radushkevich were tested to estimate biosorption capacity. Figures 6, 7 and 8 represent the linear form of Langmuir, Freundlich and Dubinin–Radushkevich isotherms. A linear plot of $1/q_e$ versus $1/C_e$ gives Langmuir parameters b and k as the intercept and slope (Fig. 6); Freundlich isotherm parameters k and n were estimated from the intercept and slope of linear plot lnq_e vs lnc_e (Fig. 7) and D–R model constants k_{DR} and q_m were evaluated from intercept and slope of a linear plot of $ln\ q_e$ versus ε^2 (Fig. 8). The characteristic parameters and correlation coefficient (R^2) obtained by fitting these three isotherms models with the experimental results are summarized in Table 1. As illustrated in Table1, the R_L value for the Langmuir isotherm equation was 0.118 which confirms the favourable biosorption of cadmium by *Moringa oleifera*. Furthermore, D–R model fits well for experimental data with a regression coefficient of 0.919. Moreover, D–R model yields mean adsorption energy (E) of 223 J/mol which is less than 8 kJ mol^{-1} showing that biosorption of cadmium by *Moringa oleifera* proceeds towards physical adsorption (Webber and Chakkravorti 1974). Based on higher regression coefficient values shown in Table 1, experimental data are well fitted for the Freundlich isotherm model with an R^2 of 0.986. Similar results were observed by Mataka et al. (2010) and Dawodu et al. (2019).

Fig. 6 Langmuir's equilibrium isotherm for cadmium (II)

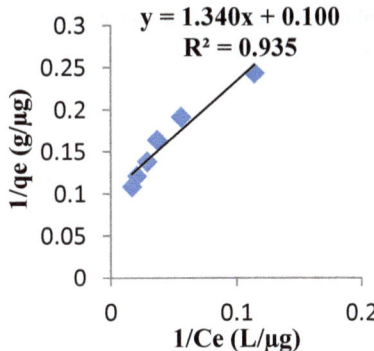

Fig. 7 Freundlich's equilibrium isotherm for cadmium (II)

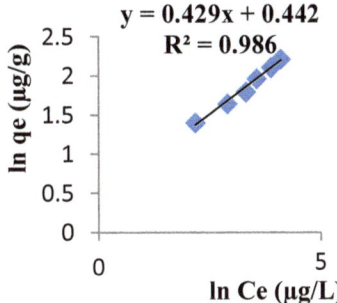

Fig. 8 The Radushkevich–Dubinin isotherm for cadmium (II)

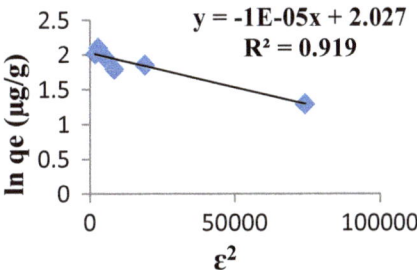

3.4 Biosorption Kinetic Models for Biosorption of Cd (II) Using Moringa Oleifera

The mechanism involved in Cadmium removal was assessed by utilizing kinetic models like Lagergren's equation (pseudo-first order) and pseudo-second order. The straight line plots of log(qe-qt) versus t shown in Fig. 9 were used to obtain k_1 (rate constant) and R^2 (correlation coefficient) for pseudo-first-order model, and the straight line plots of t/q_t against t shown in Fig. 10 were used to obtain k_2 (rate constant) and R^2 (regression coefficient) for pseudo-second-order model. It is

Table 1 Isotherm model data for cadmium (II) biosorption by *Moringa oleifera*

Models	Fitting equation	Langmuir's parameter	R^2
Langmuir	$y = 1.34x - 0.1$	$q_{max} = 10\ \mu g/g$	0.934
		$b = 0.075\ L/\mu g$	
		$R_L = 0.118$	
		Freundlich's parameter	0.986
Freundlich	$y = 4.29x + 0.442$	$n = 2.33$	
		$K_f = 1.56\ \mu g/g$	
		The Dubinin–Radushkevich parameter	0.919
Dubinin–Radushkevich	$y = -1 \times 10^{-5}x + 2.027$	$q_{max} = 7.59\ \mu g/g$	
		$k_{DR} = -1 \times 10^{-5}\ mol/J^2$	
		$E = 223\ J/mol$	

confirmed from Table 2 that pseudo- second-order model provides higher linearity with a regression coefficient of R^2 of 0.992, indicating that the rate of expulsion of cadmium (II) by *Moringa oelifera* could be better explained by pseudo-second-order model. Also, the estimated value of equilibrium uptake capacity (q_e) deduced from the pseudo-second order was 12.048 μg/g. Similar findings were observed by (Konada et al. 2017).

Fig. 9 Pseudo-first order for cadmium (II) biosorption

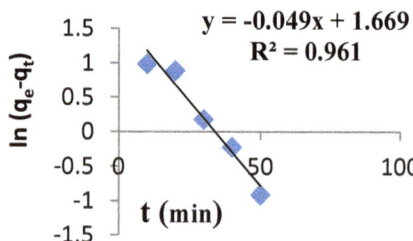

Fig.10 Pseudo-second order for cadmium (II) biosorption

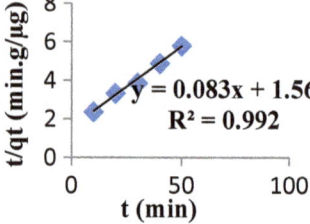

Models	Fitting equation	Corresponding parameter	R^2
Pseudo-first order	$y = -0.049x + 1.669$	$q_e = 5.296\ \mu g/g$	0.961
		$k_1 = 0.049\ \text{min}^{-1}$	
Pseudo-second order	$y = 4.29x + 0.442$	$q_e = 12.048\ \mu g/g$	0.992
		$k_2 = 4.42 \times 10^{-3}\ g / (\mu g.\ \text{min})$	

Table 2 Kinetic parameters for cadmium (II) biosorption by *Moringa oleifera*

3.5 Thermodynamics Study for Biosorption of Cd (II) Using Moringa Oleifera

Batch experiments were performed with cadmium (II) solution separately at various temperatures (293–318 K) by keeping optimum biosorbent dose (1 g), initial concentration of adsorbate (100 μg/L) and pH 7 for cadmium (II)) constant. The experimental data found from batch studies at various temperatures (293–318 K) were used to evaluate the thermodynamic parameters. The value of ΔG for cadmium (II) were -1316.6, $-3015\ 91$, -3900.75, -4690.11, -5800.36 and $-6249\ J\ \text{mol}^{-1}$ at 293, 298, 303, 308, 313 and 318 K, respectively, which confirms that the cadmium (II) expulsion utilizing *Moringa oleifera* was spontaneous in nature and thermodynamically feasible. The increase of ΔG value with decreasing temperature suggests more efficient and easier biosorption at a lower temperature. The positive value of enthalpy (ΔH) implies that biosorption cadmium (II) on *Moringa oleifera* is endothermic in nature, i.e., the biosorption of cadmium (II) increases with the rise in temperature. Moreover, the value of ΔH indicates the nature of the biosorption process to be chemical adsorption (40–120 kJ/mol), or physical adsorption (2.1–40 kJ/mol) (Vijayakumar et al. 2011). The obtained value of ΔH *is* 34851 J/mol which is less than 40 kJ mol^{-1} showing that biosorption of cadmium (II) by *Moringa oleifera* proceeds towards physical adsorption. Again, a positive ΔS value reveals an increase in the randomness of the adsorbed species and hence a better affinity of *Moringa oleifera* biomass (Fig. 11 and Table 3). Similar results were observed by Kebedea et al. (2018), Dawodu et al. (2020).

4 Conclusions

The biosorption capability of '*Moringa oleifera* seed' for the decontamination of Cd (II) from wastewater was investigated by performing batch tests. Under optimized process conditions (Moringa oleifera dose 1 g, pH 7, initial cadmium concentration 100 μg/L, time of contact of 40 and temperature 30 °C), *Moringa oleifera* could

Fig. 11 Thermodynamic
plots for cadmium (II)
biosorption

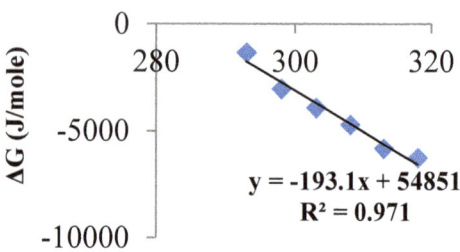

$$y = -193.1x + 54851$$
$$R^2 = 0.971$$

Table 3 Thermodynamic parameters for cadmium (II) biosorption by *Moringa oleifera*

Temperature (K)	ΔG (J/mol)	ΔH (J/mol)	ΔS (J/mol K)	R^2
293	−1316.6	34,851	193.1	0.971
298	−3015.9			
303	−3900.8			
308	−4690.1			
313	−5800.4			
318	−6249.2			

remove up to 91.4% of cadmium (II). It was concluded that experimental data were well fitted for the Freundlich isotherm model with an R^2 of 0.986. The findings from the kinetic study indicated that pseudo-second-order model provides higher linearity with a regression coefficient of 0.992. Also, the estimated value of equilibrium capacity (q_e) deduced from the pseudo-second order was 12.048 µg/g. The thermodynamic studies concluded that the values of ΔG for cadmium (II) were − 1316.6, −3015 91, −3900.75, −4690.1, −5800.36 and −6249 J mol^{-1} at 293, 298, 303, 308, 313 and 318 K, respectively, which confirms that the biosorption of Cd (II) on *Moringa oleifera* is spontaneous in nature and thermodynamically feasible. The positive value of enthalpy (ΔH) implies that biosorption cadmium on *Moringa oleifera* is endothermic in nature. The obtained value of ΔH was *34,851* J/mol for cadmium, showing that biosorption of cadmium (II) by *Moringa oleifera* proceeds towards physical adsorption. Again, a positive ΔS value reveals an increase in the randomness of the adsorbed species and hence a better affinity of *Moringa oleifera* biomass. Therefore, from the practical and economic point of view, *Moringa oleifera* seed without any chemical treatment could be considered as natural, sustainable and low-cost biosorbent for the removal of Cd (II) from drinking water for rural community households, where there are not much power sources and technical assistance.

Acknowledgements The authors gratefully acknowledge the laboratory facilities provided by R V College of Engineering, Bengaluru-560059.

References

Abdolali A, Ngo HH, Guo W, Zhou JL, Zhang J, Liang S, Chang SW, Nguyen DD, Liu Y (2017) Application of a breakthrough biosorbent for removing heavy metals from synthetic and real wastewaters in a lab-scale continuous fixed-bed column. Bioresource 229:78–87

Agrawal H (2007) Isolation of a 66 KDa protein with coagulation activity from seeds of Moringa oleifera. Glob J Biotechnol Biochem 2:36–39

Akinyeye OJ, Ibigbami TB, Odeja OO, Sosanolu OM (2020) Evaluation of kinetics and equilibrium studies of biosorption potentials of bamboo stem biomass for removal of lead (II) and cadmium (II) ions from aqueous solution. Afr J Pure Appl Chem 14:24–41

Amin F, NazTalpur F, Balouch A, Afridi HI, Khaskheli AA (2018) Efficient entrapping of toxic Pb(II) ions from aqueous system on a fixed-bed column of fungal biosorbent. Geol Ecol Landscapes 2(1):39–44

Asif Z, Chen Z (2017) Removal of arsenic from drinking water using rice husk. Appl Water Sci 7:1449–1458

Ayangbenro AS, Babalola OO (2017) A new strategy for heavy metal polluted environments: a review of microbial biosorbents. Int J Environ Res Public Health 14:94

Bibi S, Farooqi A, Yasmin A, Kamran MA, Niazi NK (2017) Arsenic and fluoride removal by potato peel and rice husk (PPRH) ash in aqueous environments. Int J Phytoremediation 19(11):1029 1036

Castagnetto JM, Hennessy SW, Roberts VA, Getzoff ED, Tainer JA, Pique ME (2002) MDB: the metalloprotein database and browser at the Scripps Research Institute. Nucleic Acids Res 30(1):379–382

Chen Z, Ma W, Han M (2008) Biosorption of nickel and copper onto treated alga (Undariapinnarlifida): application of isotherm and kinetic models. J Hazard Mater 155:327–333

Das B, Mondal NK, Bhaumik R, Roy P (2014) Insight into adsorption equilibrium, kinetics and thermodynamics of lead onto alluvial soil. Int J Environ Sci Technol 11(4):1101–1114

Dawodu FA, Onuh CU, Akpomie KG, Unuabonah EI (2019) Synthesis of silver nanoparticle from vigna unguiculata stem as adsorbent for malachite green in a batch system. SN Appl Sci 1:346

Dawodu FA, Akpan BM, Akpomie KG (2020) Sequestered capture and desorption of hexavalent chromium from solution and textile wastewater onto low cost Heinsia crinita seed coat biomass. Appl Water Sci 10:32. https://doi.org/10.1007/s13201-019-1114-6

Eman NA (2010) Production technique of natural coagulant from Moringa oleifera seeds. In: Fourteenth international water technology conference. Cairo, Egypt, pp 95–103

Farooq U, Kozinski JA, Khan MA, Athar M (2010) Biosorption of heavy metal ions using wheat based biosorbents-a review of the recent literature. Biresour. Technol. 101:5043–5053

Foroughi-Dahr M, Abolghasemi H, Esmaili M, Shojamoradi A, Fatoorehchi H (2015) Adsorption characteristics of congo red from aqueous solution onto tea waste. Chem Eng Commun 202:81–193

Giese EC (2020) Biosorption as green technology for the recovery and separation of rare earth elements. World J Microbiol Biotechnol 36:52. https://doi.org/10.1007/s11274-020-02821-6

Gupta NK, Gupta A, Ramteke P, Sahoo H, Sengupta A (2019) Biosorption-a green method for the preconcentration of rare earth elements (REE) from waste solutions: a review. J Mol Liq 274:148–164. https://doi.org/10.1016/j.molliq.2018.10.134

Hebbani AV, Anantha R, Manjunath GK, Kulkarni A, Sam R, Mishra A (2021) Evaluation of cadmium biosorption property of de-oiled palm kernel cake. Int J Phytorem 23(5):522–529. https://doi.org/10.1080/15226514.2020.1829544

Isawi H (2020) Using zeolite/polyvinyl alcohol/sodium alginate nanocomposite beads for removal of some heavy metals from wastewater. Arab J Chem 13:5691–5716

Jagaba AH, Kutty SRM, Khaw SG, Lai CL, Isa MH, Baloo L, Lawal IM, Abubakar S, Umaru I, Zango ZU (2020) Derived hybrid biosorbent for zinc (II) removal from aqueous solution by continuous-flow activated sludge system. J. Water Process Eng. 34:101152

Johri N, Jacquillet G, Unwin R (2010) Heavy metal poisoning: the effects of cadmium on the kidney. Biometals 23:783–792

Kara A, Demirbel E (2012) Kinetic, isotherm and thermodynamic analysis on adsorption of Cr(VI) ions from aqueous solutions by synthesis and characterization of magnetic-poly (divinylbenzene–vinylimidazole) microbeads. Water Air Soil Pollut 223:2387–2403

Karda A, Raj KR, Arora JK, Srivastava MM, Srivastava S (2010) Artificial neural network modeling for sorption of cadmium from aqueous system by shelled *Moringa Oleifera* seed powder as an agricultural waste. J Water Resourc Protect 2:339–344

Kebedea TG, Mengistie AA, Dube S, Nkambul TT, Nindi MM (2018) Study on adsorption of some common metal ions present in industrial effluents by Moringa stenopetala seed powder. J Environ Chem Eng 6:1378–1389

Konada RS, Reddy VK, Prasad MNV, Kumar NS (2017) *Moringa oleifera* (drumstick tree) seed coagulant protein (MoCP) binds cadmium–preparation and characterization of nanoparticles. Euro Biotech J 1(4):285–292

Li MR, Wei D, Liu T et al (2019) EDTA functionalized magnetic biochar for Pb (II) removal: adsorption performance, mechanism and SVM model prediction. Sep Purif Technol 227:115696

Madhuranthakam CMR, Thomas A, Akhter Z, Fernandes SQ, Elkamel A (2021) Removal of Chromium(VI) from contaminated water using untreated Moringa Leaves as biosorbent. Pollutants 1:51–64. https://doi.org/10.3390/pollutants1010005

Mataka LM, Sajidu SMI, Masamba WRL, Mwatseteza JF (2010) Cadmium sorption by *Moringa stenopetala* and *Moringa oleifera* seed powders: batch, time, temperature, pH and adsorption isotherm studies. Int J Water Resourc Environ Eng 2(3):50–59

Mondal NK, Samanta A, Chakraborty S, Shaikh WA (2018) Enhanced chromium (VI) removal using banana peel dust: isotherms, kinetics and thermodynamics study. Sustain Water Resour Manag 4:489

Monte Blanco SPD, Scheufele FB, Módenes AN, Schneider K, de Oliveira AP, Paraíso PR, Bergamasco R (2019) Adsorption study of heavy metals in aqueous solutions aiming at the treatment of contaminated groundwater. J Environ Sci Health Part A 54(14):1400–1411. https://doi.org/10.1080/10934529.2019.1646086

Pirkwieser P, López-López JA, Kandioller W, Keppler BK, Moreno C, Jirsa F (2018) Novel 3-hydroxy-2-naphthoate-based task-specific ionic liquids for an efficient extraction of heavy metals. Front Chem 6:172

Rajeswari M, Agrawal P, Pavithra S, Priya G, Sandhya R, Pavithra GM (2013) Continuous biosorption of cadmium by *Moringa oleifera* in a packed column. Biotechnol Bioprocess Eng 18:321–325. https://doi.org/10.1007/s12257-012-0424-4

Rajeswari M, Agrawal P, Rao NN, Sharma A, Hiremath L, Tippareddy KS (2021) Modelling and efficiency assessment of the up flow fixed bed process packed with Moringa oleifera for continuous Cd (II) removal from drinking water. J Mol Struct 1236(2021):130328

Shaikh RB, Saifullah B, Ur Rehman F, Shaikh RI (2018) Greener method for the removal of toxic metal ions from the wastewater by application of agricultural waste as an adsorbent. Water 10(10) https://doi.org/10.3390/w10101316

Sharma P, Kumari P, Srivastava MM, Srivastava S (2006) Removal of cadmium from aqueous system by shelled Moringa oleifera Lam. seed powder. Biores Technol 97:299–305

Tang JQ, Xi JB, Yu JX et al (2018) Novel combined method of biosorption and chemical precipitation for recovery of Pb2+ from wastewater. Environ Sci Pollut R 25:28705–28712

Vijayakumar G, Tamilarasan R, Dharmendra Kumar M (2011) Removal of Cd^{2+} ions from aqueous solution using live and dead *bacillus Subtilis*. Chem Eng Res Bull 15:18–24

Villarreal JS, Gándara JR, Navarrete D, Bejarano ML, Landázuri AC (2020) Lead (Pb2+) adsorption by means of pristine and prewashed residual Moringa oleifera Lam. seed husk biomass for water treatment applications. Int J Sustain Eng https://doi.org/10.1080/19397038.2020.1862350

Webber TW, Chakkravorti RK (1974) Pore and solid diffusion models for fixed—bed absorbers. AIChE J 20:228–238

Wernke G, Fagundes-Klen MR, Vieira MF, Suzaki PYR, de Souza HKS, Shimabuku QL, Bergam-asco R (2020) Mathematical modelling applied to the rate-limiting mass transfer step determination of a herbicide biosorption onto fixed-bed columns. Environ Technol 41(5):638–648. https://doi.org/10.1080/09593330.2018.1508252

Zhang R, Tian Y (2020) Characteristics of natural biopolymers and their derivative as sorbents for chromium adsorption: a review. J Leather Sci Eng 2:24. https://doi.org/10.1186/s42825-020-000 38-9

Zhu HY, Fu YQ, Jiang R, Jiang JH, Xiao L, Zeng GM, Zhao SL, Wang Y (2011) Adsorption removal of congo red on magnetic cellulose/Fe3O4/activated carbon composite: equilibrium, kinetic and thermodynamic studies. Chem Eng J 173:494–502

Green Synthesis of Zinc Oxide Nanoparticles Using *Citrus Sinensis* (Orange) Peel Extract for Achieving Ultraviolet Blocking Properties

M. Rajeswari, Nagashree N. Rao, Tanmay Agarwal, and S. Kavyasree

1 Introduction

1.1 Zinc Oxide Nanoparticles (ZnO NPs)

Nanoparticles (NPs) are particles with the size extending from 1 to 100 nm (nm). Zinc oxide Nanoparticles (ZnO NPs) are metal oxide nanoparticles that are harmless, have a high breaking point (2360 °C) and liquefying point (1975 °C), show up as a white powder and have been widely examined as a result of their great strength against brutal procedure conditions (Ramesh et al. 2014). They are multifunctional inorganic nanomaterials with one-of-a-kind properties, for example, semiconducting, antifungal, antibacterial (Geraci et al. 2017; Muthaiyan et al. 2012), UV filtering (Jayaprakash et al. 2020; Holmes et al. 2016) and piezoelectric properties (Zayed et al. 2021; Sabir et al. 2014). ZnO NPs have tremendous applications in biomedical and thin films, water purification process, beauty care products such as UV safeguard, elite nanosensors (Ma et al. 2016), sun-oriented cells, light-discharging diodes and so on. The green synthesis of ZnO NPs using citrus sinensis (orange) peel extract was investigated in the present work.

1.2 UV Blocking Properties of ZnO NPs

Consistently, around one million individuals are determined to have a skin disease. The face, head, neck and hands are the regions of the body that are most frequently exposed to the sun causing skin malignant growth (Xu and Wang 2011; Pandimurugan and Thambidurai 2017).

M. Rajeswari (✉) · N. N. Rao · T. Agarwal · S. Kavyasree
Department of Biotechnology, R.V. College of Engineering, Bengaluru 560059, India
e-mail: rajeshwarim@rvce.edu.in

The Sun Protection Factor (SPF) is a quantitative measurement of the effectiveness of a Sunscreen which is defined as the UV energy required to produce a minimal erythema dose (MED) on protected skin, divided by the UV energy required to produce a MED on unprotected skin:

$$\text{SPF} = \frac{\text{minimal erythema dose in sunscreen protected skin}}{\text{minimal erythema dose in non sunscreen protected skin}}$$

MED is characterized as the most reduced time interim or measurement of UV illumination adequate to deliver a negligible, noticeable abnormal redness on vulnerable skin (Wolf et al. 2011; Murphy 2000). The SPF of the synthesized ZnO NPs was calculated in-vitro by determining their absorbance using UV Spectrophotometry. The more the absorbance of the ZnO NPs, the more will be the SPF value. The higher the SPF value, the more effective is the synthesized product in preventing sunburn. The SPF values are calculated by the Mansur equation and these values are determined at different concentrations of ZnO NPs. UVA (320–400 nm), UVB (290–320 nm) and UVC (200–290 nm) are the three regions classified under Ultraviolet (UV) radiations of the electromagnetic spectrum. The air blocks UVC radiations before reaching the earth. The ozone layer doesn't completely block the UVB radiations and causes harm because of sunburn. The more profound layers of the epidermis are reached by the UVA radiations causing the untimely maturing of the body skin. Thus, ordinary items such as lotions, creams, moisturizers, shampoos and other hair and skin arrangements are included with Sunscreens. The proposed UV spectrophotometric technique is straightforward, fast, utilizes minimal effort reagents and is utilized in the evaluation of SPF values in numerous cosmetic formulations. It can be utilized in the examination of the sunscreens and can give data before continuing to the in-vivo tests.

2   Materials and Methods

2.1   Procedure for Synthesizing ZnO NPs

The orange peels were washed completely with distilled water to remove all the dirt. The orange peels were then grated and later dried under the sun for 4–5 days until all the moisture was evaporated. The dried peels were then grinded into powder and stored in an air-tight container. 10 g of the acquired powder was taken in a 250 ml measuring glass and 100 ml of distilled water was added to it and blended. The measuring glass was covered with aluminium foil and placed on a magnetic stirrer at 50 °C for 2 h at 850 rpm. After incubation, the extract was centrifuged at 10,000 rpm for 20 min. The supernatant was collected in a beaker and the pellet was discarded. The supernatant was then filtered utilizing Whatman's filter paper. The residue was then disposed of and the filtrate was taken. 50 ml of the filtrate was taken in a measuring glass and 5 g of Zinc acetate was added in splits of 1 g with a period interval of 5 min between each addition. The temperature was kept up at 70 °C all through in a water bath. The addition process was under continuous stirring with a glass rod to facilitate the mixing process. The heating was continued and the same temperature was maintained throughout the mixing process until a thick yellow solution was obtained. The content was dried in an hot air oven maintained at 60 °C for 30 h to obtain the desired ZnO NPs.

2.2 Characterization of the Synthesized ZnO NPs

2.2.1 FTIR Spectra Analysis

The synthesized ZnO NPs were exposed to FTIR analysis to detect the different functional groups related to it. FTIR analysis was carried out in the wavenumber range of 400–4000 cm^{-1} at room temperature. The characteristic absorption peaks of FTIR spectra of the ZnO NPs were determined by comparing them with reported FTIR analysis patterns from different papers.

2.2.2 Morphological Study

The surface morphology of the synthesized ZnO NPs was examined using SEM under different magnifications.

2.2.3 UV–VIS Spectra Analysis

UV–Vis analysis was used for the identification of the optical property of ZnO NPs.

2.2.4 XRD AND Structural Study

X-Ray Diffraction (XRD) pattern of the synthesized ZnO NPs was analysed on X-ray Diffractometer using Cu target for generating CuK alpha radiation ($\lambda = 1.54056$ Å) as the X-Ray source. XRD was carried out in the diffraction angle of 2θ in the range of 30°–80° at 40 kV, 30 mA, with the scanning rate of 2°/min. Table 1 reveals Standard data of JCPDS (Card No. 36-1451) belonging to the hexagonal Wurtzite crystal phase of ZnO.

XRD Pattern of ZnO NPs:

The crystal size of the ZnO NPs was calculated using the Debye-Scherer equation (Eq. 1) according to the highest diffraction peak in the XRD pattern:

Table 1 Standard data of JCPDS (Card No. 36-1451) belonging to hexagonal Wurtzite crystal phase of ZnO

2θ	31.78° (100)	34.43° (002)	36.28° (101)	47.57° (102)	56.26° (110)	62.86° (103)	69.13° (112)

$$D = \frac{0.9\lambda}{\beta \cos\theta} \qquad (1)$$

where

D = Crystal size,

k = Debye-Scherer constant = 0.9,

λ = X-Ray wavelength coming from CuK alpha radiation,

θ = Diffraction angle in degrees and

β = FWHM (Full width at the half maximum) of the diffraction peak in radians.

2.3 Determination of Sun Protection Factor (SPF) of the Synthesized ZnO NPs

The absorption spectra of 5 ml of various concentrations (20, 40, 60, 80 and 100 mg/ml) of the synthesized ZnO NPs (5 tests) were acquired in the scope of 290 to 450 nm utilizing 1 cm quartz cell and ethanol as a blank. The absorption data were gotten in the scope of 290–320 nm pursued by the use of the Mansur equation (Mansur et al. 1986). Mansur built up a basic scientific equation that uses the in-vitro technique found out by Sayre (Sayre et al. 1979), using UV spectrophotometry and the accompanying Mansur numerical equation (Eq. 2)

$$SPF = CF \times \sum_{290}^{320} EE(\lambda) \times I(\lambda) \times Abs(\lambda) \qquad (2)$$

3 Results and Discussion

3.1 Synthesized ZnO NPs

By non-toxic green synthesis method, 7.2 g of the desired ZnO NPs were synthesized using *Citrus Sinensis* (orange) peel extract (Fig. 1a, b).

a　　　　　　　　　　　　　　　　b

Fig. 1 **a** Crystal ZnO nanoparticles. **b** ZnO nanoparticles particles

3.2　Characterization of the Synthesized ZnO NPs

The synthesized ZnO NPs were then characterized by FTIR, XRD, SEM and UV–Vis spectroscopy.

3.2.1　FTIR Spectra Analysis

FTIR Spectra of ZnO NPs have been shown in Fig. 2. The absorption peak at 622.03 cm^{-1} corresponds to the characteristic signal of the ZnO bond which confirms the presence of ZnO NPs. The absorption peak at 3156.89 cm^{-1} corresponds to the stretching vibration of hydroxyl groups. The absorption peaks at 1565.85 cm^{-1} and 2507.51 cm^{-1} correspond to the stretching vibration of $C = O$ functional groups and alkanes groups, respectively. The absorption peaks at 1021.64 cm^{-1}, 1054.46 cm^{-1}, 1445.61 cm^{-1} and 2353.27 cm^{-1} correspond to the $S = O$ stretching, amines, C-N stretching, carboxylic acids, respectively. FTIR Spectra of ZnO NPs have been shown in Fig. 2 (Chithra et al. 2015).

3.2.2　Morphological Study

The SEM images in Fig. 3 showed that most of the ZnO NPs were in various irregular shapes of agglomerated particles.

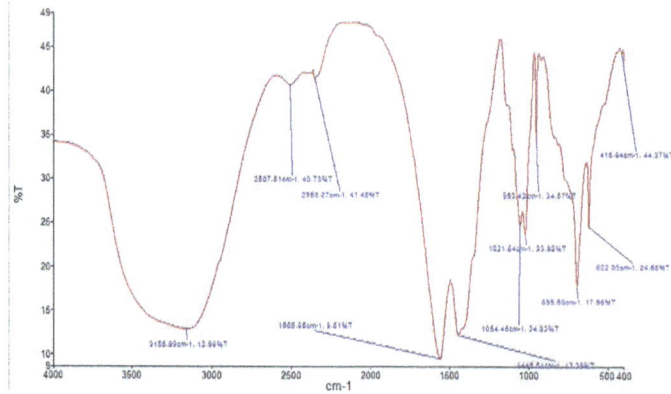

Fig. 2 FTIR spectra of ZnO NPs

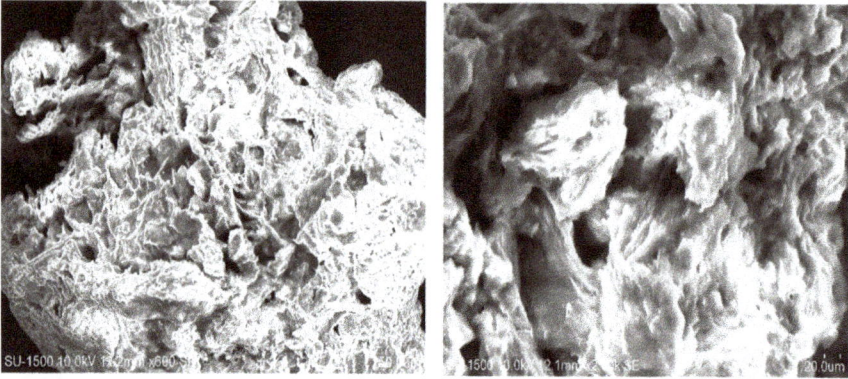

Fig. 3 SEM image of ZnO NPs

3.2.3 UV–VIS Spectra Analysis

UV–Vis absorption spectrum for orange peel extract in Fig. 4 showed the peak at 344.4 nm which correlated with the UV-spectrum range of ZnO that is between 320 and 390 nm (Sangeetha et al. 2011).

3.2.4 XRD AND Structural Study

XRD pattern of the synthesized ZnO NPs in Fig. 5 showed strong intensity diffraction peaks at $2\theta = 31.6°, 33.465°, 37.4°, 42.23°, 57.62°, 61.9233°$ and $69.63°$ which correspond to the lattice planes (100), (002), (101), (102), (110), (103) and (112), respectively. These diffraction peaks resembled the standard data of JCPDS belonging to the Hexagonal Wurtzite crystal phase of ZnO. Therefore, the obtained XRD pattern

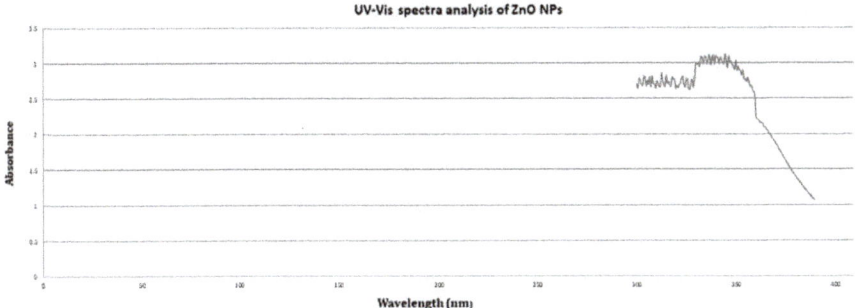

Fig. 4 UV–Vis spectra analysis of ZnO NPs

Fig. 5 XRD pattern of ZnO NP

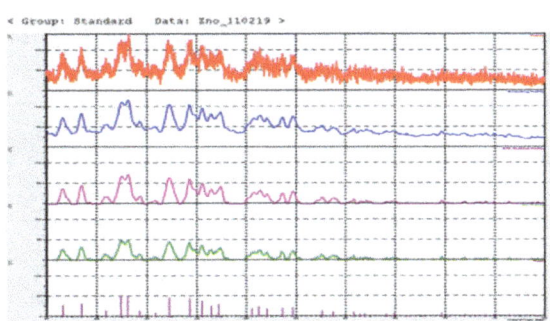

confirms the formation of ZnO NPs with a Hexagonal Wurtzite crystal structure (El-Naggar et al. 2017; Ibrahim et al. 2017; Ibrahim et al. 2018).

3.3 Evaluation of Sun Protection Factor (SPF) of the Synthesized ZnO NPs

The results showed that the SPF values calculated for the synthesized ZnO NPs (5 samples) were 4.228, 5.545, 5.93, 6.428 and 7.866, respectively. The 5 samples (Tables 2, 3, 4, 5 and 6) have shown increasing UV blocking properties as the concentrations of the ZnO NPs increased. Therefore, the synthesized ZnO NPs can be employed in Sunscreens to protect the skin from harmful effects of the sun's UV radiations.

$$SPF = CF \times \sum_{290}^{320} EE(\lambda) \times I(\lambda) \times Abs(\lambda)$$

Table 2 Sample 1 (20 mg/ml)

Wavelength (nm)	Abs	EE × I	Abs × (EE × I)
290	0.438	0.0150	0.0065
295	0.434	0.0817	0.0354
300	0.429	0.2874	0.1232
305	0.423	0.3278	0.1386
310	0.417	0.1864	0.0777
315	0.398	0.0839	0.0333
320	0.452	0.0180	0.0081
Total			0.4228

Table 3 Sample 2 (40 mg/ml)

Wavelength (nm)	Abs	EE × I	Abs × (EE × I)
290	0.572	0.0150	0.0085
295	0.567	0.0817	0.0463
300	0.560	0.2874	0.1609
305	0.554	0.3278	0.1816
310	0.548	0.1864	0.1021
315	0.531	0.0839	0.0445
320	0.594	0.0180	0.0106
Total			0.5545

Table 4 Sample 3 (60 mg/ml)

Wavelength (nm)	Abs	EE × I	Abs × (EE × I)
290	0.612	0.0150	0.0091
295	0.605	0.0817	0.0494
300	0.599	0.2874	0.1721
305	0.593	0.3278	0.1943
310	0.585	0.1864	0.1090
315	0.573	0.0839	0.0480
320	0.620	0.0180	0.0111
Total			0.593

$$SPF = CF \times \sum_{290}^{320} EE(\lambda) \times I(\lambda) \times Abs(\lambda)$$

$$SPF = CF \times \sum_{290}^{320} EE(\lambda) \times I(\lambda) \times Abs(\lambda)$$

Table 5 Sample 4 (80 mg/ml)

Wavelength (nm)	Abs	EE × I	Abs × (EE × I)
290	0.692	0.0150	0.0103
295	0.680	0.0817	0.0555
300	0.666	0.2874	0.1914
305	0.637	0.3278	0.2088
310	0.613	0.1864	0.1142
315	0.594	0.0839	0.0498
320	0.712	0.0180	0.0128
Total			0.6428

Table 6 Sample 5 (100 mg/ml)

Wavelength (nm)	Abs	EE × I	Abs × (EE × I)
290	0.830	0.0150	0.0124
295	0.814	0.0817	0.0665
300	0.808	0.2874	0.2322
305	0.785	0.3278	0.2573
310	0.759	0.1864	0.1414
315	0.735	0.0839	0.0616
320	0.847	0.0180	0.0152
Total			0.7866

$$SPF = CF \times \sum_{290}^{320} EE(\lambda) \times I(\lambda) \times Abs(\lambda)$$

$$SPF = CF \times \sum_{290}^{320} EE(\lambda) \times I(\lambda) \times Abs(\lambda)$$

4 Conclusions

The present study dealt with the synthesis of zinc nanoparticles which was carried out using *citrus sinensis* (orange) peel extract by the green synthesis method. Characterization was done by FTIR spectroscopy, XRD, SEM and UV–visible spectroscopy. FTIR results showed relative peaks for green synthesized ZnO at 622.03 cm^{-1}. XRD confirmed ZnO structure by showing peaks at $2\theta = 31.6°$, 33.465°, 37.4°, 42.23°, 57.62°, 61.9233° and 69.63°. The SEM image showed that most of the ZnO NPs were in various irregular shapes of agglomerated particles. UV-spectroscopy confirmed the presence of ZnO in the extract by a strong peak at 344.4 nm.

The SPF of the synthesized ZnO NPs was calculated in-vitro by determining their absorbance using UV Spectrophotometry. The SPF values were calculated by the Mansur equation and these values were determined at different concentrations of ZnO NPs. The results showed that the SPF values calculated for the synthesized ZnO NPs (5 tests) were 4.228, 5.545, 5.93, 6.428 and 7.866, respectively. The 5 samples have shown increasing UV blocking properties as the concentrations of the ZnO NPs increased. Therefore, the synthesized ZnO NPs can be employed in Sunscreens to protect the skin from harmful effects of the sun's UV radiations.

References

Chithra MJ, Sathya M, Pushpanathan K (2015) Effect of pH on crystal size and photoluminescence property of ZnO nanoparticles prepared by chemical precipitation method. Acta Metallurgica Sinica (English Letters) 28(3):394–404

El-Naggar ME, Hassabo AG, Mohamed AL, Shaheen TI (2017) Surface modification of SiO_2 coated ZnO nanoparticles for multifunctional cotton fabrics. J Colloid Interface Sci 498:413–422. https://doi.org/10.1016/j.jcis.2017.03.080

Geraci A, Di Stefano V, Di Martino E, Schillaci D, Schicchi R (2017) Essential oil components of orange peels and antimicrobial activity. Nat Prod Res 31(6):653–659

Holmes AM, Song Z, Moghimi HR et al (2016) Relative penetration of zinc oxide and zinc ions into human skin after application of different zinc oxide formulations. ACS Nano 10:1810–1819

Ibrahim NA, Nada AA, Eid BM, Al-Moghazy M, Hassabo AG, Abou-Zeid NY (2018) Nano-structured metal oxides: synthesis, characterization and application for multifunctional cotton fabric. Adv Nat Sci Nanosci Nanotechnol 9:035014

Ibrahim NA, Nada AA, Hassabo AG, Eid BM, Noor El-Deen AM, Abou-Zeid NY (2017) Effect of different capping agents on physicochemical and antimicrobial properties of ZnO nano-particles. Chem Pap 71:1365–1375. https://doi.org/10.1007/s11696-017-0132-9

Jayaprakash N, Suresh R, Rajalakshmi S, Raja S, Sundaravadivel E, Gayathri M, Sridharan M (2020) One-step synthesis, characterisation, photocatalytic and bio-medical applications of ZnO nanoplates. Mater Technol 35(2):112–124. https://doi.org/10.1080/10667857.2019.1659533

Ma LT, Fan HQ, Tian HL et al (2016) The n-ZnO/n-In_2O_3 heterojunction formed by a surface-modification and their potential barrier-control in methanal gas sensing. Sens Actuators B-Chem 222:508–516

Mansur JS, Breder MNR, Mansur MCA, Azulay RD (1986) Determinação do fator de proteção solar por espectrofotometria An Bras Dermatol Rio de Janeiro 61:121–124

Murphy WC (2000) Sunscreens efficacy Glob. Cosmet Ind Duluth 167:38–44

Muthaiyan A, Biswas D, Crandall PG, Wilkinson BJ, Ricke SC (2012) Application of orange essential oil as an antistaphylococcal agent in a dressing model. BMC Complement Altern Med 12(1):125

Pandimurugan R, Thambidurai S (2017) UV protection and antibacterial properties of seaweed capped ZnO nanoparticles coated cotton fabrics. Int J Biol Macromol 105:788–795

Ramesh P, Rajendran A, Meenakshisundaram M (2014) Green synthesis of zinc oxide nanoparticles using flower extract Cassia auriculata. J Nanosci Nanotechnol 2(1):41–45

Sabir S, Arshad M, Chaudhari SK (2014) Zinc oxide nanoparticles for revolutionizing agriculture: synthesis and applications. Sci World J 1–8

Sangeetha G, Rajeshwari S, Venckatesh R (2011) Green synthesis of zinc oxide nanoparticles by aloe barbadensis miller leaf extract: structure and optical properties. Mater Res Bull 46(12):2560–2566

Sayre RM, Agin PP, Levee GJ, Marlowe E (1979) Comparison of in vivo and in vitro testing of sunscreening formulas. Photochem. Photobiol Oxford 29:559–566

Wolf R, Wolf D, Morganti P, Ruocco V (2011) Sunscreens. Clinic Dermatol N Y 19:452–459

Xu S, Wang ZL (2011) One-dimensional ZnO nanostructures: solution growth and functional properties. Nano Res 4:1013–1098

Zayed M, Ghazal H, Othman HA, Hassabo AG (2021) Synthesis of different nanometals using Citrus Sinensis peel (orange peel) waste extraction for valuable functionalization of cotton fabric. Chemical papers. https://doi.org/10.1007/s11696-021-01881-8

Biohydrogen Production from Food and Beverage Processing: A Promising Strategy for Wastewater Management

Pragnesh N. Dave, Shalini Chaturvedi, and Lokesh Kumar Sahu

1 Introduction

The biomass was an important resource of fuels and energy in the beginning of 1800s. Beginning of the fossil fuel epoch has nearly phased out its continuation from the industrialized nations waiting until the *"First oil Shock"* shuddered the planet in mid-1970s, merely then biomass revival started to commence. The consequence of the biomass was recognized by the governments and policy makers so as to use it to salvage the equilibrium of reducing natural non-renewable resources. The inflated global demand for energy over and above the declining of fossil fuel deposits has required a world search for another energy sources. One of the best options to tone down the troubles coupled with climate change and global warming is to decrease burning of fossil fuels and look for new alternative energy resources. Biohydrogen gas has absorbed much consideration as an expectant option of energy, credit to its zero greenhouse emissions, vis-a-vis petroleum fuels (Arimi et al. 2015). The leading advantage of biohydrogen is that it can be shaped from renewable raw materials such as organic wastes, contributory to human-derived waste disposal and their resurgence in an environmentally pleasant way (Khanal et al. 2004).

P. N. Dave (✉)
Department of Chemistry, Sardar Patel University, Vallabh Vidynagar, Anand, Gujarat 388 120, India
e-mail: pragnesh7@yahoo.com

S. Chaturvedi
Samarpan Science and Commerce College, Gandhinagar, Gujarat, India

L. K. Sahu
Physical Research Laboratory, Space and Atmospheric Sciences Division, Ahmedabad 380 009, India
e-mail: lokesh@prl.res.in

© The Author(s), under exclusive license to Springer Nature Singapore Pte Ltd. 2023   287
A. K. Mishra and C. M. Hussain (eds.), *Biobased Materials*,
https://doi.org/10.1007/978-981-19-6024-6_14

The fossil fuels are depleting and ensuing in grave environment issues. For a future energy economy, hydrogen gas is regarded as a potential candidate. As water is its final combustion product, hydrogen is the only carbon-free fuel. For the energy-related environmental issues, such as greenhouse emission or acid rain therefore the application of hydrogen will greatly contribute. Biohydrogen is defined as hydrogen produced biologically, most commonly by bacteria, algae and archaea from both cultivation and from waste organic materials. Most biologically produced hydrogen in the biosphere is evolved during microbial fermentation processes. These organisms decay organic matter to carbon dioxide and hydrogen.

For biohydrogen generation, a number of potentially suitable residual substrates have been assessed through dark fermentation, thus far (Arimi et al. 2015; Khanal et al. 2004; Poggi-Varaldo et al. 2014) a number of researchers worked with the industrial wastewater from beverage industry, sugar industry, palm oil effluent, chemical industry, distillery industry effluent as substrate for the biohydrogen production. The hydrogen is produced as a spin-off and the yield is very little in acidogenic phase of an anaerobic digestion. To enhance the production of hydrogen the dark fermentation methods were recognized, whereas it is a challenge towards marketable use. Among these, the organic part of municipal solid wastes or food wastes represents a zero-cost, suitable source for hydrogen production, mainly due to its high carbohydrate content and its wide availability (Gioannis et al. 2013; Xiao et al. 2013; Yasin et al. 2013). In the recent years, numerous studies have been published on fermentative hydrogen production of food waste, using pure (Xiao et al. 2013; Hu et al. 2013; Hu et al. 2014) or mixed microbial cultures (Han et al. 2015; Sivagurunthan et al. 2014), using several reactor configurations, both batch (Sivagurunthan et al. 2014; Cappai et al. 2014) and continuous (Karadag et al. 2014; Sivagurunthan et al. 2015).

2 Major Applications of Hydrogen

Hydrogen is the latest buzzword when it comes to dealing with energy issues. The type of "hydrogen" is listed below

Type of Hydrogen

Green Hydrogen	Brown/Grey Hydrogen	Blue Hydrogen
• Derived from electrolysis of water using renewable sources like solar, wind and biomass for generating electricity . • Today, less than 1% hydrogen of the world 's is green .	• Derived from fossil fuels depending upon whether it is coal (brown) or gas (grey). • Being produced all over the world.	• Produced from fossil fuels, but a part of the carbon is absorbed using carbon capture utilization and storage technology .

The hydrogen is considered as carbon-free fuel as the only by-product H_2 energy after its combustion, is water. Providing of low-carbo energy for heat, balancing of electricity at national grid, and application in combustive engine favored hydrogen over other hydrocarbon-based gaseous biofuels. The petroleum and fertilizers companies are considered to be the leading users of H_2, from the statistical point of view, which nearly account for 50 and 37% correspondingly (Nath and Das 2003). By way of the demanding of H_2 engine, the H_2 fuel cells requirement is amplified by six percent in the past five years (Dicks et al. 2004). The H_2 energy will be used alongside many other forms of energy; therefore, it is important to understand and enhance the large effectiveness of delivering clean fuel for a diversify applications. How H_2 will be used in the future depends on the explicit wants of the public. There are some other foremost applications of present H_2 exploitation proposed as follows:

- H_2 is applied for hydration of considerable oils for fuel production, alkali hydration, and foods hydration for fertilizers production.
- Since H_2 is the ideal electron donor, it is used for diminishing nitrate, perchlorate, selenite, and a set of other oxidized water contaminations.
- At present, the industrial application of hydrogen is corresponding to merely 3% of the entire energy consumption, and it is predictable to develop considerably in the years to come (Das and Veziroğlu 2001).

Hydrogen is a energy dense and zero-carbon alternative energy delivery service with clean burning properties and biohydrogen creation by microalgae can trim down manufacturing connected greenhouse gas (GHG) releases to a great degree. Biohydrogen can be formed during dark fermentation by means of starch, sugars, or cellulosic materials. In recent times, microalgae-based biohydrogen production branded a silver lining for biohydrogen manufacturing by way of photolysis or being a substrate for anaerobic fermentation. This chapter writes the lists of methods of hydrogen production by different techniques.

3 Biomass as an Alternative of Fossil Fuels for Hydrogen Production

Biomass, which is the most pliable renewable sources, is the product of photosynthesis, can be valuable for sustainable biohydrogen production. The biomass resources comprise aquatic plants, wood waste urban garbage and household effluents and agricultural wastes with their lignocellulogic products. Since biomass utilizes atmospheric CO_2 for its augmentation, consequently it possesses an insignificant net CO impact as compared to the fossil fuels. The overall yearly generation of lignocellulosic biomass residues is nearly 220 billion tons, which is tantamount to 60–80 billion tons of crude, and was considered as a huge environmental load primarily due to absence of its proper management percepts. Usually, biomass, used as a fuel to harness heat, comprised of hydrocarbon materials like combustion of wood, dried plant, etc. The potential applications for microbial-assisted H_2 production of biomass are shown in Fig. 1. Bio-masses like energy crops, forestry waste, agricultural waste, and industrial and municipal waste are used. Microbial ability in terms of maximum hydrogen production to decompose biomass has been extensively studied (Crutzen and Andreae 1990).

H_2 production methods, utilizing biomass as a feedstock, can be divided into two categories Table 1 (Asif and Muneer 2007).

Table 1. H_2 production methods.

Fig. 1 Potential biomass as renewable energy

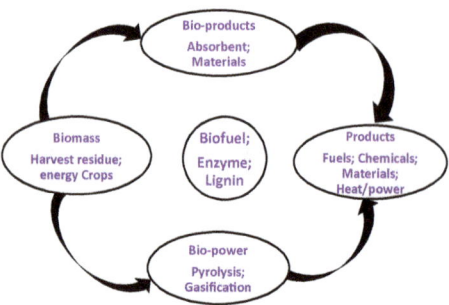

Table 1 Fermentative hydrogen production from various biomass (Mishra et al. 2017)

Substrate	Inoculum	Operating parameter		Mode	Maximum H_2 yield	References
		Temperature	pH			
Glucose	Clostridium sp. YM1	37 °C	6.5	Batch	1.7 mol H_2/mol glucose 3821 ml/L (CUH)	Hiligsmann et al. (2014)
Glucose monohydrate	Clostridium strains CWBI1009	30 °C	5.2	Sequenced Batch	302 mL/g glucose consumed, i.e., 2.4 mol/mol	Noblecourt et al. (2018)
Lactate	Microbial consortium	35 °C	6	Batch culture	0.4 mol $H_2 \cdot$mollactate^{-1}	Mishra et al. (2017)
Mannose monohydrous	Bacillus anthracis strain PUNAJAN 1	35 °C	6.5	Batch	236 ml H_2/g chemical oxygen demand (COD)	Hamilton et al. (2018)
Glucose monohydrate	Clostridium butyricum CWBI1009	30 °C	7.3	Batch	1.43 mol H2.molglucose^{-1}	Han et al. (2015)
Enzymatic hydrolyzed food waste	Immobilized sludge	–	–	Continuous	85.6 ml/g food waste	Pachapur et al. (2015)
Crude glycerol and apple pomace hydrolysate	Co-culture of Enterobacter aerogenes and Clostridium butyricum	30 °C	6.5	Batch	26.07 ± 1.57 mmol H_2/L of medium	Dhar et al. (2015)
Sugar beet juice	Sludge was preheated	37 °C	5.5 ± 0.2	Batch	3.2 mol H_2/mol hexose	Patel et al. (2017)
Mixed bio–wastes	Mixed cultures	35 °C	7.0	Batch	84 L/kg TS	Wang et al. (2018)

(continued)

Table 1 (continued)

Substrate	Inoculum	Operating parameter		Mode	Maximum H_2 yield	References
		Temperature	pH			
Concentrated sludge	Alkaline pre-treated sludge	37 °C	9.5	Batch test	15.6 mL per gram/VSS	Argun and Dao (2017)
Waste peach pulp	Anaerobic sludge	37 °C	5.9	Batch	123.27 mLH$_2$/gTOC	Silva-Illanes et al. (2017)
Glycerol	Enriched mixed microflora	70 °C	5.0	Continuous	$(0.58 \pm 0.13 \text{ molH}_2 \text{ mol}^{-1}$ glycerol	Kongjan and Angelidaki (2010)
Hydrosate of wheatstraw	Mixed extreme thermophiles	37 °C	6.5	Batch mode	212.0 ± 6.6 mL–H$_2$/g-sugars	Sydney et al. (2018)
Vinasse-based medium supplemented with sugarcane juice	Microbial consotorium > 50% Oxalobacteraceae	70 °C	7.0	Batch	1.59 ± 0.21 molH$_2$/molglucos	Singh et al. (2013)
Palm oil mill effluent	Immobilize Clostridium butyricum EB6	37 °C	5.5	Batch operation	5.35 LH$_2$/L-POME	Wang et al. (2003)
Palm oil mill effluent	Bacillus anthracis PUNAJAN 1	35 °C	6.5	Batch	2.42 mol H$_2$/mol mannose	Mishra et al. (2017)

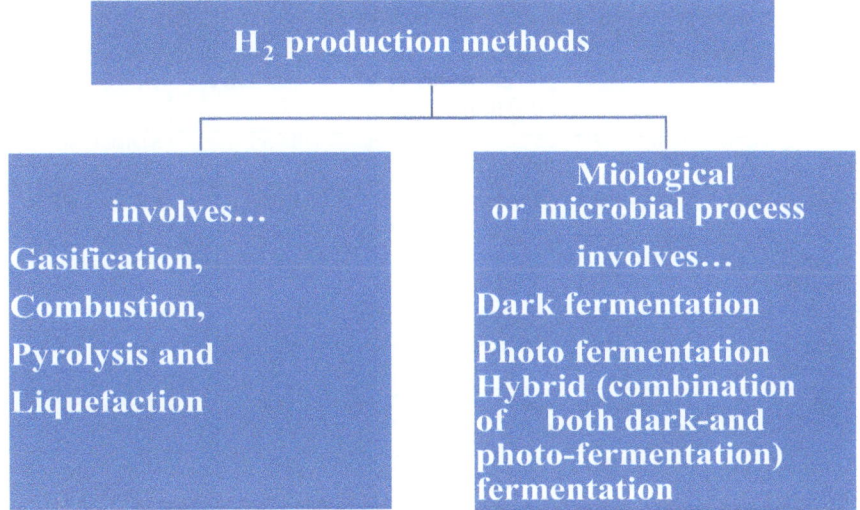

Thermochemical process customarily has an exorbitant design arrangement and wants high temperature, high pressure, and electricity for thermochemical process. That exhilarated investigators to catch on sustainable and flexible H_2 manufacturing methods which make practice of the biomass as a substrate (Patel et al. 2016). Microbial H_2 production has advantages over the thermochemical process method. Anaerobic fermentation by anaerobic bacteria has two different phases, (i) first phase known as *acidogenic phase* which is categorized by high hydrogen production and the speedy growth along with acetic and butyric acid production, while (ii) second phase known as *solventogenic phase* is categorized by slower growth with low hydrogen and solvent production.

Whether a solvent or acidic generating pathway is operated by anaerobic bacteria reliant on ATP and NADH levels. With the recognition of anaerobic digestion processes of wastewater are eased of hydrogen production as a result of the progress of dark-fermentation approach. By using a variety of biomass resources such as glucose, molasses, sugar cane, potato, starch residues, and so on as a feedstock shown in Table 2 (Hallenbeck 2012). Long-term and stable H_2 production for practical wastewater treatment has not been adequately achieved and are yet to be broadly explored further at molecular level though immense advancement has already been made in the diverse aspects of fermentative H_2 production like biomass immobilization, microbial community control, and bioreactor design (Abdeshahian et al. 2014).

This book chapter provides here and now an overview of approaches of fermentative hydrogen production. This includes the following:

Table 2. Fermentative hydrogen production from various biomass.

Table 2 Hydrogen production by dark fermentation (Khabirul Islam et al. 2021)

Effluent source	Inoculum source	pH	Temp (°C)	Substrate concentration (gCOD/L)	Operation mode	H_2 production rate (L/L/d)	References
Beverage WW	EMC	6.5	37.0	5.0	Batch	1.75	Kumar et al. (2017)
Beverage WW	AM	4–6	28	2.4–4.7		0.03	Kumar et al. (2015)
Cheese processing WW	ADS	4.8	35–38	5–7		1.0	Yang et al. (2007)
Cheese whey WW	ADS	5.5	55	21–47		1.5	Azbar et al. (2009)
Distillery WW	AS	5.5	37.0	34.8		2.88	Wicher et al. (2013)
Plastic industry WW	AS	5.5	36.0	3		0.28	Moreno-Andrade et al. (2015)
Olive mill WW	AS	7.0	37.0	50		0.42	Gonçalves et al. (2014)
Textile WW	–	7.0	37.0	20		4.32	Li et al. (2012)
Sugary WW	AS	4.5	35	6	Continuous	3.45	Wang et al. (2013)
Molasses WW	AS	4.4	35	8		7.47	Wang et al. (2013)
Olive mill WW	AS	7.0	35	39		7.00	Kumar et al. (2017)
Cheese whey WW	ADS	5.9	22–25	20		8.64	Azbar et al. (2009)

Legend WW: wastewater; AS: anaerobic sludge; ADS: anaerobic digester sludge; EMC: enriched mix culture; AM: anaerobic mixed microflora

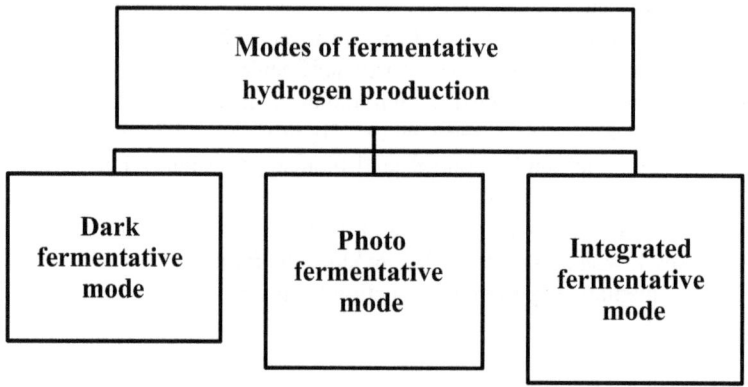

3.1 Dark Fermentative Mode (DFM)

Hydrogen-producing microorganisms and microalgae can produce hydrogen, in the absence of light and oxygen in dark fermentative mode, from a wide variety of substrates, including the organic fraction of wastewater (Fig. 2). The glucose considered to be ideal substance for fermentation, during glycolysis, is changed to pyruvate through various pathways and H_2 can be produced (Eqs. (1)–(4)).

The major soluble and bio-available organic products, in the case of wastewater, include organic acids like acetic, propionic, and butyric besides short-chain alcohols (Singh et al. 2015).

1 mole of glucose ≡ Produce 12 mole of hydrogen (Eq. (1)),

$$C_6H_{12}O_6 + 6H_2O \rightarrow 12H_2 + 6CO_2(\Delta G_o = +3.2\,kJ) \tag{1}$$

but practically, the production rate is much lower usually 3.47 mol H_2/mol glucose through the acetic acid pathway (Ginkel and Logan 2005). An array of operational parameters, viz., volume and rate of production can influence hydrogen production in batch production, including the activity and growth rate of the anaerobic microorganisms (Ortigueira et al. 2015), with continuous operation mode being more complex as the microbial activity is very sensitive to pH and toxic upset (Won and Lau 2011). The content and bio-availability of organic and inorganic substrates also significantly influence process performance (Kumar et al. 2016).

$$C_6H_{12}O_6 + 2H_2O \rightarrow 2CH_3COOH + 2CO_2 + 4H_2 \text{ (acetic acid pathway)} \tag{2}$$

$$C_6H_{12}O_6 \rightarrow CH_3CH_2CH_2COOH + 2CO_2 + 2H_2 \text{ (butyric acid pathway)} \tag{3}$$

$$C_6H_{12}O_6 \rightarrow CH_3COOH + CH_3CH_2COOH + CO_2 + H_2 \text{ (propionic acid pathway)} \tag{4}$$

Table 2 compares the DF hydrogen production rates from a range of industrial effluents by batch and continuous mode processes. DF operates in acidic pH conditions, varies from 4.5 to 7.5, at low operational temperature. In continuous mode,

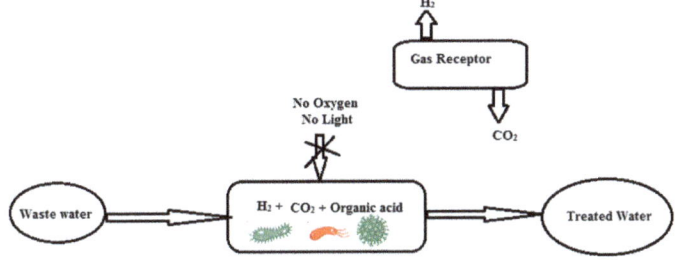

Fig. 2 Schematic diagram of H_2 production process by dark fermentative mode

Table 3 Key parameters influencing dark fermentative H_2 production (Khabirul Islam et al. 2021)

Parameter	Description	References
Inoculum	The choice of organisms is critical. Soil, wastewater sludge, compost, manure, digester sludge, and solid waste can all be used	Wang and Wan (2009)
Pre-treatment	Thermal, mechanical, chemical, microwave, and biological pre-treatment enhance the bio-availability of the substrate and the hydrogen yield from both waste and wastewater	Leaño and Babel (2012)
Temperature	Typically, mesophilic conditions (25–49 °C) produce a higher H_2 yield with mixed cultures. Effective H_2 yields are possible with increase in temperature to 60 °C	Moreno-Andrade et al. (2015)
pH	pH is a critical factor significantly determining the growth and metabolic activities of microbes. Optimum pH ranges from 4.5 to 9	Wang and Wan (2009)
HRT	Depending on reactor conditions and inoculum, the optimum HRT for hydrogen production ranges between hours and days	Wang et al. (2013)

wastewater was reported to yield higher rates of H_2 production than using batch mode; but, close process control is necessary to support steady gas production in the continuous mode (Azbar et al. 2009). The substrate COD is a key element in the production of H_2, and DF is a multi-parametric process.

Process Parameters in DF:

The mode of operation, substrate, and microorganism culture/inoculums source play a significant role in DF-mediated H_2 production (Table 2). The control reactor parameters include pH, temperature, and HRT, with organisms particularly sensitive to high-strength effluents. The key parameters investigated in the literature are described in Table 3.

Strengths, Weaknesses, and H_2 Enhancement Strategy:

Dark fermentation is a comparatively low-tech, low-cost process producing moderate rates of H_2 and organic removal (Nikolaidis and Poullikkas 2017). Added reward comprises the following***:

- Compounded forms of organic substrate can be utilized by anaerobic microorganisms or microalgae—Unassuming reactor assembly.
- The likelihood of producing value-added by-products.
- No need for outside energy such as light or electrical bias input.
- Incessant, all-day process is thinkable.

The challenge of anaerobic systems is, the effluent gases to guarantee that methane, hydrogen sulfide and carbon monoxide are separated from H_2 (Kumar et al. 2015).

The pre-treatment of feedstock, process optimization, co-fermentation, the supplementation of additives such as metal and their ions Ni, NiO and CoO, and improving inoculum specificity for H_2 production are avenues for sustained research (Manish and Banerjee 2008; Akkerman et al. 2002).

3.2 PFM

A fermentative conversion of organic substrates into hydrogen and carbon dioxide is achieved by using sunlight as an energy source (which is limiting process efficiency), and the reactions shown in Eqs. (5) and (6) portray the energetics of the process for two model compounds, viz., acetic acid and glucose, confirming the non-spontaneous reaction. By photosynthetic bacteria, the organic content is converted to hydrogen, primarily via nitrogenase and hydrogenase enzyme systems. Batch mode photoreactors ensure the absence of oxygen, authorizing purple non-sulfur (PNS) bacteria (Abo-Hashesh et al. 2011) to anaerobically break down organic compounds producing hydrogen (Fig. 3) (Li and Fang 2009) (see Fig. 4).

$$2CH_3COOH + 2H_2O \rightarrow 4H_2 + 2CO_2, (\Delta G^\circ = +104\,kJ) \tag{5}$$

$$C_6H_{12}O_6 + 6H_2O \rightarrow 12H_2 + 6CO_2 \,(\Delta G^\circ = +3.2\ kJ) \tag{6}$$

Using light as the energy source, the organic acid substrates, for example, purple non-sulfur bacteria (PNS) are oxidized using the tricarboxylic acid cycle (TCA), producing electrons, protons, and carbon dioxide (Manish and Banerjee 2008; Akkerman et al. 2002).

Advantages:

- In removal of environmental pollutants, use of industrial waste and use of organic acids produced from dark fermentation.

Disadvantages:

- Need nitrogen to limit condition and pretreatment of industrial effluent as it may be toxic (Mathews and Wang 2009).

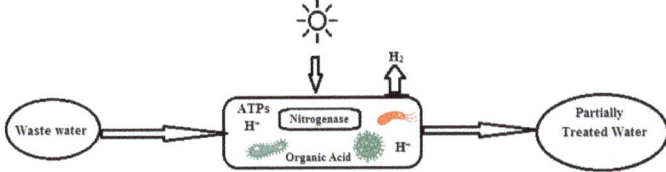

Fig. 3 Schematic diagram of photo-fermentation process

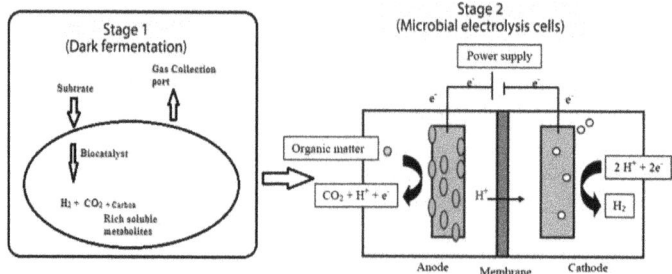

Fig. 4 Microbial electrolysis cells (MECs) integrated with the dark fermentation process for higher H$_2$ yield. *Legends* (A: anode; C: cathode; Biofilm: electrochemically active mixed microbial population) (Chandrasekhar et al. 2015)

In the presence of molecular nitrogen (N$_2$), nitrogenase catalyzes the formation of ammonium (NH$_4^+$) and H$_2$ (Eq. (7)), with the reaction also sustainable in the absence of N$_2$ (Eq. (8)).

$$N_2 + 8\,e^- + 10\,H^+ + 16\,MgATP \rightarrow 2\,NH_4^+ + H_2 + 16\,MgADP + 16\,Ph \quad (7)$$

$$8\,e^- + 8\,H^+ + 16\,MgATP \rightarrow 4\,H_2 + 16\,MgADP + 16\,Ph \quad (8)$$

where Ph signifies orthophosphate.

3.3 IF Mode

Many integrated approaches recently were proposed to conquer the limitations of several processes to increase the production of H$_2$ in dark fermentation. Mainly when in the form of an integrated two-stage energy producing process, the usage of the residual acid-rich organic substances from the fermentation effluents as carbon-rich substrates for further energy recovery is a viable and novel idea. With the primary dark fermentative process of H$_2$ production, numerous secondary processes were integrated, like methanogenesis for methane, acidogenic fermentation for H$_2$, photobiological processes for H$_2$ (Chandrasekhar et al. 2015), MECs for H$_2$, anoxygenic nutrient-limiting processes for bioplastics, cultivation of heterotrophic algae for lipids, and MFCs for bioelectricity generation. With these integrated approaches, the primary process uses these further substrates for additional energy production, and therefore, the entire process is more economically viable and practically applicable than without the integration.

With these integrated approaches, the primary process uses further substrates for additional energy production, and therefore, the entire process is more economically viable and practically applicable than without the integration.

The physiology and metabolic activities of bacteria vary significantly under dark- and photo-fermentative conditions. The efficiency of bacteria depends primarily on the types of enzymes involved in H_2 evolution. Under dark-fermentative conditions, hydrogenase and nitrogenase are the major enzymes responsible for this process (Kalia and Purohit 2008; Das et al. 2006). The efficiency of the dark-fermentative H_2 evolution process is governed by VFAs (Eqs. 9–12), such that acetic acid generation can lead to an additional 4 mol of H_2, whereas butyric acid is expected to generate 2 mol of H_2/mol of substrate. Lactic acid and ethanol are considered to be counter-productive to H_2 evolution process (Kalia and Purohit 2008; Patel et al. 2012). The intermediates of the dark-fermentative BHP, such as acetic and butyric acid, can be taken up by photosynthetic organisms to generate additional H_2 (Eqs. 13–14) (Afsar et al. 2011; Androga et al. 2012; Fang et al. 2005; Özgür et al. 2010).

$$C_6H_{12}O_6 (\text{Hexose}) + 2H_2O \rightarrow 2CH_3COOH \text{ (Acetate)} + 4H_2 + 2CO_2 \quad (9)$$

$$C_6H_{12}O_6 \rightarrow 2CH_3CH_2CH_2COOH \text{ (Butyrate)} + 2H_2 + 2CO_2 \quad (10)$$

$$C_6H_{12}O_6 \circledR 2CH_3CH(OH)COOH \text{ (Lactate)} \quad (11)$$

$$C_6H_{12}O_6 \rightarrow 2CH_3CH_2OH \text{ (Ethanol)} + 2CO_2 \quad (12)$$

$$CH_3COOH + 2H_2O \rightarrow 4H_2 + 2CO_2 \quad (13)$$

$$CH_3CH_2CH_2COOH + 6H_2O \circledR 10H_2 + 4CO_2 \quad (14)$$

It has been possible to achieve 2.86–6.07 mol H_2/mol hexose, using the organisms present in activated sludge enriched for dark-fermentative H_2-producers, along with photosynthetic organisms such as Rhodobacter Sphaeroides, Rhodopseudomonas palustris, and undefined hotosynthetic bacteria (Cheng et al. 2011; Su et al. 2009), over an incubation period ranging from 1 to 6 days of dark-fermentation followed by 5–14 days of photo-fermentative phase (Argun et al. 2008; Cheng et al. 2011). In most of the cases, the temperature of 31–37 °C has been found to be optimal during the dark phase and 30 °C during the light phase.

4 Techno Economical Challenges

To make the H_2 production process economically more practicable were endeavored during the last two decades (Hallenbeck 2011). However, some key technical challenges remain, and if these challenges are overcome, the overall H_2 production efficiency will increase through the biological pathways described below (Chandrasekhar et al. 2015) (Table 4).

Table 4 Biological pathways for H_2 production and the technical limitations (Chandrasekhar et al. 2015)

Type of bioprocess	Technical challenges
Dark fermentation	• low substrate conversion efficiency • low H_2 yield • thermodynamic limitations • mixture of H_2 and CO_2 gases as products, which require separation
Photofermentation	• requirement of an external light source • the process is limited by day and night cycles, with sunlight as the light source • low H_2 yield caused by extremely low light conversion efficiency
Direct biophotolysis	• O_2 generation caused by the activity of PS II • requirement for customized photobioreactors • low H_2 yield caused by extremely low light conversion efficiency
Indirect biophotolysis	• lower H_2 yield caused by hydrogenase(s) • requirement of an external light source • total light conversion efficiency was very low

For successful bioreactor design and the determination of the ideal process parameters, to get better yield of H_2 the efficient design of H_2 producing bioreactors, process modifications, selection of appropriate feedstocks, and with the selection of suitable and efficient microbial strains are required (Venkata Mohan et al. 2013; Das 2009). Similarly, hybrid processes that include subsequent methane production or electro-fermentation are also being considered to increase the energy recovery of the process. Although there is an abundance of research in the past and at present, these specific areas must be investigated to further enhance the production of H_2 via the biological pathways (Hallenbeck and Ghosh 2009). The integration of the H_2 production process with a conventional wastewater treatment process has several advantages, such as waste remediation with simultaneous generation of clean energy. In the imminent time, carbon-rich organic wastes may be targeted as suitable feedstocks for H_2 production because of their natural abundance. The use of cheaper raw material substrates would increase the H_2 yield from the biological processes, which would help significantly to make the process more economically viable and cost effective.

5 Strategies to Enhance the Efficiency of the Process

The major deterrents to the conventional biological H_2 production from any of the processes described above are the low substrate conversion efficiency and the accumulation of VFAs. Because of these deterrents, the overall yield of H_2 is far too low for the process to be economically feasible and commercially applicable (Chandrasekhar and Venkata Mohan 2014; Vijaya Bhaskar et al. 2008; Abo-Hashesh and Hallenbeck 2012). Additionally, after dark fermentation, significant amounts of residual organic substances such as VFAs or solvent remain in the effluent. Thus, additional treatments are necessary before disposal into the environment. The reuse of the residual carbon fraction of the fermentative effluents for further energy generation together with proper environmental treatment would be wise considering the environmental and economic factors (Mohanakrishna et al. 2010). Moreover, the design and fabrication of photobioreactors that use the internal light supply efficiently remains a challenge in photofermentation (Gadhamshetty et al. 2011; Keskin et al. 2011).

6 Future Perspectives

In the developing countries accounting from the biomass, 38% of the primary energy consumption is the largest source of energy in the world. The increased interest in biological hydrogen production represents an electrifying area of developed technology for energy generation using various biomasses. The organic rich biomass has great potential as the substrate for dark, photo, and integrated dark-photo fermentative hydrogen production. The integrated approach of dark-photo fermentation has many advantages, though single-stage dark and photo fermentative hydrogen, including maximum conversion of substrate to the hydrogen. The fermentative hydrogen production is not only restricted to the key operational parameters including pH and temperature. Significant improvements to succeed in sustainable hydrogen production at a commercial level can be expected through the devolvement of genetically engineered fermentative microorganism, advancement in bioreactor designing, etc.

The multidisciplinary fermentation processes used for the production of H_2 were numerous and a variety of substrates were also examined. The inherent limitations, such as low substrate conversion efficiency, accumulation of VFAs as carbon-rich acid intermediates, and change in system redox conditions and buffering capacity. Thus, to overcome the potential limiting factors and to improve the efficiency of the H_2 production process, an understanding of the mechanisms of H^+ reduction, functional roles of membrane components, composition of the communities, development of cultures, and design and development of competent bioreactors are the critical areas for both photo- and dark fermentation processes. The optimization of the process parameters is necessary to scale up the technology. The integrated approach that use the acid-rich reactor effluents with the simultaneous recovery of energy must be efficient and completely established for the commercialization of the process to be economically feasible. Although various novel approaches are anticipated in future years to overcome some of the persistent problems, biological H_2 production technology requires a diverged method for the process to be eco-friendly and economically feasible.

7 Way Forward

Research in microbial biomass conversion (MEC) for hydrogen production has progressed in recent years; key areas of research and development are working to address a number of challenges, including

- Through a number of methods such as microbial strain improvements, reactor system optimization, and identifying feedstock sources and processing methods with the highest yields improving the rates and yields of hydrogen production from fermentation processes.
- Developing MEC systems that can be scaled up to commercially relevant sizes while maintaining the production rates and system efficiencies seen at the bench scale and minimizing the costs of the reactor components.

References

(a)Mishra P, Thakur S, Singh L, Krishnan S, Sakinah M, Ab Wahid Z (2017) Fermentative hydrogen production from indigenous mesophilic strain Bacillus anthracis PUNAJAN 1 newly isolated from palm oil mill effluent. Int J Hydrogen Energy 42:16054–16063. (b) Mishra P, Krishnan S, Rana S, Singha L, Sakinah M, Ab Wahid Z (2019) Outlook of fermentative hydrogen production techniques: an overview of dark, photo and integrated dark-photo fermentative approach to biomass. Energ Strat Rev 24:27–37

Abdeshahian P, Al-Shorgani NKN, Salih NK, Shukor H, Kadier A, Hamid AA, Kalil MS (2014) The production of biohydrogen by a novel strain Clostridium sp. YM1 in dark fermentation process. Int J Hydrogen Energy 39:12524–12531

Abo-Hashesh M, Ghosh D, Tourigny A, Taous A, Hallenbeck PC (2011) Single stage photofermentative hydrogen production from glucose: an attractive alternative to two stage photofermentation or co-culture approaches. Int J Hydrog Energy 36:13889–13895

Abo-Hashesh M, Hallenbeck PC (2012) Microaerobic dark fermentative hydrogen production by the photosynthetic bacterium, rhodobacter capsulatus jp91. Int J Low Carbon Technol 7:97–103

Afsar N, Özgür E, Gürgan M, Akköse S, Yücel M, Gündüz U, Eroglu I (2011) Hydrogen productivity of photosynthetic bacteria on dark fermenter effluent of potato steam peels hydrolysate. Int J Hydrogen Energy 36:432–438

Akkerman I, Janssen M, Rocha J, Wijffels RH (2002) Photobiological hydrogen production: photochemical efficiency and bioreactor design. Int J Hydrogen Energy 27:1195–1208

Androga DD, Ozgur E, Eroglu I, Gunduz U, Yucel M (2012) Amelioration of photofermentative hydrogen production from molasses dark fermenter effluent by zeolite-based removal of ammonium ion. Int J Hydrogen Energy 37(21):16421–16429

Argun H, Dao S (2017) Bio-hydrogen production from waste peach pulp by dark fermentation: effect of inoculum addition. Int J Hydrogen Energy 42:2569–2574

Argun H, Kargi F, Kapdan IK, Oztekin R (2008) Light fermentation of dark fermentation effluent for bio-hydrogen production by different Rhodobacter species at different initial volatile fatty acid (VFA) concentrations. Int J Hydrogen Energy 33:7405–7412

Arimi MM, Knodel J, Kiprop A, Namango SS, Zhang Y, Geissen S-U (2015) Strategies for improvement of biohydrogen production from organic-rich wastewater: a review. Biomass Bioenerg 75:101–118

Asif M, Muneer T (2007) Energy supply, its demand and security issues for developed and emerging economies. Renew Sustain Energy Rev 11:1388–1413

Azbar N, Çetinkaya Dokgöz FT, Keskin T, Korkmaz KS, Syed HM (2009) Continuous fermentative hydrogen production from cheese whey wastewater under thermophilic anaerobic conditions. Int J Hydrogen Energy 34:7441–7447

Cappai G, De Gioannis G, Friargriu M, Massi E, Muntoni A, Polettini A, Pomi R, Spiga D (2014) An experimental study on fermentative H_2 from food waste as affected by pH. Waste Manage 34:1510–1519

Chandrasekhar K, Lee Y-J, Lee D-W, Production B (2015) Strategies to improve process efficiency through microbial routes. Int J Mol Sci 16:8266–8293

Chandrasekhar K, Venkata Mohan S (2014) Bio-electrohydrolysis as a pretreatment strategy to catabolize complex food waste in closed circuitry: function of electron flux to enhance acidogenic biohydrogen production. Int J Hydrog Energy 39:11411–11422

Cheng J, Su H, Zhou J, Song W, Cen K (2011) Hydrogen production by mixed bacteria through dark- and photo-fermentation. Int J Hydrogen Energy 36:450–457

Cheng J, Su H, Zhou J, Song W, Cen K (2011) Microwaveassisted alkali pretreatment of rice straw to promote enzymatic hydrolysis and hydrogen production in dark- and photo-fermentation. Int J Hydrogen Energy 36:2093–2101

Crutzen PJ, Andreae MO (1990) Biomass burning in the tropics: impact on atmospheric chemistry and biogeochemical cycles. Science 250:1669–1678

Das D (2009) Advances in biohydrogen production processes: an approach towards commercialization. Int J Hydrog Energy 34:7349–7357

Das D, Veziroğlu TN (2001) Hydrogen production by biological, processes: a survey of literature. Int J Hydrogen Energy 26:13–28

Das D, Dutta T, Nath K, Kotay SM, Das AK, Veziroglu TN (2006) Role of Fe-hydrogenase in biological hydrogen production. Curr Sci 90:1627–1637

De Gioannis G, Muntoni A, Polettini A, Pomi R (2013) A review of dark fermentative hydrogen production from biodegradable municipal waste fractions. Waste Manage 33:1345–1361

Dhar BR, Elbeshbishy E, Hafez H, Lee H-S (2015) Hydrogen production from sugar beet juice using an integrated biohydrogen process of dark fermentation and microbial electrolysis cell. Bioresour Technol 198:223–230

Dicks A, da Costa JD, Simpson A, McLellan B (2004) Fuel cells, hydrogen and energy supply in Australia. J Power Sour 13:1–12

Fang HHP, Liu H, Zhang T (2005) Phototrophic hydrogen production from acetate and butyrate in wastewater. Int J Hydrogen Energy 90:785–793

Gadhamshetty V, Sukumaran A, Nirmalakhandan N (2011) Review: Photoparameters in photofermentative biohydrogen production. J Biosci Bioeng 41:1–51

Van Ginkel S, Logan BE (2005) Inhibition of biohydrogen production by undissociated acetic and butyric acids. Environ Sci Technol 39:9351–9356

Gonçalves MR, Costa JC, Pereira MA, Abreu AA, Alves MM (2014) On the independence of hydrogen production from methanogenic suppressor in olive mill wastewater. Int J Hydrog Energy 39:6402–6406

Hallenbeck PC (2011) Microbial paths to renewable hydrogen production. Biofuels 2:285–302

Hallenbeck PC, Ghosh D (2009) Advances in fermentative biohydrogen production: the way forward? Trends Biotechnol 27:287–297

Hallenbeck PC (2012) Hydrogen production by cyanobacteria, microbial technologies in advanced biofuels, production. Springer, pp 15–28

Hamilton C, Calusinska M, Baptiste S, Masset J, Beckers L, Thonart P, Hiligsmann S (2018) Effect of the nitrogen source on the hydrogen production metabolism and hydrogenases of Clostridium butyricum CWBI1009. Int J Hydrogen Energy 43:5451–5462

Han W, Liu DN, Shi YW, Tang JH, Li YF, Ren NQ (2015) Biohydrogen production from food waste hydrolysate using continuous mixed immobilized sludge reactors. Bioresour Technol 180:54–58

Han W, Liu DN, Shi YW, Tang JH, Li YF, Ren NQ (2015) Biohydrogen production from food waste hydrolysate using continuous mixed immobilized sludge reactors. Biores Technol 180:54–58

Hiligsmann S, Beckers L, Masset J, Hamilton C, Thonart P (2014) Mesophilic biohydrogen production by Clostridium butyricum CWBI1009 in trickling biofilter reactor. Int J Hydrogen Energy 39:6899–6911

Hu CC, Giannis A, Chen C-L, Qi W, Wang J-Y (2013) Comparative study of biohydrogen production by four dark fermentative bacteria. Int J Hydrogen Energy 38:15686–15692

Hu CC, Giannis A, Chen C-L, Wang L-Y (2014) Evaluation of hydrogen producing cultures using pretreated food waste. Int J Hydrogen Energy 39:19337–19342

Kalia VC, Purohit HJ (2008) Microbial diversity and genomics in aid of bioenergy. J Ind Microbiol Biotechnol 35:403–419

Karadag D, Koroglu OE, Ozkaya B, Cakmakci M, Heaven S, Banks C (2014) A review on fermentative hydrogen production from dairy industry wastewater. J Chem Technol Biotechnol 89:1627–1636

Keskin T, Abo-Hashesh M, Hallenbeck PC (2011) Photofermentative hydrogen production from wastes. Bioresour Technol 102:8557–8568

Khabirul Islam AKM, Dunlop PSM, Hewitt NJ, Lenihan R, Brandoni C (2021) Bio-hydrogen production from wastewater: a comparative, study of low energy intensive production processes. Clean Technol 3:156–182

Khanal SK, Chen W-H, Li L, Sung S (2004) Biological hydrogen production: effects of pH and intermediate products. Int J Hydrogen Energy 29:1123–1131

Kongjan P, Angelidaki I (2010) Extreme thermophilic biohydrogen production from wheat straw hydrolysate using mixed culture fermentation: effect of reactor configuration. Bioresour Technol 101:7789–7796

Kumar G, Bakonyi P, Sivagurunathan P, Kim SH, Nemestóthy N, Bélafi-Bakó K, Lin CY (2015) Enhanced biohydrogen production from beverage industrial wastewater using external nitrogen sources and bioaugmentation with facultative anaerobic strains. J Biosci Bioeng 120:155–160

Kumar G, Sivagurunathan P, Park JH, Park JH, Park HD, Yoon JJ, Kim SH (2016) HRT dependent performance and bacterial community population of granular hydrogen-producing mixed cultures fed with galactose. Bioresour Technol 206:188–194

Kumar G, Sivagurunathan P, Pugazhendhi A, Thi NBD, Zhen G, Chandrasekhar K, Kadier A (2017) A comprehensive overview on light independent fermentative hydrogen production from wastewater feedstock and possible integrative options. Energy Convers Manag 14:390–402

Leaño EP, Babel S (2012) Effects of pretreatment methods on cassava wastewater for biohydrogen production optimization. Renew Energy 39:339–346

Li YC, Chu CY, Wu SY, Tsai CY, Wang CC, Hung CH, Lin CY (2012) Feasible pretreatment of textile wastewater for dark fermentative hydrogen production. Int J Hydrog Energy 37:15511–15517

Li RY, Fang HHP (2009) Heterotrophic photo fermentative hydrogen production. Crit Rev Environ Sci Technol 39:1081–1108

Manish S, Banerjee R (2008) Comparison of biohydrogen production processes. Int J Hydrogen Energy 33(1):279–286

Mathews J, Wang G (2009) Metabolic pathway engineering for enhanced biohydrogen production. Int J Hydrogen Energy 34:7404–7416

Mohanakrishna G, Venkata Mohan S, Sarma PN (2010) Utilizing acid-rich effluents of fermentative hydrogen production process as substrate for harnessing bioelectricity: an integrative approach. Int J Hydrogen Energy 35:3440–3449

Moreno-Andrade I, Moreno G, Kumar G, Buitrón G (2015) Biohydrogen production from industrial wastewaters. Water Sci Technol 71:105–110

Nath K, Das D (2003) Hydrogen from biomass. Curr Sci 265–271

Nikolaidis P, Poullikkas A (2017) A comparative overview of hydrogen production processes. Renew Sustain Energy Rev 67:597–611

Noblecourt A, Christophe G, Larroche C, Fontanille P (2018) Hydrogen production by dark fermentation from pre-fermented depackaging food wastes. Bioresour Technol 247:864–870

Ortigueira J, Alves L, Gouveia L, Moura P (2015) Third generation biohydrogen production by Clostridium butyricum and adapted mixed cultures from Scenedesmus obliquus microalga biomass. Fuel 153:128–134

Özgür E, Afsar N, de Vrije T, Yücel M, Gündüz U, Claassen PA, Eroglu I (2010) Potential use of thermophile dark fermentation effluents in photofermentative hydrogen production by Rhodobacter capsulatus. J Clean Prod 18:S23–S28

Pachapur VL, Sarma SJ, Brar SK, Le Bihan Y, Buelna G, Verma M (2015) Biohydrogen production by co-fermentation of crude glycerol and apple pomace hydrolysate using co-culture of Enterobacter aerogenes and Clostridium butyricum. Bioresour Technol 193:297–306

Patel SKS, Singh M, Kumar P, Purohit HJ, Kalia VC (2012) Exploitation of defined bacterial cultures for production of hydrogen and polyhydroxybutyrate from pea-shells. Biomass Bioenergy 36:218–225

Patel M, Zhang X, Kumar A (2016) Techno-economic and life cycle assessment on lignocellulosic biomass thermochemical conversion technologies: a review. Renew Sustain Energy Rev 53:1486–1499

Patel SK, Lee J-K, Kalia VC (2017) Dark-fermentative biological hydrogen production from mixed biowastes using defined mixed cultures, Indian. J Microbiol 57:171–176

Poggi-Varaldo HM, Munoz-Paez KM, Escamilla-Alvarado C, Robledo-Narvaez PN, Ponce-Noyola MT, Calva-Calva G, Rios-Leal E, Galindez-Mayer J, Estrada-Vazquez CE, Ortega-Clementa A, Rinderknecht-Seijas NF (2014) Biohydrogen, biomethane and bioelectricity as crucial components of biorefinery of organic wastes: a review. Waste Manage Res 32:353–365

Silva-Illanes F, Tapia-Venegas E, Schiappacasse MC, Trably E, Ruiz-Filippi G (2017) Impact of hydraulic retention time (HRT) and pH on dark fermentative hydrogen production from glycerol. Energy 141:358–367

Singh L, Wahid ZA, Siddiqui MF, Ahmad A, Rahim MHA, Sakinah M (2013) Biohydrogen production from palm oil mill effluent using immobilized Clostridium butyricum EB6 in polyethylene glycol. Process Biochem 48:294–298

Singh A, Sevda S, Abu,Reesh IM, Vanbroekhoven K, Rathore D (2015) Pant D. Biohydrogen production from lignocellulosic biomass: technology and sustainability. Energies 8:13062–13080

Sivagurunthan P, Sen B, Lin C-Y (2014) Batch fermentative hydrogen production by enriched mixed culture: combination strategy and their microbial composition. J Biosci Bioeng 117(2):222–228

Sivagurunthan P, Sen B, Lin C-Y (2015) High rate fermentative hydrogen production from beverage wastewater. Appl Energy 147:1–9

Su H, Cheng J, Zhou J, Song W, Cen K (2009) Improving hydrogen production from cassava starch by combination of dark and photo fermentation. Int J Hydrogen Energy 34:1780–1786

Sydney EB, Novak AC, Rosa D, Medeiros ABP, Brar SK, Larroche C, Soccol CR (2018) Screening and bioprospecting of anaerobic consortia for biohydrogen and volatile fatty acid production in a vinasse based medium through dark fermentation. Process Biochem 67:1–7

Venkata Mohan S, Chandrasekhar K, Chiranjeevi P, Babu PS (2013) Biohydrogen production from wastewater. In: Pandey A, Chang J-S, Hallenbeck PC, Larroche C (eds) Biohydrogen. Amsterdam, The Netherlands, Elsevier, pp 223–257

Vijaya Bhaskar Y, Venkata Mohan S, Sarma PN (2008) Effect of substrate loading rate of chemical wastewater on fermentative biohydrogen production in biofilm configured sequencing batch reactor. Bioresour Technol 99:6941–6948

Wang C, Chang C, Chu C, Lee D, Chang B-V, Liao C (2003) Producing hydrogen from wastewater sludge by Clostridium bifermentans. J Biotechnol 102:83–92

Wang D, Duan Y, Yang Q, Liu Y, Ni B-J, Wang Q, Zeng G, Li X, Yuan Z (2018) Free ammonia enhances dark fermentative hydrogen production from waste activated sludge. Water Res 133:272–281

Wang B, Li Y, Ren N (2013) Biohydrogen from molasses with ethanol-type fermentation: effect of hydraulic retention time. Int J Hydrog Energy 38:4361–4367

Wang J, Wan W (2009) Factors influencing fermentative hydrogen production: a review. Int J Hydrog Energy 34:799–811

Wicher E, Seifert K, Zagrodnik R, Pietrzyk B, Laniecki M (2013) Hydrogen gas production from distillery wastewater by dark fermentation. Int J Hydrog Energy 38:7767–7773

Won SG, Lau AK (2011) Effects of key operational parameters on biohydrogen production via anaerobic fermentation in a sequencing batch reactor. Bioresour Technol 102:6876–6883

Xiao L, Deng Z, Fung KY, Ng KM (2013) Biohydrogen generation from anaerobic digestion of food waste. Int J Hydrogen Energy 38:13907–13913

Yang P, Zhang R, McGarvey JA, Benemann JR (2007) Biohydrogen production from cheese processing wastewater by anaerobic fermentation using mixed microbial communities. Int J Hydrog Energy 32:4761–4771

Yasin NHM, Mumtaz T, Hassan MA, Rahman NA (2013) Food waste and food processing waste for biohydrogen production: a review. J Environ Manage 130:375–385

Milton Keynes UK
Ingram Content Group UK Ltd.
UKHW021936251023
431302UK00003B/10